Statistical Estimation of Epidemiological Risk

STATISTICS IN PRACTICE

Statistics in Practice is an important international series of texts, which provide detailed coverage of statistical concepts, methods and worked case studies in specific fields of investigation and study.

With sound motivation and many worked practical examples, the books show in down-to-earth terms how to select and use an appropriate range of statistical techniques in a particular practical field within each title's special topic area.

The books provide statistical support for professionals and research workers across a range of employment fields and research environments. Subject areas covered include medicine and pharmaceutics; industry, finance and commerce; public services; the earth and environmental sciences, and so on.

The books also provide support to students studying statistical courses applied to the above areas. The demand for graduates to be equipped for the work environment has led to such courses becoming increasingly prevalent at universities and colleges.

It is our aim to present judiciously chosen and well-written workbooks to meet everyday practical needs. The feedback of views from readers will be most valuable to monitor the success of this aim.

A complete list of titles in this series appears at the end of the volume.

Statistical Estimation of Epidemiological Risk

Kung-Jong Lui
Department of Mathematics and Statistics
San Diego State University, USA

Telephone (+44) 1243 779777

Email (for orders and customer service enquiries): cs-books@wiley.co.uk
Visit our Home Page on www.wileyeurope.com or www.wiley.com

Other Wiley Editorial Offices

John Wiley & Sons Inc., 111 River Street, Hoboken, NJ 07030, USA

Jossey-Bass, 989 Market Street, San Francisco, CA 94103-1741, USA

Wiley-VCH Verlag GmbH, Boschstr. 12, D-69469 Weinheim, Germany

John Wiley & Sons Australia Ltd, 33 Park Road, Milton, Queensland 4064, Australia

John Wiley & Sons (Asia) Pte Ltd, 2 Clementi Loop #02-01, Jin Xing Distripark, Singapore 129809

John Wiley & Sons Canada Ltd, 22 Worcester Road, Etobicoke, Ontario, Canada M9W 1L1

Wiley also publishes its books in a variety of electronic formats. Some content that appears in print may not be available in electronic books.

British Library Cataloguing in Publication Data

A catalogue record for this book is available from the British Library

ISBN 0-470-85071-X

Typeset in 10/12pt Photina by Laserwords Private Limited, Chennai, India

This book is printed on acid-free paper responsibly manufactured from sustainable forestry in which at least two trees are planted for each one used for paper production.

In memory of my parents
Shung-Wu and Li-Ching

Contents

About the Author

KUNG-JONG LUI is a professor in the Department of Mathematics and Statistics at San Diego State University. Since he obtained his Ph.D. in biostatistics from UCLA in 1982, he has published more than 100 papers in peer-reviewed journals, including *Biometrics, Statistics in Medicine, Biometrical Journal, Psychometrika, Communications in Statistics: Theory and Methods, Science, Proceedings of National Academy of Sciences, Controlled Clinical Trials, Journal of Official Statistics, IEEE Transactions on Reliability, Environmetrics, Test, Computational Statistics and Data Analysis, American Journal of Epidemiology, American Journal of Public Health*, etc. He is a Fellow of the American Statistical Association, a life member of the International Chinese Statistical Association, and a member of the Western North American Region of the International Biometric Society.

Preface

The estimation of epidemiological indices plays an important role in epidemiological investigations. One aim of this book is to provide biostatisticians, epidemiologists, and medical researchers with a useful resource on the different estimators of the most commonly used measures of risk in a variety of designs. Through a systematic presentation and discussion, it is hoped that the reader will appreciate better the use and limitations of, and the relationships among, these indices. Because the material in each chapter is generally self-contained, readers may choose chapters according to their own interests without the need to read through all the preceding chapters. This may increase the utility of the book, although I must admit that some definitions are repeated between chapters to avoid ambiguities in the formulae.

This book is intended for postgraduates and researchers who have one year of training in biostatistics and possess some basic knowledge of epidemiological terms, such as prevalence, risk difference, odds ratio, relative risk, and attributable risk. It is also intended for students of biostatistics and epidemiology as a one-semester graduate course, focusing on statistical estimation of risk in epidemiology. Because research on estimation of epidemiological risk has been quite intensive in the last two decades, to provide readers with up-to-date information I have included many recently developed estimators and relevant references. Thus, this book may also be used as a desk reference for established researchers. Although the book is mainly directed at biostatisticians and epidemiologists, because measures such as the risk difference, relative difference (or relative risk reduction), and number needed to treat are often used to report clinical findings, the book should be useful for statisticians and clinicians working in pharmaceutical areas as well.

When the underlying disease is rare, the probability of obtaining only a few or zero cases in a sample under binomial sampling can be large or non-negligible. To ensure that a reasonable number of cases are obtained, we may consider use of inverse sampling, a fact which has not been widely familiar among practicing biostatisticians or epidemiologists. This may be the first book to attempt to systematically introduce in a unified manner statistical methods relevant to inverse sampling in epidemiology. In contrast to binomial sampling, we show that the bias of estimators for the relative risk or the odds ratio in paired-sample data can easily be avoided by using inverse sampling. Furthermore, when the sample size is small, asymptotic interval estimators for the relative difference, the attributable risk in case–control studies when the underlying disease is rare, or the odds ratio in paired-sample data may be inappropriate. We note that under inverse sampling the derivation of exact confidence intervals for these indices is straightforward. The results and discussions on inverse sampling presented in this book can provide readers with an alternative way to design their studies.

When the response variable is on an ordinal scale with more than two categories, the odds ratio is inapplicable without arbitrarily collapsing the data. This book also includes a chapter (Chapter 6) focusing on the generalized odds ratio. This measure has an easy interpretation and should be useful for epidemiologists and clinicians when they wish to provide a quantitative measure of the strength of association for ordinal data between two comparison groups without assuming any parametric models.

The attributable risk (AR), representing the proportion of cases that may be prevented if the underlying risk factor under investigation is completely eliminated, is probably one of the most important indices for public health administrators to rank the relative importance of risk factors for intervention. Although there have been numerous recent publications that focus estimation on this useful measure in a variety of designs, many textbooks have touched this topic superficially by considering only the simplest cases in which there are no confounders. I discuss estimation of the AR from the simplest case – no confounders under a variety of designs – to the more complicated case with confounders. I also discuss estimation of the AR for paired-sample data. I further consider the situation in which the exposure variable has multiple levels, and the situation in which one applies the logistic regression model to adjust for the effects of confounding variables in case–control studies. A brief discussion on estimation of the AR under inverse sampling has also been included. The discussions on the AR presented in this book should be useful for researchers working in public health administration by providing relatively complete information on recent developments.

Upon the request of an anonymous reviewer, I have also included a chapter that discusses the use of the 'number needed to treat' (NNT). Because it can be easily understood by clinicians, this index has frequently been employed in randomized trials and evidence-based medicine. However, it has been subject to criticism by statisticians due to misuse and misunderstanding. For example, there are published papers that report the union of two disjoint open intervals as a confidence interval for the NNT, or provide a confidence interval that does not even contain the NNT point estimate. I have tried to present this index in such a way that these criticisms can be avoided. I sincerely hope that readers find the discussion presented here useful in clarifying the limitations of the NNT and in computing interval estimators for it.

I wish to express my indebtedness to my colleagues Drs. Duane Steffey, Colleen Kelly, and Richard Levine at San Diego State University, as well as to the three anonymous reviewers who generously provided valuable comments on an early draft of the manuscript. I also wish to thank the particular reviewer who spent valuable time on the revised draft and provided additional suggestions which led to improvements in the content of this book. I would like to thank Dr. N. Breslow at the University of Washington, the International Agency for Research on Cancer Center, the International Biometric Society, the American Medical Association, and Oxford University for their permission to include the data sets used to illustrate the methods discussed here. I would also like to express my gratitude to Drs. William G. Cumberland, A. A. Afifi, Sander Greenland, Frank Massey, Jr., Olive Dunn, Charles Stone, Robert Jennrich, Potter Chang, and Donald Ylvisaker for their teaching in biostatistics, epidemiology, and mathematical statistics when I was a student at UCLA, as well as Dr. Thomas Ferguson at UCLA and Dr. Daniel McGee at Florida State University for their encouragement and advice in the past. I wish especially to thank Mr. Rob Calver, Editor of Statistics and Mathematics at John Wiley & Sons, for his help and time during the preparation of this book. I also want to thank my wife Jen-Mei, whose patience and understanding have endured throughout so many years and made the work much more pleasant than it otherwise would have been. Finally, I want to express my deepest appreciation to my parents Shung-Wu and Li-Ching for their endless love, support, and guidance, which will live forever in my memory.

Kung-Jong Lui
San Diego, California

1

Population Proportion or Prevalence

To quantify the impact of a given disease on public health in a community, or in studying the variation of a disease distribution between geographical regions to locate the potential causes, we may wish to first estimate the prevalence of the disease, defined as the population proportion of subjects who have it. In this chapter, we start by discussing the estimation of population prevalence under the most commonly assumed case – binomial sampling, in which we take a random sample of n subjects and obtain X cases. For example, to estimate the prevalence of HIV-infected subjects, we may take a random sample of ($n =$) 1000 subjects in a local community and obtain ($x =$) 5 subjects with positive results from an HIV-antibody test. In practice, however, a complete list of the sampling population needed to employ binomial sampling may not be available. We therefore discuss estimation under cluster sampling, in which the sampled unit is the cluster itself rather than the individual subject. As an example, we take a random sample of households and estimate the proportion of people who went to see a doctor in the last 12 months (Cochran, 1977). In this case, the sampled units are households rather than individuals. Other examples of the use of cluster sampling include the study of the effect of an educational intervention program on the use of solar protection among children (Mayer *et al.*, 1997) and the effect of vitamin A supplementation on child mortality (Herrera *et al.*, 1992). As noted by Cochran (1977), the estimate of the population prevalence can be subject to a large relative error when the underlying population prevalence is small under binomial sampling. Furthermore, when the disease is rare, we may even obtain 0 cases in the sample. To alleviate these concerns, we discuss the use of inverse sampling (Haldane, 1945), in which we continue sampling subjects until we obtain a predetermined number x of cases. For example, we may decide to sample subjects until we obtain, say, 5 HIV-infected cases when estimating the prevalence of HIV-infected subjects in a community. In contrast to binomial sampling, the number of cases x under inverse sampling is fixed, but the total number of sampled subjects N needed to obtain these x cases is random. Except

Statistical Estimation of Epidemiological Risk K-J. Lui
 ISBN: 0-470-85071-X (HB)

for specifically referring to the incidence rate, calculated as the number of events divided by the number of person-years of follow-up time, we will generally use the terms probability, proportion, risk, and rate synonymously in this book (Fleiss, 1981). An excellent discussion on explicit definitions of these terms as used in epidemiology appears elsewhere (Selvin, 1996).

1.1 BINOMIAL SAMPLING

Suppose that a random sample of size n is taken from a very large population so that we can reasonably assume that the probability of a randomly selected subject being a case equals a constant π and the events for each randomly selected subject of being a case or a non-case are all mutually independent. Let X denote the random number of cases among these n sampled subjects. The random variable X then follows the binomial distribution with parameters n and π:

$$P(X = x|\pi) = \binom{n}{x} \pi^x (1-\pi)^{n-x}, \tag{1.1}$$

where $x = 0, 1, \ldots, n$, $0 < \pi < 1$, and π denotes the underlying population proportion of cases. The most commonly used point estimator of the parameter π is simply the sample proportion of cases:

$$\hat{\pi} = X/n. \tag{1.2}$$

Note that under distribution (1.1), the point estimator $\hat{\pi}$ (1.2) has the expectation $E(\hat{\pi}) = \pi$ (i.e., $\hat{\pi}$ is an unbiased estimator of the population proportion π) and the variance $\text{Var}(\hat{\pi}) = \pi(1-\pi)/n$ (**Exercise 1.1**). In fact, the estimator $\hat{\pi}$ is the uniformly minimum variance unbiased estimator (UMVUE) of π under (1.1). By the central limit theorem, the random quantity $(\hat{\pi} - \pi)/\sqrt{\text{Var}(\hat{\pi})}$ has the asymptotic standard normal distribution as $n \to \infty$. Thus, by Slutsky's theorem (Casella and Berger, 1990), we obtain an asymptotic $100(1-\alpha)$ percent confidence interval for π using Wald's statistic (Agresti and Coull, 1998),

$$[\max\{\hat{\pi} - Z_{\alpha/2}\sqrt{\hat{\pi}(1-\hat{\pi})/n}, 0\}, \quad \min\{\hat{\pi} + Z_{\alpha/2}\sqrt{\hat{\pi}(1-\hat{\pi})/n}, 1\}]. \tag{1.3}$$

Note that when $\hat{\pi} = 0$ or $\hat{\pi} = 1$, the estimated variance $\hat{\pi}(1-\hat{\pi})/n$ equals 0. Obviously, this underestimates the true variance. Therefore, whenever $\hat{\pi} = 0$ or $\hat{\pi} = 1$, we recommend use of $\hat{\pi}^*(1-\hat{\pi}^*)/n$ to estimate the variance, where $\hat{\pi}^* = (X + 0.5)/(n + 1)$. Note also that although interval estimator (1.3) is easy to use, it is well known that when n is not so large that both $n\hat{\pi} \geq 5$ and $n(1-\hat{\pi}) \geq 5$ hold, (1.3) is not expected to perform well due to the possibly skewed sampling distribution of $\hat{\pi}$. To improve the performance of (1.3), we consider the probability $P([(\hat{\pi} - \pi)/\sqrt{\text{Var}(\hat{\pi})}]^2 \leq Z^2_{\alpha/2}) \doteq 1 - \alpha$ as n is large. This leads us to obtain the following quadratic equation (Wilson, 1927; Fleiss, 1981; Casella and Berger, 1990; Newcombe, 1998):

$$A\pi^2 - 2B\pi + C \leq 0, \quad (1.4)$$

where $A = 1 + Z^2_{\alpha/2}/n$, $B = \hat{\pi} + Z^2_{\alpha/2}/(2n)$, and $C = \hat{\pi}^2$. Because $A > 0$, (1.4) is always convex. Furthermore, we can show that $B^2 - AC > 0$ (**Exercise 1.2**) and hence the two distinct roots of $A\pi^2 - 2B\pi + C = 0$ always exist. Thus, an asymptotic $100(1 - \alpha)$ percent confidence interval, which can also be derived from the score test (Wilson, 1927; Agresti and Coull, 1998; Casella and Berger, 1990; Newcombe, 1998; see also the Appendix), is given by

$$[(B - \sqrt{B^2 - AC})/A, \quad (B + \sqrt{B^2 - AC})/A]. \quad (1.5)$$

Note that an asymptotic confidence interval similar to (1.5) but with a continuity correction can be found elsewhere (Fleiss, 1981; Newcombe, 1998). Using a continuity correction can always increase the coverage probability through an increase in the length of the resulting interval estimate, but may produce a conservative confidence interval (Agresti and Coull, 1998). Note also that although interval estimator (1.5) generally outperforms (1.3), both of these confidence intervals are derived from large-sample theory. When n is small, for $X = x > 0$, we may consider using the confidence interval derived on the basis of the exact distribution (1.1) (Casella and Berger, 1990; Clopper and Pearson, 1934; Jowett, 1963):

$$[x/\{x + (n - x + 1)F_{2(n-x+1),2x,\alpha/2}\},$$
$$\{(x + 1)F_{2(x+1),2(n-x),\alpha/2}\}/\{(n - x) + (x + 1)F_{2(x+1),2(n-x),\alpha/2}\}], \quad (1.6)$$

where $F_{f_1,f_2,\alpha}$ is the upper 100αth percentile of the central F distribution with f_1 and f_2 degrees of freedom. If $x = 0$, then we would define the lower limit of (1.6) to be 0. Similarly, if $x = n$, then we would define the upper limit of (1.6) to be 1. Applying interval estimator (1.6) can always guarantee the coverage probability to be larger than or equal to the desired confidence level $100(1 - \alpha)$ percent for any positive integer n. Details of the derivation of confidence limits (1.6) are given in **Exercises 1.3** and **1.4**. However, it is well known that (1.6) is likely to be conservative, especially when n is not large. Blyth and Still (1983) propose another exact binomial confidence interval that satisfies a few desirable statistical properties. To facilitate the use of their interval estimator, Blyth and Still (1983) tabulate the 95% and 99% confidence limits for $n \leq 30$. They note that in some cases the interval estimate they propose can actually be contained in the resulting estimate using (1.6). Vollset (1993), Agresti and Coull (1998), and Newcombe (1998) all provide good systematic discussions comparing the performance of different interval estimators for a binomial proportion. Other closed-form interval estimators using transformations of $\hat{\pi}$ appear in **Exercises 1.5** and **1.6**.

Example 1.1 We are interested in estimating the prevalence π of subjects with hypertension in a city. Suppose that a random sample of size 200 is taken, and 35 of these 200 sampled subjects are identified to be cases. Given these data, the point estimate $\hat{\pi}$ (1.2) of the hypertension prevalence π is 0.175. The interval estimators (1.3), (1.5), and (1.6) give 95% confidence intervals for π of [0.122, 0.228], [0.129, 0.234], and [0.125, 0.235], respectively. Because both the estimates $n\hat{\pi}$ and $n(1-\hat{\pi})$ are reasonably large (at least 5), these resulting interval estimates are similar to one another; they are all appropriate for use.

Example 1.2 In a pilot study of a rare disease, suppose that we obtain only a single case with exposure to a risk factor of interest out of a random sample of 10 cases. We are interested in estimating the exposure prevalence π in the case population. Employing (1.3), (1.5), and (1.6), the corresponding 95% confidence intervals for π are [0, 0.286], [0.018, 0.404], and [0.003, 0.445]. Note that the interval estimate using (1.3) tends to shift to the left as compared with the those using (1.5) and (1.6) and therefore may not appropriate for use in this situation. It may come as no surprise that the interval estimate obtained using (1.6) is the longest of the three. This is because the coverage probability of (1.6) can be larger than the desired confidence level when n is small.

1.2 CLUSTER SAMPLING

Because of the practical difficulty of obtaining a complete list of subjects in a population, it will often be convenient to employ cluster sampling to collect data. In fact, in many circumstances clustering is unavoidable; it may even occur by study design. For example, in a study concerned with an educational intervention program on behavior change, the data are grouped into small classes (Mayer *et al.*, 1997; Lui *et al.*, 2000) and hence it is natural to treat the classes as the sampled units. When any two subjects are randomly selected from the same class, the events that these two subjects have the outcome of interest are likely to be positively correlated. Thus, the interval estimators (1.3), (1.5), and (1.6) of π, in which the intraclass correlation is not taken into account, will tend to overestimate the precision of the resulting estimate, so that the actual coverage probability of these estimators under cluster sampling will likely be less than the desired confidence level. The results presented in this section can also be useful in the situation where the measurement of the underlying response on subjects is unreliable or the cost of obtaining a new subject is much higher than obtaining a measurement from someone who is already a sampled subject (Lui, 1991). In this case, we may consider taking more than one measurement per subject to increase the efficiency or reduce the expense of a study. The number of repeated measurements taken from each subject then forms a cluster.

Suppose that a random sample of n clusters with varying cluster size $m_i (i = 1, 2, \ldots, n)$ is taken. Define $X_{ij} = 1$ if the jth $(j = 1, 2, \ldots, m_i)$ subject in the

ith cluster is a case, and $X_{ij} = 0$ otherwise. Let p_i denote the probability that a randomly selected subject from cluster i is a case; that is $P(X_{ij} = 1) = p_i$ and $P(X_{ij} = 0) = 1 - p_i$, where $0 < p_i < 1$. To account for the intraclass correlation between the outcomes of subjects within clusters, we assume that the p_i independently and identically follow a beta distribution beta(α, β) with mean $\pi = \alpha/T$ and variance $\pi(1 - \pi)/(T + 1)$, where $T = \alpha + \beta$, because this family is rich in shapes and is commonly used to model Bernoulli data (Johnson and Kotz, 1969). On the basis of the above model assumptions, we can easily show that the intraclass correlation between the outcomes X_{ij} and $X_{ij'}$, $j \neq j'$, within cluster i is $\rho = 1/(T + 1)$, which is always positive (**Exercise 1.7**). We can further show that the probability of a randomly selected subject being a case under the above model assumption is simply $E(X_{ij}) = E(E(X_{ij}|p_i)) = E(p_i) = \pi$.

Given p_i fixed, the conditional distribution of $X_{i.} = \sum_j X_{ij}$ follows the binomial distribution with m_i and p_i. Define

$$\hat{\pi} = \sum_i X_{i.}/m_{.}, \tag{1.7}$$

where $m_{.} = \sum_i m_i$ is the total number of sampled subjects. Note that $\hat{\pi}$ is simply the sample proportion of subjects who are the cases. We can easily show that $\hat{\pi}$ is an unbiased estimator of π under cluster sampling as well. Furthermore, we can show that the variance $\mathrm{Var}(\hat{\pi})$ (**Exercise 1.8**) is equal to

$$\mathrm{Var}(\hat{\pi}) = \pi(1 - \pi)f(\mathbf{m}, \rho)/m_{.}, \tag{1.8}$$

where $\mathbf{m}' = (m_1, m_2, \ldots, m_n)$ and $f(\mathbf{m}, \rho)$ is the variance inflation factor due to the intraclass correlation ρ and equals $\sum_i m_i[1 + (m_i - 1)\rho]/m_{.}$, which is always greater than or equal to 1. The larger the value of ρ, the larger is the value of $f(\mathbf{m}, \rho)$. When the intraclass correlation ρ between the outcomes of all subjects within clusters equals 0, $f(\mathbf{m}, 0) = 1$ and hence the variance $\mathrm{Var}(\hat{\pi})$ reduces to $\pi(1 - \pi)/m_{.}$. On the other hand, when ρ equals 1, $f(\mathbf{m}, \rho)$ reaches the maximum $\sum m_i^2/m_{.}$. For a given total number of subjects $m_{.}$, using equal cluster size m_i will minimize the inflation factor $f(\mathbf{m}, \rho)$. To estimate ρ, we can apply the traditional intraclass correlation estimator (Fleiss, 1986; Lui *et al.*, 1996; Elston, 1977; Yamamoto and Yanagimoto, 1992)

$$\hat{\rho} = (\mathrm{BMS} - \mathrm{WMS})/[\mathrm{BMS} + (m^* - 1)\mathrm{WMS}],$$

where

$$\mathrm{BMS} = \left[\sum_i (X_{i.}^2/m_i) - \left(\sum_i X_{i.}\right)^2 /m_{.}\right] \Big/ (n - 1) \text{ and}$$

$$\mathrm{WMS} = \left[\sum_i X_{i.} - \sum_i (X_{i.}^2/m_i)\right] \Big/ \left[\sum_i (m_i - 1)\right]$$

are the between mean-squared and within mean-squared errors, respectively, and

$$m^* = \left[\left(\sum_i m_i\right)^2 - \sum_i m_i^2\right] \Big/ \left[(n-1)\sum_i m_i\right].$$

Note that under the common correlation model (Mak, 1988), the variance formula (1.8) and the traditional intraclass correlation $\hat{\rho}$ as given above are still valid (**Exercise 1.9**).

On the basis of the above results, an asymptotic $100(1-\alpha)$ percent confidence interval for π is

$$\begin{aligned}&[\max\{\hat{\pi} - Z_{\alpha/2}\sqrt{\hat{\pi}(1-\hat{\pi})f(\mathbf{m}, \hat{\rho})/m_{.}}, 0\},\\ &\min\{\hat{\pi} + Z_{\alpha/2}\sqrt{\hat{\pi}(1-\hat{\pi})f(\mathbf{m}, \hat{\rho})/m_{.}}, 1\}]. \end{aligned} \quad (1.9)$$

Note that when the cluster size $m_i = 1$ for all i, interval estimator (1.9) reduces to (1.3). Thus, when the number of subjects $m_.$ is small, (1.9), although simple to use, is unlikely to perform well. Following ideas similar to those for deriving interval estimator (1.5), we consider the following quadratic equation in π:

$$\mathcal{A}\pi^2 - 2\mathcal{B}\pi + \mathcal{C} \leq 0 \quad (1.10)$$

where $\mathcal{A} = [1 + Z^2_{\alpha/2} f(\mathbf{m}, \hat{\rho})/m_.]$, $\mathcal{B} = [\hat{\pi} + Z^2_{\alpha/2} f(\mathbf{m}, \hat{\rho})/(2\,m_.)]$, and $\mathcal{C} = \hat{\pi}^2$. Because $\mathcal{A} > 0$, (1.10) is always convex. Furthermore, we can show that $\mathcal{B}^2 - \mathcal{AC} > 0$ and hence an asymptotic $100(1-\alpha)$ percent confidence interval for π is given by

$$[(\mathcal{B} - \sqrt{\mathcal{B}^2 - \mathcal{AC}})/\mathcal{A},\ (\mathcal{B} + \sqrt{\mathcal{B}^2 - \mathcal{AC}})/\mathcal{A}]. \quad (1.11)$$

When $m_i = 1$ for all i, as expected, interval estimator (1.11) reduces to (1.5).

In an effort to improve the performance of (1.9), we consider use of the logarithmic transformation to improve the normal approximation of $\hat{\pi}$ (**Exercise 1.5**). By the delta method (Agresti, 1990; Casella and Berger, 1990; see also the Appendix), we can show that the asymptotic variance of $\log(\hat{\pi})$ is $\text{Var}(\hat{\pi}) = (1-\pi)f(\mathbf{m}, \rho)/(m_.\pi)$. Therefore, we obtain an asymptotic $100(1-\alpha)$ percent confidence interval for π to be

$$[\hat{\pi}\exp(-Z_{\alpha/2}\sqrt{(1-\hat{\pi})f(\mathbf{m}, \hat{\rho})/(m_.\hat{\pi})}),\ \hat{\pi}\exp(Z_{\alpha/2}\sqrt{(1-\hat{\pi})f(\mathbf{m}, \hat{\rho})/(m_.\hat{\pi})})]. \quad (1.12)$$

Note that when $\hat{\pi} = 0$, $\log(\hat{\pi})$ is not defined, and when $\hat{\pi} = 1$, the estimated variance of $\log(\hat{\pi})$ is 0. In these cases, we may apply a commonly used *ad hoc* adjustment procedure for sparse data by substituting $\left(\sum X_{i.} + 0.5\right)/(m_. + 1)$ for $\hat{\pi}$ in (1.12).

Example 1.3 Consider the study of an educational intervention program on behavior change with regard to solar protection (Mayer *et al.*, 1997). There are 29 classes with sizes ranging from 1 to 6 in the intervention group and 29 classes with sizes ranging from 1 to 4 in the control groups (Lui *et al.*, 2000). Suppose that we are only interested in estimating the prevalence rate π of children who do not have an adequate level of solar protection in the intervention group. The class size and the corresponding number of children not possessing an adequate level of solar protection in this group are given in Table 1.1. The point estimate $\hat{\pi}$ in the intervention group is 0.422. Applying (1.9), (1.11), and (1.12), we obtain 95% confidence intervals for π of [0.273, 0.570], [0.286, 0.571], [0.297, 0.600], respectively. As seen for the binomial sampling (i.e., $m_i = 1$ for all i), interval estimate (1.9) using Wald's statistic tends to shift to the left as compared with the other two estimates.

Example 1.4 A simple random sample of 30 households of size m_i ranging from 1 to 6 persons is drawn from a census taken in 1947 in wards 5 and 6 of the Eastern Health District of Baltimore (Cochran, 1977, p. 67). For each of these 30 sampled households, we ask how many persons went to see a doctor in the last 12 months. We summarize the data in Table 1.2. Suppose that we want to estimate the proportion π of people who consulted a doctor. From Table 1.2, the point estimate $\hat{\pi}$ is 0.288. The 95% confidence intervals for π obtained from (1.9), (1.11), and (1.12) are [0.148, 0.429], [0.177, 0.469], and [0.172, 0.442], respectively. Note that if we employed the ratio estimator discussed elsewhere (Cochran, 1977) to estimate π under cluster sampling, we would obtain a 95% confidence interval for π of [0.147, 0.430], which is almost the same as that obtained using (1.9), but is less preferable to interval estimates using (1.11) or (1.12).

Table 1.1 The class size and (in parentheses) the observed number of children with an inadequate level of solar protection in the intervention and control groups.

Intervention group
3(1), 2(1), 2(1), 5(0), 4(1), 3(2), 1(1), 2(2), 2(2), 2(1), 1(1), 3(2), 1(1), 3(2), 2(2), 2(0), 6(0), 2(0), 4(0), 2(1), 2(2), 2(1), 2(1), 1(1), 1(1), 1(0), 1(0), 1(0), 1(0)
Control group
2(0), 4(0), 3(2), 2(2), 3(0), 4(4), 4(2), 2(1), 2(1), 3(3), 2(2), 2(1), 4(1), 3(3), 2(2), 3(3), 1(1), 1(0), 2(1), 2(2), 2(1), 3(1), 3(2), 4(4), 1(1), 1(1), 1(1), 1(0), 1(0)

Source: Lui *et al.* (2000).

Table 1.2 Household size and (in parentheses) the observed number of people who consulted a doctor in the last 12 months for a random sample of 30 households.

5(5), 6(0), 3(2), 3(3), 2(0), 3(0), 3(0), 3(0), 4(0), 4(0), 3(0), 2(0), 7(0), 4(4), 3(1), 5(2), 4(0), 4(0), 3(1), 3(3), 4(2), 3(0), 3(0), 1(0), 2(2), 4(2), 3(0), 4(2), 2(0), 4(1)

Source: Cochran (1977).

1.3 INVERSE SAMPLING

When the underlying disease is rare (i.e., $\pi \doteq 0$), the coefficient of variation $\sqrt{(1-\pi)/(n\pi)}$ for estimator $\hat{\pi}$ (1.2) under binomial sampling (1.1) is large. Furthermore, when π is extremely small, the probability of obtaining 0 cases in our sample under (1.1) is no longer negligible for a small or even moderate sample size n. To alleviate this practical concern, we may apply inverse sampling (Haldane, 1945), in which we continue sampling subjects until we obtain a predetermined number x of cases. Let Y denote the number of non-cases before we obtain exactly x cases. The random variable Y then follows the negative binomial distribution with parameters x and π:

$$P(Y=y|\pi) = \binom{x+y-1}{y} \pi^x (1-\pi)^y, \qquad y = 0, 1, 2, \ldots. \tag{1.13}$$

Under distribution (1.13), we can show that the maximum likelihood estimator (**MLE**) of π is given by

$$\hat{\pi} = x/N, \tag{1.14}$$

where $N = x + Y$. Note that (1.14) is actually a biased estimator of π. The asymptotic variance $\mathrm{Var}(\hat{\pi})$ can be shown to equal $\pi^2(1-\pi)/x$ (**Exercise 1.11**). Thus, the asymptotic coefficient of variation of $\hat{\pi}$ under distribution (1.13) is $\sqrt{(1-\pi)/x}$, which is approximately equal to $\sqrt{1/x}$ as the underlying prevalence rate $\pi \doteq 0$. In contrast to binomial sampling, we can ensure that the relative error is smaller than a given precision by simply increasing the predetermined number x of cases. Furthermore, an asymptotic $100(1-\alpha)$ percent confidence interval for π using Wald's statistic is given by

$$[\max\{\hat{\pi} - Z_{\alpha/2}\sqrt{\hat{\pi}^2(1-\hat{\pi})/x}, 0\},\ \min\{\hat{\pi} + Z_{\alpha/2}\sqrt{\hat{\pi}^2(1-\hat{\pi})/x}, 1\}]. \tag{1.15}$$

As noted before, $\hat{\pi}$ is a biased estimator of π. To alleviate this concern, for $x > 1$ we may consider use of the unbiased estimator

$$\hat{\pi}^{(u)} = (x-1)/(N-1), \tag{1.16}$$

which is, in fact, the UMVUE of π (**Exercise 1.12**). Best (1974) derives a closed-form expression for the variance of this estimator:

$$\mathrm{Var}(\hat{\pi}^{(u)}) = (x-1)(1-\pi)\left[\sum_{k=2}^{x-1}(-\pi/(1-\pi))^k/(x-k) - (-\pi/(1-\pi))^x \log(\pi)\right] - \pi^2. \tag{1.17}$$

As shown in **Exercise 1.13**, an unbiased estimator of $\mathrm{Var}(\hat{\pi}^{(u)})$ (1.17) for $x > 2$ is given by (Finney, 1949)

$$\widehat{\mathrm{Var}}(\hat{\pi}^{(u)}) = \hat{\pi}^{(u)}(1 - \hat{\pi}^{(u)})/(N - 2). \quad (1.18)$$

When $x > 2$, (1.16) and (1.18) lead to an asymptotic $100(1-\alpha)$ percent confidence interval for π given by

$$[\max\{\hat{\pi}^{(u)} - Z_{\alpha/2}\sqrt{\hat{\pi}^{(u)}(1 - \hat{\pi}^{(u)})/(N-2)}, 0\},$$
$$\min\{\hat{\pi}^{(u)} + Z_{\alpha/2}\sqrt{\hat{\pi}^{(u)}(1 - \hat{\pi}^{(u)})/(N-2)}, 1\}]. \quad (1.19)$$

Note that both interval estimators (1.15) and (1.19) may not be appropriate for use when N is not large. When N $(= x + y)$ is small, we may consider using the exact $100(1-\alpha)$ percent confidence interval $[\pi_{\mathrm{l}}^{(e)}, \pi_{\mathrm{u}}^{(e)}]$ on the basis of distribution (1.13), where $\pi_{\mathrm{l}}^{(e)}$ and $\pi_{\mathrm{u}}^{(e)}$ are the solutions of the following two equations: $\sum_{y'=0}^{y} \mathrm{P}(Y = y'|\pi_{\mathrm{l}}^{(e)}) = \alpha/2$ and $\sum_{y'=y}^{\infty} \mathrm{P}(Y = y'|\pi_{\mathrm{u}}^{(e)}) = \alpha/2$ (Casella and Berger, 1990). From **Exercises 1.3** and **1.14**, for $Y = y > 0$ we obtain an exact $100(1-\alpha)$ percent confidence interval for π (**Exercise 1.15**; Casella and Berger, 1990), given by

$$[x/\{x + (y+1)F_{2(y+1),2x,\alpha/2}\}, \quad xF_{2x,2y,\alpha/2}/\{xF_{2x,2y,\alpha/2} + y\}]. \quad (1.20)$$

When $y = 0$, we define the upper limit of (1.20) to be 1 for convenience. Note that the confidence limits proposed by George and Elston (1993) are actually a special case of (1.20) when $x = 1$. Lui (1995) discusses the expected length of (1.20) as a function of x and the relationship between (1.20) and the confidence limits on the expected number of trials in reliability studies previously discussed by Clemans (1959). When the underlying disease is rare (i.e., π is small), Bennett (1981) proposes an approximate $100(1-\alpha)$ percent confidence interval for π on the basis of the χ^2 distribution. Details of this can be found in **Exercise 1.20**.

Example 1.5 Suppose that we employ inverse sampling and collect 100 non-cases before obtaining exactly 20 cases. Applying interval estimators (1.15), (1.19), and (1.20), we obtain 95% confidence intervals for π of [0.100, 0.233], [0.094, 0.226], and [0.105, 0.238]. Given such an adequate number of cases, these resulting interval estimates are all similar to one another.

Example 1.6 Under inverse sampling, suppose that we decide to continue sampling subjects until we obtain exactly 2 cases. Suppose we obtain 10 non-cases in our sample. Applying interval estimators (1.15), (1.19), and (1.20), we obtain 95% confidence intervals for π of [0.000, 0.378], [0.000, 0.269], and [0.021, 0.413]. As compared with the exact 95% confidence interval, interval estimators (1.15) and (1.18), derived from large-sample theory, tend to shift to the left and are probably inadequate for use in this case.

Note that the sum $\sum_i X_i$ of independent random variables X_i, each following the binomial distribution (1.1) with parameters n_i and π, follows the binomial distribution with parameters $\sum n_i$ and π. Furthermore, the sum $\sum_i Y_i$ of independent negative binomial random variables Y_i, each following the negative binomial distribution (1.13) with parameters x_i and π, follows the negative binomial with parameters $\sum_i x_i$ and π (Hoel *et al.*, 1971). Thus, in practice, we can simultaneously send several surveyors to a homogeneous population and ask each surveyor to sample a desired number of subjects n_i under binomial sampling (or continue sampling subjects until he/she has obtained a predetermined number x_i of cases under inverse sampling). We can then combine all these samples into a single database and calculate the confidence limits by simply substituting $\sum_i n_i$ for n and $\sum_i X_i$ for X under the binomial distribution (1.1) (or $\sum_i x_i$ for x and $\sum_i Y_i$ for Y under the negative binomial distribution (1.13)), respectively. All the results derived here can then be employed. Note also that when studying a rare disease in a follow-up study, we often assume that the number of cases follows a Poisson distribution. We present some useful results on estimation of the disease incidence rate under this distribution in **Exercise 1.21**. We will discuss the use of Poisson sampling in much more detail in Chapters 2 and 4.

EXERCISE

1.1. Suppose that the random variable X follows the binomial distribution (1.1) with n and π.
(a) Show that $\mathrm{E}(\hat{\pi}) = \pi$ and $\mathrm{Var}(\hat{\pi}) = \pi(1-\pi)/n$.
(b) Find an unbiased estimator of $\mathrm{Var}(\hat{\pi})$.

1.2. Suppose that the random variable X follows the binomial distribution (1.1) with parameters n and π. Show that the inequality $B^2 - AC > 0$ always holds, where $A = 1 + Z_{\alpha/2}^2/n$, $B = \hat{\pi} + Z_{\alpha/2}^2/(2n)$, $C = \hat{\pi}^2$, and $\hat{\pi} = X/n$, and that the two distinct roots of $A\pi^2 - 2B\pi + C = 0$ always fall between 0 and 1 when $\hat{\pi} > 0$.

1.3. (a) Prove

$$\sum_{k=0}^{x} \binom{n}{k} \pi^k (1-\pi)^{n-k} = (n-x)\binom{n}{x} \int_0^{1-\pi} t^{n-x-1}(1-t)^x \, dt.$$

(Hint: use a similar principle to mathematical induction.)
(b) Show that if F follows the F distribution with p and q degrees of freedom, then $(p/q)F/[1 + (p/q)F]$ follows the beta distribution beta$(p/2, q/2)$.
(c) On the basis of the results in (a) and (b), show that

$$\mathrm{P}(X \le x) = \mathrm{P}\left(F > \frac{(n-x)\pi}{(x+1)(1-\pi)}\right),$$

where X is binomial with parameters n and π, and $F \sim F_{2(x+1),2(n-x)}$, respectively.

1.4. Based on the result in part (c) of **Exercise 1.3**, derive the confidence limits (1.6).

1.5. Using the delta method (Agresti, 1990), show that the asymptotic variance of $\log(\hat{\pi})$ is $(1-\pi)/(n\pi)$ under distribution (1.1) and discuss how to apply this result to derive an asymptotic $100(1-\alpha)$ percent confidence interval for π.

1.6. Using the delta method, show that the asymptotic variance of $2\sin^{-1}\sqrt{\hat{\pi}}$ is $1/n$ under distribution (1.1) and discuss how to apply this result to derive an asymptotic $100(1-\alpha)$ percent confidence interval for π.

1.7. Suppose that the Bernoulli random variable X_{ij} has the probability mass function $P(X_{ij}=1)=p_i$ and $P(X_{ij}=0)=1-p_i$, where p_i follows the beta distribution with mean $E(p_i)=\pi$ and variance $\pi(1-\pi)/(T+1)(i=1,2,..,n, j=1,2,\ldots,m_i)$. Suppose further that, given p_i fixed, X_{ij} and $X_{ij'}$ are conditionally independent for $j\neq j'$. Show that the intraclass correlation between X_{ij} and $X_{ij'}$ (where $j\neq j'$) within cluster i is $\rho=1/(T+1)$.

1.8. Show that the variance of $\hat{\pi}$ $(=\sum_{i=1}^{n} X_{i.}/m_.$, where $X_{i.}=\sum_{j=1}^{m_i} X_{ij}$ and $m_. = \sum_i m_i)$ under the model assumption in **Exercise 1.7** is $\text{Var}(\hat{\pi}) = \pi(1-\pi)f(\mathbf{m},\rho)/m_.$, where $\mathbf{m}'=(m_1,m_2,\ldots,m_n)$ and $f(\mathbf{m},\rho)=\sum_i m_i[1+(m_i-1)\rho]/m_.$.

1.9. Under the common correlation model, we assume that the joint probabilities of any two different dichotomous responses X_{ij} and $X_{ij'}$ within a given cluster i are defined as follows:

$$P(X_{ij}=1, X_{ij'}=1)=\pi^2+\rho\pi(1-\pi),$$

$$P(X_{ij}=0, X_{ij'}=0)=(1-\pi)[(1-\pi)+\rho\pi],$$

$$P(X_{ij}=0, X_{ij'}=1)=P(X_{ij}=1, X_{ij'}=0)=\pi(1-\pi)(1-\rho).$$

(a) Show that the intraclass correlation between X_{ij} and $X_{ij'}$ is equal to ρ.
(b) Show that the variance $\text{Var}(\hat{\pi})$ (where $\hat{\pi}=\sum_{i=1}^{n} X_{i.}/m_.$) is equal to $\pi(1-\pi)f(\mathbf{m},\rho)/m_.$, where $f(\mathbf{m},\rho)=\sum_i m_i[1+(m_i-1)\rho]/m_.$. This is actually the same as that given under the beta-binomial model for $\rho=1/(T+1)$.
(c) Show that the expectation $E(\text{WMS})=\pi(1-\pi)(1-\rho)$ and $E(\text{BMS})=\pi(1-\pi)(1-\rho)+m^*\pi(1-\pi)\rho$, where $m^*=\left[\left(\sum_i m_i\right)^2-\sum_i m_i^2\right]\Big/\left[(n-1)\sum_i m_i\right]$. Thus, we may apply the traditional intraclass correlation estimator $\hat{\rho}=(\text{BMS}-\text{WMS})/[\text{BMS}+(m^*-1)\text{WMS}]$ to estimate ρ.

1.10. Consider the data for the control group in Table 1.1. (a) What is the MLE $\hat{\pi}$ of the prevalence of children with an inadequate level of solar protection in the control group? (b) What are the corresponding 95% confidence intervals for π using (1.9), (1.11), and (1.12)?

1.11. Show that the asymptotic variance $\mathrm{Var}(\hat{\pi})$ under distribution (1.13) is $\pi^2(1-\pi)/x$, where $\hat{\pi}$ is given in (1.14). Thus, the coefficient of variation of $\hat{\pi}$ is $\sqrt{(1-\pi)/x}$.

1.12. Show that $\hat{\pi}^{(u)}(=(x-1)/(N-1))$ is an unbiased estimator of π under distribution (1.13). Note that because $\hat{\pi}^{(u)}$ is a function of complete sufficient statistic, $\hat{\pi}^{(u)}$ is the UMVUE of π (Casella and Berger, 1990).

1.13. Show that for $x > 2$, we have $\mathrm{E}[\hat{\pi}^{(u)}(1-\hat{\pi}^{(u)})/(N-2)] = \mathrm{Var}(\hat{\pi}^{(u)})$, where the expectation is taken with respect to distribution (1.13) (Finney, 1949).

1.14. Show that the cumulative distribution $\sum_{y'=0}^{y} \mathrm{P}(Y = y'|\pi)$, where Y follows the negative binomial distribution (1.13) with parameters x and π, equals $\sum_{x'=0}^{y} \mathrm{P}(X = x'|1-\pi)$, where X has the binomial distribution with parameters $n = x + y$ and $1 - \pi$ (Morris, 1963).

1.15. On the basis of the results found in **Exercises 1.3** and **1.14**, for $y > 0$ derive the $100(1-\alpha)$ percent confidence limits $[\pi_l^{(e)}, \pi_u^{(e)}]$ given in (1.20) where $\pi_l^{(e)} = x/[x + (y+1)F_{2(y+1),2x,\alpha/2}]$, and $\pi_u^{(e)} = xF_{2x,2y,\alpha/2}/[xF_{2x,2y,\alpha/2} + y]$, respectively.

1.16. Consider an experiment consisting of x randomly selected devices (where x is fixed), each subject to a series of independent and identical trials until it fails. Suppose that the failure probability at each trial equals a constant π and all these failures between trials are mutually independent. Discuss how we can apply formula (1.20) to derive a $100(1-\alpha)$ percent confidence interval for the expected number of trials $(1-\pi)/\pi$ before the failure of a given device. (Hint: $f(\pi) = (1-\pi)/\pi$ is a monotonically decreasing function of π.)

1.17. Using the delta method, show that $2\sinh^{-1}(\sqrt{Y/x})$, where Y follows the negative binomial distribution (1.13) has the asymptotic variance $1/x$. Thus, the transformation $2\sinh^{-1}(\sqrt{Y/x})$ can be used to stabilize the variance of the negative binomial random variable Y.

1.18. Suppose that a random sample of size 1000 subjects is taken. Suppose further that we find 5 cases in this sample. What are the 95% confidence intervals for the prevalence of cases using (1.3), (1.5), and (1.6)?

1.19. Let Y_i denote the number of trials before failure for device $i(i = 1, 2, \ldots, 5)$ in **Exercise 1.16**. Suppose that $\sum Y_i = 100$. What is the 95% confidence interval for the expected number of trials $(1-\pi)/\pi$ before failure for a given device.

1.20. When the underlying prevalence rate π is small, show that $2(x+Y)\pi$, where Y follows the negative binomial distribution (1.13), follows approximately the χ^2 distribution with $2x$ degrees of freedom (Bennett, 1981). How can we apply this result to derive an approximate $100(1-\alpha)$ percent confidence interval for π?

1.21. When the underlying disease incidence rate λ is small in cohort studies, the number X of cases is often assumed to follow a Poisson distribution: $\exp(-\lambda n^*)(\lambda n^*)^X/X!$, where n^* is a known total of follow-up time in person-years and $X = 0, 1, 2, \ldots$.
(a) Show that $\hat{\lambda} = X/n^*$ is the MLE and an unbiased estimator of λ with variance λ/n^*.
(b) Show that an asymptotic $100(1-\alpha)$ percent confidence interval for λ is given by $\hat{\lambda} \pm Z_{\alpha/2}\sqrt{\hat{\lambda}/n^*}$.
(c) Show that an asymptotic $100(1-\alpha)$ percent confidence interval for λ can be given by solving the two distinct roots of the following quadratic equation: $A^\dagger\lambda^2 - 2B^\dagger\lambda + C^\dagger = 0$, $A^\dagger = 1$, $B^\dagger = \hat{\lambda} + Z^2_{\alpha/2}/(2n^*)$, and $C^\dagger = \hat{\lambda}^2$.
(d) Using the fact that

$$\sum_{X=0}^{x_0} \exp(-\lambda n^*)(\lambda n^*)^X/X! = P(\chi^2_{2(x_0+1)} > 2n^*\lambda),$$

where $\chi^2_{2(x_0+1)}$ is a chi-squared random variable with $2(x_0+1)$ degrees of freedom (Casella and Berger, 1990), show that an exact $100(1-\alpha)$ percent confidence interval for λ is $[\chi^2_{2x_0,1-\alpha/2}/(2n^*), \chi^2_{2(x_0+1),\alpha/2}/(2n^*)]$, where $\chi^2_{f,\alpha}$ is the upper 100αth percentile of the central χ^2 distribution with f degrees of freedom. Note that if $x_0 = 0$, we define the lower limit to be 0.)

1.22. Suppose that in **Exercise 1.21**, we follow a group of 25 subjects for 2 years and obtain $X = 10$ cases.
(a) What is the MLE $\hat{\lambda}$ of the disease incidence rate λ?
(b) What is the 95% confidence interval for λ using Wald's statistic?
(c) What is the 95% confidence interval for λ using the quadratic equation in **Exercise 1.21**?
(d) What is the exact 95% confidence interval for λ?

REFERENCES

Agresti, A. (1990) *Categorical Data Analysis*. Wiley, New York.
Agresti, A. and Coull, B. A. (1998) Approximate is better than 'exact' for interval estimation of binomial proportions. *American Statistician*, **52**, 119–126.
Bennett, B. M. (1981) On the use of the negative binomial in epidemiology. *Biometrical Journal*, **23**, 69–72.
Best, D. J. (1974) The variance of the inverse binomial estimator. *Biometrika*, **67**, 385–386.
Blyth, C. R. and Still, H. A. (1983) Binomial confidence intervals. *Journal of the American Statistical Association*, **78**, 108–116.
Casella, G. and Berger, R. L. (1990) *Statistical Inference*. Duxbury, Belmont, CA.
Clemans, K. G. (1959) Confidence limits in the case of the geometric distribution. *Biometrika*, **46**, 260–264.
Clopper, C. J. and Pearson, E. S. (1934) The use of confidence or fiducial limits illustrated in the case of the binomial. *Biometrika*, **26**, 404–413.
Cochran, W. G. (1977) *Sampling Techniques*, 3rd edition. Wiley, New York.

Elston, R. C. (1977) Response to query: Estimating 'heritability' of a dichotomous trait. *Biometrics*, **33**, 232–233.
Finney, D. J. (1949) On a method of estimating frequencies. *Biometrika*, **36**, 233–234.
Fleiss, J. L. (1981) *Statistical Methods for Rates and Proportions*. Wiley, New York.
Fleiss, J. L. (1986) *The Design and Analysis of Clinical Experiments*, Wiley, New York.
George, V. T. and Elston R. C. (1993) Confidence limits based on the first occurrence of an event. *Statistics in Medicine*, **12**, 685–690.
Haldane, J. B. S. (1945) On a method of estimating frequencies. *Biometrika*, **33**, 222–225.
Herrera, M. G., Nestel, P., Amin, A. E., Fawzi, W. W., Mohamed, K. A. and Weld, L. (1992) Vitamin A supplementation and child survival. *Lancet*, **340**, 267–271.
Hoel, P. G., Port, S. C. and Stone, C. J. (1971) *Introduction to Probability Theory*. Houghton Mifflin, Boston.
Johnson, N. L. and Kotz, S. (1969) *Distributions in Statistics: Discrete Distributions*. Wiley, New York.
Jowett, G. H. (1963) The relationship between the binomial and *F* distributions. *The Statistician*, **13**, 55–57.
Lui, K. -J. (1991) Sample size for repeated measurements in dichotomous data. *Statistics in Medicine*, **10**, 463–472.
Lui, K. -J. (1995) Confidence limits for the population prevalence rate based on the negative binomial distribution. *Statistics in Medicine*, **14**, 1471–1477.
Lui, K. -J., Cumberland, W. G. and Kuo, L. (1996) An interval estimate for the intraclass correlation in beta-binomial sampling. *Biometrics*, **52**, 412–425.
Lui, K. -J., Mayer, J. A. and Eckhardt, L. (2000) Confidence intervals for the risk ratio under cluster sampling based on the beta-binomial model. *Statistics in Medicine*, **19**, 2933–2942.
Mak, T. K. (1988) Analysing intraclass correlation for dichotomous variables. *Applied Statistics*, **37**, 344–352.
Mayer, J. A., Slymen, D. J., Eckhardt, L., *et al.* (1997) Reducing ultraviolet radiation exposure in children. *Preventive Medicine*, **26**, 516–522.
Morris, K. W. (1963) A note on direct and inverse binomial sampling. *Biometrika*, **50**, 544–545.
Newcombe, R. G. (1998) Two-sided confidence intervals for the single proportion: comparison of seven methods. *Statistics in Medicine*, **17**, 857–872.
Selvin, S. (1996) *Statistical Analysis of Epidemiologic Data*. Oxford University Press, New York.
Vollset, S. E. (1993) Confidence intervals for a binomial proportion. *Statistics in Medicine*, **12**, 809–824.
Wilson, E. B. (1927) Probable inference, the law of succession, and statistical inference. *Journal of the American Statistical Association*, **22**, 209–212.
Yamamoto, E. and Yanagimoto, T. (1992) Moment estimators for the beta-binomial distribution. *Journal of Applied Statistics*, **19**, 273–283.

2

Risk Difference

When we determine the relative order of importance of diseases in terms of public health issues, it is appealing to use an index that can reflect the magnitude of the excess mortality attributed to each disease. This leads us to consider the use of the mortality risk difference (RD) between the two groups determined by the presence or the absence of the disease. In fact, Berkson (1958) strongly criticizes the use of the ratio of two rates as a measure of association by noting that the level of rate can be lost when a ratio is used. For example, a tenfold increase with respect to a rate of one per million is the same as a tenfold increase with respect to a rate of one per thousand, even though the latter increase might have far more serious implications than the former, especially if the rates are deaths from some disease (Fleiss, 1981). Therefore, Berkson contends that the simple difference between two rates should be used as the index to measure the association. Similarly, when allocating limited resources to reduce the occurrence of a disease, we may want to search for risk factors which, if eliminated, can result in the largest possible reduction in the underlying disease rate. To help investigators locate these risk factors, we may compare the morbidity RDs among the potential risk factors under investigation. When we calculate the RD from a cohort of subjects who are originally free of the disease, RD represents the incidence RD. When we calculate the RD from a cross-sectional study, in which we simultaneously determine the status of each sampled subject with respect to the exposure and outcome, RD represents the prevalence RD. Because the methods of estimation in the following discussion are generally applicable to both of these RDs, we use RD for brevity unless there is a need to distinguish between them. By definition, the range of RD between two population proportions is $-1 < \text{RD} < 1$. When there is no association between the risk factor and the disease, $\text{RD} = 0$.

We first discuss estimation under the simplest case – independent binomial sampling. We then extend this to accommodate the situation in which we employ pre-stratified sampling in multicenter studies or meta-analysis. Also, because clustered data commonly occur in health-related and epidemiological studies, we discuss estimation of RD while accounting for the intraclass correlation under independent beta-binomial sampling. Furthermore, we may

Statistical Estimation of Epidemiological Risk K-J. Lui
 ISBN: 0-470-85071-X (HB)

often encounter paired-sample data in surveys (Agresti, 1990; Fleiss, 1981), we consider estimation of RD in this situation as well. When the underlying disease is rare, to ensure that we can obtain an appropriate number of cases we include a brief discussion on estimation of RD under independent negative binomial (or inverse) sampling. Finally, when assessing the effect of a risk factor on the incidence rate of a rare chronic disease in cohort studies, to account for the possible difference in the follow-up time in person-years between two comparison groups, we discuss estimation of the incidence RD under Poisson sampling.

2.1 INDEPENDENT BINOMIAL SAMPLING

Suppose that we independently sample n_i subjects from a population with exposure ($i = 1$) and a population with non-exposure ($i = 0$) to a risk factor. Suppose further that we obtain X_i cases from group i. The random variable X_i then follows the binomial distribution (1.1) with parameters n_i and π_i, where π_i denotes the probability that a randomly selected subject from population i is a case. The RD is defined as $\Delta = \pi_1 - \pi_0$. Because $\mathrm{E}(\hat{\pi}_i) = \pi_i$ under (1.1), where $\hat{\pi}_i = X_i/n_i$, the simple difference between the two sample proportions,

$$\hat{\Delta} = \hat{\pi}_1 - \hat{\pi}_0, \tag{2.1}$$

is an unbiased estimator of RD (**Exercise 2.1**). Furthermore, the variance of $\hat{\Delta}$ is equal to

$$\mathrm{Var}(\hat{\Delta}) = \pi_1(1-\pi_1)/n_1 + \pi_0(1-\pi_0)/n_0. \tag{2.2}$$

By the central limit theorem, we may claim that $(\hat{\Delta} - \Delta)/\sqrt{\mathrm{Var}(\hat{\Delta})}$ has the asymptotic standard normal distribution as $n \to \infty$. Thus, by Slutsky's theorem (Casella and Berger, 1990), an asymptotic $100(1-\alpha)$ percent confidence interval for Δ using Wald's statistic is given by

$$\begin{aligned}&[\max\{\hat{\Delta} - Z_{\alpha/2}\sqrt{\hat{\pi}_1(1-\hat{\pi}_1)/n_1 + \hat{\pi}_0(1-\hat{\pi}_0)/n_0}, -1\},\\ &\min\{\hat{\Delta} + Z_{\alpha/2}\sqrt{\hat{\pi}_1(1-\hat{\pi}_1)/n_1 + \hat{\pi}_0(1-\hat{\pi}_0)/n_0}, 1\}],\end{aligned} \tag{2.3}$$

where Z_α is the upper 100αth percentile of the standard normal distribution. Note that $\hat{\pi}_i(1-\hat{\pi}_i)/(n_i-1)$ is an unbiased estimator of $\pi_i(1-\pi_i)/n_i$ (**Exercise 2.2**), and substituting $\hat{\pi}_i(1-\hat{\pi}_i)/(n_i-1)$ for $\hat{\pi}_i(1-\hat{\pi}_i)/n_i$ in (2.3) may improve its performance when n_i is not large. Note also that a continuity correction is commonly applied to (2.3) by subtracting $(1/n_1 + 1/n_0)/2$ from the lower limit and adding the same quantity to the upper limit of $\hat{\Delta} \mp Z_{\alpha/2}\sqrt{\hat{\pi}_1(1-\hat{\pi}_1)/n_1 + \hat{\pi}_0(1-\hat{\pi}_0)/n_0}$, subject to Δ remaining strictly between -1 and 1 (Fleiss, 1981). When both n_i are large, such continuity correction is unnecessary as it has little or no effect on the interval estimates. Readers who are interested in use of the continuity correction may refer to Hauck and Anderson

(1986), who use simulation methods to provide a systematic evaluation of various continuity correction schemes.

Recall that the probability $P([(\hat{\Delta} - \Delta)/\sqrt{\mathrm{Var}(\hat{\Delta})}]^2 \leq Z^2_{\alpha/2}) \doteq 1 - \alpha$ as both n_i are large. Define $\mathcal{T} = \pi_1 + \pi_0$. Then $\pi_1 = (\mathcal{T} + \Delta)/2$ and $\pi_0 = (\mathcal{T} - \Delta)/2$. Thus, we can express the variance $\mathrm{Var}(\hat{\Delta})\left(= \sum_{i=0}^{1} \pi_i(1 - \pi_i)/n_i\right)$ in terms of parameters $\mathcal{T}$ and Δ. These results lead us to consider the following quadratic equation in Δ (**Exercise 2.3**):

$$A\Delta^2 - 2B\Delta + C \leq 0, \tag{2.4}$$

where

$$A = 1 + (1/n_1 + 1/n_0)Z^2_{\alpha/2}/4,$$

$$B = \hat{\pi}_1 - \hat{\pi}_0 + (1 - \mathcal{T})(1/n_1 - 1/n_0)Z^2_{\alpha/2}/4,$$

$$C = (\hat{\pi}_1 - \hat{\pi}_0)^2 - \mathcal{T}(2 - \mathcal{T})(1/n_1 + 1/n_0)Z^2_{\alpha/2}/4.$$

Because $A > 0$, (2.4) is always convex. If $B^2 - AC > 0$, then an asymptotic $100(1 - \alpha)$ percent confidence interval for Δ for a given $\mathcal{T}$ would be $[\Delta_l(\mathcal{T}), \Delta_u(\mathcal{T})]$, where $\Delta_l(\mathcal{T}) = \max\{(B - \sqrt{B^2 - AC})/A, -1\}$ and $\Delta_u(\mathcal{T}) = \min\{(B + \sqrt{B^2 - AC})/A, 1\}$ are the two distinct roots such that the equality in (2.4) holds, subject to Δ lying in the range $(-1, 1)$. But since the nuisance parameter $\mathcal{T}$ is often unknown in practice, the interval estimator $[\Delta_l(\mathcal{T}), \Delta_u(\mathcal{T})]$ cannot be used. In the following discussion, we consider a few methods for estimating $\mathcal{T}$, each of which produces a slightly different interval estimator for Δ.

First, we consider the simplest unbiased estimator $\hat{\mathcal{T}}_1 = \hat{\pi}_1 + \hat{\pi}_0$ of $\mathcal{T}$. Hence, we obtain an asymptotic $100(1 - \alpha)$ percent confidence interval for Δ from (2.4):

$$[\Delta_l(\hat{\mathcal{T}}_1), \Delta_u(\hat{\mathcal{T}}_1)]. \tag{2.5}$$

Note that if π_i were extremely small (or extremely close to 1), the probability that the resulting estimate $\widehat{\mathrm{Var}}(\hat{\pi}_i)(= \pi_i(1 - \pi_i)/n_i)$ is 0 would be non-negligible. This obviously underestimates the variance $\mathrm{Var}(\hat{\pi}_i)$, and hence estimators (2.3) and (2.5) are not likely to perform well in this case. Thus, in application of (2.3), whenever either of $\hat{\pi}_0$ or $\hat{\pi}_1$ equals 0 or 1, we recommend use of the *ad hoc* adjustment procedure of adding 0.5 for sparse data and applying $\hat{\pi}_i^*(1 - \hat{\pi}_i^*)/n_i$ to estimate $\mathrm{Var}(\hat{\pi}_i)$, where $\hat{\pi}_i^* = (X_i + 0.5)/(n_i + 1)$. Similarly, when either of $\hat{\pi}_0$ or $\hat{\pi}_1$ equals 0 or 1, we may use $\hat{\mathcal{T}}_2 = \hat{\pi}_1^* + \hat{\pi}_0^*$ to estimate $\mathcal{T}$ in application of (2.5). In fact, Beal (1987) also notes that interval estimator (2.5) without use of the above adjustment procedure may occasionally result in too short an interval estimate. Beal (1987) proposes use of $\hat{\mathcal{T}}_2$ to estimate $\mathcal{T}$ regardless of whether $\hat{\pi}_i = 0$ or 1. Thus, our asymptotic $100(1 - \alpha)$ percent confidence interval for Δ is

$$[\Delta_l(\hat{\mathcal{T}}_2), \Delta_u(\hat{\mathcal{T}}_2)]. \tag{2.6}$$

Note that interval estimator (2.6), in which the parameter T is estimated by $\hat{T}_2$, is slightly different from (2.5), in which T is estimated by $\hat{T}_1$ when neither $\hat{\pi}_0$ nor $\hat{\pi}_1$ equals 0 or 1, and by $\hat{T}_2$ when either $\hat{\pi}_0$ or $\hat{\pi}_1$ equals 0 or 1. Note also that Beal (1987) refers to (2.5) and (2.6) as Haldane and Jeffreys–Perks intervals, respectively.

The derivation of interval estimators (2.5) and (2.6) does not incorporate the condition $\pi_1 - \pi_0 = \Delta$ into estimation of T. Conditional upon $\pi_1 - \pi_0 = \Delta$, Mee (1984) proposes an estimator of T that requires use of a doubly iterative procedure. To simplify Mee's calculation procedure, conditional upon $\Delta = \pi_1 - \pi_0$, Wallenstein (1997) proposes the weighted least-squares (WLS) estimators $\hat{\pi}_{1|\Delta} = \bar{\pi} + \Delta n_0/n_.$ and $\hat{\pi}_{0|\Delta} = \bar{\pi} - \Delta n_1/n_.$, where $\bar{\pi} = (n_1\hat{\pi}_1 + n_0\hat{\pi}_0)/n_.$ and $n_. = n_1 + n_0$, for estimating π_1 and π_0 (**Exercises 2.4** and **2.5**). However, applying Wallenstein's procedure not only involves a tedious *ad hoc* adjustment procedure, but also loses efficiency with respect to the average length of the confidence interval (Lui, 2001a). Newcombe (1998) provides a systematic discussion comparing the performance of 11 methods, including (2.3), (2.5), and (2.6). Newcombe also notes that interval estimators (2.5) and (2.6) are generally preferable to (2.3). While referring readers to this paper for details, we note that Newcombe recommends the following asymptotic $100(1-\alpha)$ percent confidence interval for Δ:

$$[\max\{\hat{\Delta} - Z_{\alpha/2}\sqrt{l_1(1-l_1)/n_1 + u_0(1-u_0)/n_0}, -1\},$$
$$\min\{\hat{\Delta} + Z_{\alpha/2}\sqrt{u_1(1-u_1)/n_1 + l_0(1-l_0)/n_0}, 1\}], \qquad (2.7)$$

where l_i and u_i are the smaller and larger roots of $\pi_i : |\hat{\pi}_i - \pi_i| = Z_{\alpha/2}\sqrt{\pi_i(1-\pi_i)/n_i}$. Note that interval estimators (2.3) and (2.5)–(2.7) are all derived on the basis of large-sample theory. Santner and Snell (1980) develop small-sample interval estimators for Δ. Except for the conditional method, however, the approaches proposed by Santner and Snell are quite sophisticated, and we do not present these procedures here to save space. Discussions on interval estimation of RD for small-sample cases are given by Coe and Tamhane (1993) and Santner and Yamagani (1993).

Example 2.1 Consider the study of mortality due to heart disease for two treatments, tolbutamide and placebo, among patients with adult-onset diabetes (Miettinen, 1985, p. 142). The data show that 19 patients died of heart disease among 204 patients taking tolbutamide, while only 5 died out of 205 patients taking the placebo. Given these data, the estimate $\hat{\Delta}$ (2.1) is 0.069. Applying interval estimators (2.3), and (2.5)–(2.7), we obtain 95% confidence intervals for Δ of [0.024, 0.114], [0.023, 0.113], [0.022, 0.114], and [0.023, 0.118], respectively. Because the sample sizes in both groups are reasonably large, all these interval estimates are similar to one another. Since all the lower limits are above 0, we may conclude that the risk of dying from heart disease in the tolbutamide treatment is significantly higher than that in the placebo group at the 5% level.

Example 2.2 Consider the double-blind, placebo-controlled study of topiramate in patients with refractory partial epilepsy (Sharief *et al.*, 1996), in which 47 patients are randomly assigned to either topiramate ($n_1 = 23$) or placebo ($n_0 = 24$) treatment for a 3-week titration and an 8-week stabilization period. We obtain 8 patients in the topiramate group versus 2 patients in the placebo group who have a 50% or greater reduction from baseline in seizure rate during the double-blind phase. On the basis of these data, we estimate $\hat{\Delta}$ (2.1) to be 0.264. Applying interval estimators (2.3), and (2.5)–(2.7), we obtain 95% confidence intervals for Δ of [0.041, 0.488], [0.029, 0.461], [0.025, 0.466], and [0.027, 0.476] respectively. The interval estimates from (2.5)–(2.7) are similar to one another, while that from (2.3) is shifted to the right. Because the number of subjects with 50% reduction is only 2 in the placebo group, use of interval estimator (2.3) in this case can be misleading. Since all the lower limits are above 0, there is significant evidence at the 5% level that use of topiramate tends to increase the proportion of patients with 50% reduction in seizures.

2.2 A SERIES OF INDEPENDENT BINOMIAL SAMPLING PROCEDURES

Suppose that we employ a pre-stratified sampling procedure with S strata. The strata can be formed by centers or hospitals in a multicenter study, or different studies in a meta-analysis. From stratum s ($s = 1, 2, \ldots, S$), we independently sample n_{is} subjects from the exposed ($i = 1$) and non-exposed populations ($i = 0$). Suppose we obtain X_{is} cases among n_{is} subjects. Let π_{is} denote the probability that a randomly selected subject from the ith population in the sth stratum is a case. Define $\Delta_s = \pi_{1s} - \pi_{0s}$ and $\mathcal{T}_s = \pi_{1s} + \pi_{0s}$, so that $\pi_{1s} = (\Delta_s + \mathcal{T}_s)/2$ and $\pi_{0s} = (\mathcal{T}_s - \Delta_s)/2$. Under the above assumptions, the joint probability mass function of the random vector $\mathbf{X}' = (\mathbf{X}'_1, \mathbf{X}'_0)$, where $\mathbf{X}'_i = (X_{i1}, X_{i2}, \ldots, X_{iS})$, is given by

$$f_{\mathbf{X}}(\mathbf{x}|\mathbf{n}, \boldsymbol{\Delta}, \boldsymbol{\mathcal{T}}) = \prod_{s=1}^{S} \binom{n_{1s}}{x_{1s}} \left(\frac{\Delta_s + \mathcal{T}_s}{2}\right)^{x_{1s}} \left(\frac{2 - \Delta_s - \mathcal{T}_s}{2}\right)^{n_{1s}-x_{1s}} \times \binom{n_{0s}}{x_{0s}} \left(\frac{\mathcal{T}_s - \Delta_s}{2}\right)^{x_{0s}} \left(\frac{2 - \mathcal{T}_s + \Delta_s}{2}\right)^{n_{0s}-x_{0s}}, \quad (2.8)$$

where $x_{is} = 0, 1, 2, \ldots, n_{is}$, $\mathbf{n}' = (n_{11}, n_{12}, \ldots, n_{1S}, n_{01}, n_{02}, \ldots, n_{0S})$, $\boldsymbol{\Delta}' = (\Delta_1, \Delta_2, \ldots, \Delta_S)$ and $\boldsymbol{\mathcal{T}}' = (\mathcal{T}_1, \mathcal{T}_2, \ldots, \mathcal{T}_S)$.

2.2.1 Summary interval estimators

Under distribution (2.8), the MLEs of Δ_s and $\mathcal{T}_s$ are simply $\hat{\Delta}_s = \hat{\pi}_{1s} - \hat{\pi}_{0s}$ and $\hat{\mathcal{T}}_s = \hat{\pi}_{1s} + \hat{\pi}_{0s}$, respectively, where $\hat{\pi}_{is} = X_{is}/n_{is}$. As given in (2.2), the variance

$\text{Var}(\hat{\Delta}_s)$ is $\pi_{1s}(1-\pi_{1s})/n_{1s}+\pi_{0s}(1-\pi_{0s})/n_{0s}$. In this subsection, we assume that Δ_s is constant across all strata and we denote this common value by Δ_c. Note that the restricted MLE of Δ_c subject to the condition $\Delta_1=\Delta_2=\ldots=\Delta_S$ under (2.8) is not in closed form and requires the use of an iterative numerical procedure. On the other hand, if we took a reasonably large sample size from each stratum and if the variance $\text{Var}(\hat{\Delta}_s)$ were known, we could consider the WLS estimator of stratum-specific estimators, $\sum_s W_s\hat{\Delta}_s/\sum_s W_s$, where $W_s=1/\text{Var}(\hat{\Delta}_s)$ (**Exercise 2.6**). On the other hand, if $\text{Var}(\hat{\Delta}_s)$ were unknown, we might substitute the unbiased estimator $\widehat{\text{Var}}(\hat{\Delta}_s)=\hat{\pi}_{1s}(1-\hat{\pi}_{1s})/(n_{1s}-1)+\hat{\pi}_{0s}(1-\hat{\pi}_{0s})/(n_{0s}-1)$ for $\text{Var}(\hat{\Delta}_s)$ and obtain the commonly used point estimator

$$\hat{\Delta}_{\text{WLS}}=\sum_s\hat{W}_s\hat{\Delta}_s/\sum_s\hat{W}_s, \tag{2.9}$$

where $\hat{W}_s=1/\widehat{\text{Var}}(\hat{\Delta}_s)$. Furthermore, note that the variance $\text{Var}((\sum_s W_s\hat{\Delta}_s)/(\sum_s W_s))=1/\sum_s W_s$. This leads us to obtain an asymptotic $100(1-\alpha)$ percent confidence interval for Δ_c given by

$$\left[\max\left\{\frac{\sum_s\hat{W}_s\hat{\Delta}_s}{\sum_s\hat{W}_s}-\frac{Z_{\alpha/2}}{\sqrt{\sum_s\hat{W}_s}},-1\right\},\ \min\left\{\frac{\sum_s\hat{W}_s\hat{\Delta}_s}{\sum_s\hat{W}_s}+\frac{Z_{\alpha/2}}{\sqrt{\sum_s\hat{W}_s}},1\right\}\right]. \tag{2.10}$$

To ensure that estimators (2.9) and (2.10) perform well we need to take a reasonably large sample size from each stratum (Greenland and Robins, 1985). When the stratum sizes are not large, Greenland and Robins propose a Mantel–Haenszel type estimator of Δ_c,

$$\hat{\Delta}_{\text{MH}}=\frac{\sum_s(X_{1s}n_{0s}-X_{0s}n_{1s})/n_{.s}}{\sum_s n_{1s}n_{0s}/n_{.s}}, \tag{2.11}$$

which is actually a weighted average of $\hat{\Delta}_s$ with weights proportional to $n_{1s}n_{0s}/n_{.s}$. This estimator is consistent for sparse data. Greenland and Robins further derive an asymptotic variance of $\hat{\Delta}_{\text{MH}}$. However, as noted by Sato (1989), this variance is consistent only for large strata. Sato develops the following estimated asymptotic variance of $\hat{\Delta}_{\text{MH}}$, which is consistent for both sparse data and large strata:

$$\widehat{\text{Var}}(\hat{\Delta}_{\text{MH}})=\frac{\hat{\Delta}_{\text{MH}}\sum_s P_s+\sum_s Q_s}{\left(\sum_s n_{1s}n_{0s}/n_{.s}\right)^2}, \tag{2.12}$$

where

$$\mathcal{P}_s = \frac{n_{1s}^2 X_{0s} - n_{0s}^2 X_{1s} + (n_{1s} n_{0s} (n_{0s} - n_{1s})/2)}{n_{.s}^2}$$

$$\mathcal{O}_s = \frac{X_{1s}(n_{0s} - X_{0s})/n_{.s} + X_{0s}(n_{1s} - X_{1s})/n_{.s}}{2}.$$

On the basis of (2.11) and (2.12), we obtain an asymptotic $100(1-\alpha)$ percent confidence interval for Δ_c:

$$[\max\{\hat{\Delta}_{\text{MH}} - Z_{\alpha/2}\sqrt{\widehat{\text{Var}}(\hat{\Delta}_{\text{MH}})}, -1\}, \quad \min\{\hat{\Delta}_{\text{MH}} + Z_{\alpha/2}\sqrt{\widehat{\text{Var}}(\hat{\Delta}_{\text{MH}})}, 1\}]. \tag{2.13}$$

2.2.2 Test for the homogeneity of risk difference

Before employing a summary estimator as discussed in Section 2.2.1, we may wish to examine whether the assumption that the RD is constant across strata is satisfied by our data. That is, we wish to test $H_0 : \Delta_1 = \Delta_2 = \ldots = \Delta_S$. When all n_{is} are large, we may apply the WLS test statistic (Fleiss, 1981)

$$T_{\text{WLS}} = \sum_{s=1}^{S} \hat{W}_s(\hat{\Delta}_s - \hat{\Delta}_{\text{WLS}})^2. \tag{2.14}$$

As all $n_{is} \to \infty$, we can approximate the sampling distribution of test statistic (2.14) by the chi-squared distribution with $S-1$ degrees of freedom. We reject H_0 at level 2 when $T_{\text{WLS}} > \chi^2_{S-1,\alpha}$, where $\chi^2_{S-1,\alpha}$ is the upper 100αth percentile of the χ^2 distribution with $S-1$ degrees of freedom.

As an alternative to the above chi-squared approximation, when S is also large, Lipsitz *et al.* (1998) proposes the test statistic

$$Z_{\text{WLS}} = [T_{\text{WLS}} - (S-1)]/\sqrt{2(S-1)}, \tag{2.15}$$

which asymptotically follows the standard normal distribution under H_0. Note that when the assumption of RD homogeneity does not hold, the T_{WLS} test statistic (2.14) is expected to be larger than that under H_0. In other words, we should reject H_0 only when T_{WLS} is large. Thus, we should do a one-sided test and reject H_0 at level α when the test value Z_{WLS} (2.15) is larger than Z_α.

To try to improve the normal approximation of a chi-squared distribution, following Fisher (1928), we may consider use of a logarithmic transformation of T_{WLS}. This leads us to consider the test statistic (Lui and Kelly, 2000)

$$Z_{\text{LWLS}} = \{\log(T_{\text{WLS}}/(S-1))/2 + 1/[2(S-1)]\}/\sqrt{1/[2(S-1)]}. \tag{2.16}$$

Again, we reject the null hypothesis H_0 at level α when Z_{LWLS} is larger than Z_α. Note that when $\hat{\pi}_{is} = 0$ or 1 in applications of test statistics (2.14)–(2.16), the estimated variance $\widehat{\text{Var}}(\hat{\Delta}_s)$ is certainly inappropriate, and hence we recommend use of the *ad hoc* procedure for sparse data involving adding 0.50 to each cell of the sth stratum. In fact, on the basis of simulation, Lui and Kelly (2000) note that using this adjustment procedure can substantially improve the performance of test statistic (2.14), especially when the stratum size is moderate. Lui and Kelly (2000) also note that test procedure using (2.16) is generally preferable to that using (2.14) in a variety of situations. However, this is not true for test procedure using (2.15), which often has Type I error much larger than the nominal level α. In an effort to improve the performance of test statistics (2.14)–(2.16), Lipsitz *et al.* (1998) propose three other weighted test statistics as well. However, test procedures using these statistics are generally conservative when the stratum sizes are small (Lui and Kelly, 2000).

Example 2.3 Consider the all-cause mortality data from six trials comparing aspirin ($i = 1$) with placebo ($i = 0$) in post-myocardial infarction patients (Canner, 1987). These include the two trials ($s = 1, 2$) carried out in the United Kingdom (Elwood *et al.*, 1974; Elwood and Sweetnam, 1979), the Coronary Drug Project (1976) aspirin study ($s = 3$), the German–Austrian multicenter study (Breddin *et al.*, 1979) ($s = 4$), the Persantine-Aspirin Reinfarction Study (1980) ($s = 5$), and the Aspirin Myocardial Infarction Study (1980) ($s = 6$). We summarize these data in Table 2.1. Applying test procedures (2.14)–(2.16) to test the homogeneity of the RD over these six trials, we obtain p-values 0.089, 0.075, and 0.090; there is weak evidence against the homogeneity of the RD. Calculating point estimates $\hat{\Delta}_s$ for $s = 1, \ldots, 6$, we obtain $-0.028, -0.026, -0.025, -0.018, -0.023$, and 0.011. Except for the last trial, the resulting estimates $\hat{\Delta}_s$ are similar to one another. In fact, after a careful comparison of these trials, Canner (1987) notes that the baseline imbalance of medical conditions between the aspirin and placebo groups in the sixth trial may partially explain the apparent difference of mortality between that trial and the other five. Applying test procedures (2.14)–(2.16) to test the homogeneity of the RD over the first five trials again, we obtain p-values 0.998, 0.915, and 1.00. This suggests that the assumption that RD_s for the first five trials is constant should be reasonable. Applying (2.9) and (2.11) to the data for these five trials ($s = 1, 2, \ldots, 5$), we obtain -0.025 for both point estimates of Δ. Furthermore, we obtain 95% confidence intervals for Δ_c of $[-0.040, -0.010]$ using both (2.10) and (2.13). Since the upper limit is below 0, there is significant evidence at the

Table 2.1 All-cause mortality data: the number of deaths/the number of patients in the aspirin and placebo groups in ($S =$)6 clinical trials.

$s =$	1	2	3	4	5	6
Aspirin	49/615	102/832	44/758	27/317	85/810	246/2267
Placebo	67/624	126/850	64/771	32/309	52/406	219/2257

5% level to support the hypothesis that taking aspirin may reduce the all-cause mortality as compared with the placebo group.

Example 2.4 Consider the data from a Cancer and Leukemia Group B randomized trial comparing two chemotherapy treatments with respect to survival rate by the end of the study in patients with multiple myeloma (Cooper *et al.*, 1993; Lipsitz *et al.*, 1998, p. 149). We summarize these data in Table 2.2. There are 21 institutions with the number of eligible patients accrued n_{is} ranging from 2 to 12. Applying test statistics (2.14)–(2.16) with the *ad hoc* adjustment procedure for sparse data for testing the homogeneity of the RD, we obtain p-values of 0.634, 0.666, and 0.619. Therefore, we may reasonably assume that the RD of survival rates between these two chemotherapy treatments is constant over these 21 institutions. When using $\hat{\Delta}_{WLS}$ (2.9) and $\hat{\Delta}_{MH}$ (2.11), we obtain summary point estimates 0.024 and 0.057, respectively. Because the number of patients n_{is} in each institution is small, this resulting estimate $\hat{\Delta}_{WLS}$ can be misleading. Applying interval estimator (2.13), we obtain a 95% confidence interval for Δ_c of

Table 2.2 Survival data from a Cancer and Leukemia Group B randomized trial: the number X_{is} of surviving patients and the number n_{is} of patients assigned to each of two chemotherapy treatments over 21 institutions.

Institution no.	Treatment I		Treatment II	
S	n_{1s}	X_{1s}	n_{0s}	X_{0s}
1	4	3	3	1
2	4	3	11	8
3	2	2	3	2
4	2	2	2	2
5	2	2	3	0
6	3	1	3	2
7	2	2	3	2
8	5	1	4	4
9	2	2	3	2
10	2	0	3	2
11	3	3	3	3
12	2	2	2	0
13	4	1	5	1
14	3	2	4	2
15	4	2	6	4
16	12	4	9	3
17	2	1	3	2
18	3	3	4	1
19	4	1	3	2
20	3	0	2	0
21	4	2	5	1

Source: Lipsitz *et al.* (1998).

[−0.099, 0.214]. This result suggests that the RD of survival rates between the two chemotherapy treatments is not significant at the 5% level.

2.3 INDEPENDENT CLUSTER SAMPLING

In practice, it is quite common not to be able to obtain a complete list of subjects for the entire population $i(i = 1, 0)$ and so not to employ binomial sampling to estimate π_i as considered in Section 2.1. Indeed, we may encounter many situations in which subjects are naturally grouped into clusters. For example, suppose that we want to study the relationship between goiter and iodine deficiency and collect the data from two geographical regions. One region is rich in iodine and the other is not. Because it is much easier to obtain a complete list of villages than a complete list of inhabitants in a region, we may wish to apply cluster sampling, in which we first take a random sample of villages from a region and then draw a random sample of individual subjects from each sampled village. Since the outcomes of subjects within villages (or clusters) are likely positively correlated, it is important to account for this intraclass correlation to avoid overestimating the precision of our inference. Similarly, for administrative convenience, cluster randomization trials, in which the unit of randomization is a school, a class, or a household are also widely used for evaluating an intervention for the prevention of disease (Cornfield, 1978; Donner *et al.*, 1981; Klar and Donner, 2001; Lui *et al.*, 2000; Herrera *et al.*, 1992). As an example, consider a study in which spouse pairs are randomly assigned to either a control group or an experimental group receiving a reduced amount of dietary sodium (Donner *et al.*, 1981). The clusters in this study are spouse pairs, in which the responses of husband and wife are likely to be positively correlated.

Suppose that we independently sample n_i clusters of varying size m_{ij} from a population with exposure $(i = 1)$ and a population with non-exposure $(i = 0)$ to a risk factor of interest, respectively. We define the random variable $X_{ijk} = 1$ if the kth $(k = 1, 2, \ldots, m_{ij})$ subject in the jth $(j = 1, \ldots, n_i)$ cluster from the ith population $(i = 1, 0)$ is a case, and $X_{ijk} = 0$ otherwise. We assume that the probability $P(X_{ijk} = 1) = p_{ij}$ and $P(X_{ijk} = 0) = 1 - p_{ij}$, respectively, where $0 < p_{ij} < 1$. Because the outcomes of subjects within clusters are likely correlated, we assume that the p_{ij} independently and identically follow a beta distribution beta(α_i, β_i) with mean $\pi_i = \alpha_i/T_i$ and variance $\pi_i(1 - \pi_i)/(T_i + 1)$, where $T_i = \alpha_i + \beta_i$. As noted by Johnson and Kotz (1970), the beta family is rich and flexible in shapes and includes the uniform distribution over (0, 1) as a special case for $\pi_i = 0.50$ and $T_i = 2$. On the basis of the model assumptions, the intraclass correlation between X_{ijk} and $X_{ijk'}$ for $k \neq k'$ is $\rho_i = 1/(T_i + 1)$ (**Exercise 1.7**). Note that under the model assumptions, the probability of a randomly selected subject being a case in population i is equal to $E(X_{ijk}) = \pi_i$.

Given p_{ij} fixed, the conditional distribution of $X_{ij.} = \sum_k X_{ijk}$ is the binomial distribution with parameters m_{ij} and p_{ij}. Define $\hat{\pi}_i = \sum_j X_{ij.}/m_{i.}$, where $m_{i.} =$

$\sum_j m_{ij}$ is the total number of sampled subjects from population i. The sample proportion of cases $\hat{\pi}_i$ is an unbiased estimator of π_i with variance $\text{Var}(\hat{\pi}_i)$ equal to $\pi_i(1-\pi_i)f(\mathbf{m}_i, \rho_i)/m_{i.}$, where $\mathbf{m}_i' = (m_{i1}, m_{i2}, \ldots, m_{in_i})$, and $f(\mathbf{m}_i, \rho_i)$ is the variance inflation factor due to the intraclass correlation ρ_i and equals $\sum_j m_{ij}[1+(m_{ij}-1)\rho_i]/m_{i.}$ (**Exercise 1.8**). As noted in Chapter 1, we can easily see that $f(\mathbf{m}_i, \rho_i)$ is an increasing function of ρ_i and reaches its maximum when $\rho_i = 1$.

First, note that the estimator

$$\hat{\Delta} = \hat{\pi}_1 - \hat{\pi}_0 \tag{2.17}$$

is an unbiased estimator of $\Delta (= \pi_1 - \pi_0)$ under the beta-binomial distribution with variance equal to

$$\text{Var}(\hat{\Delta}) = \pi_1(1-\pi_1)f(\mathbf{m}_1, \rho_1)/m_{1.} + \pi_0(1-\pi_0)f(\mathbf{m}_0, \rho_0)/m_{0.}. \tag{2.18}$$

We can simply substitute $\hat{\pi}_i$ for π_i and $\hat{\rho}_i$ for ρ_i in (2.18) to obtain the estimated variance $\widehat{\text{Var}}(\hat{\Delta})$, where

$$\hat{\rho}_i = (\text{BMS}_i - \text{WMS}_i)/[\text{BMS}_i + (m_i^* - 1)\text{WMS}_i]; \tag{2.19}$$

here

$$\text{BMS}_i = \left[\sum_j (X_{ij.}^2/m_{ij}) - \left(\sum_j X_{ij.}\right)^2 \Big/ \left(\sum_j m_{ij}\right)\right] \Big/ (n_i - 1) \text{ and}$$

$$\text{WMS}_i = \left[\sum_j X_{ij.} - \sum_j (X_{ij.}^2/m_{ij})\right] \Big/ \left[\sum_j (m_{ij} - 1)\right]$$

are the between mean-squared and within mean-squared errors, respectively, and

$$m_i^* = \left[\left(\sum_j m_{ij}\right)^2 - \sum_j m_{ij}^2\right] \Big/ \left[(n_i - 1)\sum_j m_{ij}\right]$$

(Fleiss, 1986; Elston, 1977; Lui *et al.*, 1996).

Therefore, an asymptotic $100(1-\alpha)$ percent confidence interval for Δ is

$$[\max\{\hat{\Delta} - Z_{\alpha/2}\sqrt{\widehat{\text{Var}}(\hat{\Delta})}, -1\}, \min\{\hat{\Delta} + Z_{\alpha/2}\sqrt{\widehat{\text{Var}}(\hat{\Delta})}, 1\}], \tag{2.20}$$

where $\widehat{\text{Var}}(\hat{\Delta}) = \hat{\pi}_1(1-\hat{\pi}_1)f(\mathbf{m}_1, \hat{\rho}_1)/m_{1.} + \hat{\pi}_0(1-\hat{\pi}_0)f(\mathbf{m}_0, \hat{\rho}_0)/m_{0.}$. Note that since the sampling distribution of $\hat{\Delta}$ can be skewed, interval estimator (2.20) may not perform well, especially when the expected number of cases in either of the two comparison groups is small. To improve the performance of (2.20), we generalize interval estimators (2.5)–(2.7) to accommodate the correlated data under cluster sampling.

First, recall that the probability $P([(\hat{\Delta} - \Delta)/\sqrt{\text{Var}(\hat{\Delta})}]^2 \le Z_{\alpha/2}^2) \doteq 1 - \alpha$ as both $m_{i.}$ are large. Following ideas similar to those for binomial sampling, we consider the following quadratic equation in Δ:

$$A^\dagger\Delta^2 - 2B^\dagger\Delta + C^\dagger \leq 0, \tag{2.21}$$

where

$$A^\dagger = 1 + [f(\mathbf{m}_1, \hat{\rho}_1)/m_{1.} + f(\mathbf{m}_0, \hat{\rho}_0)/m_{0.}]Z^2_{\alpha/2}/4,$$
$$B^\dagger = \hat{\Delta} + (1 - \mathcal{T})[f(\mathbf{m}_1, \hat{\rho}_1)/m_{1.} - f(\mathbf{m}_0, \hat{\rho}_0)/m_{0.}]Z^2_{\alpha/2}/4,$$
$$C^\dagger = \hat{\Delta}^2 - \mathcal{T}(2 - \mathcal{T})[f(\mathbf{m}_1, \hat{\rho}_1)/m_{1.} + f(\mathbf{m}_0, \hat{\rho}_0)/m_{0.}]Z^2_{\alpha/2}/4.$$

Equation (2.21) is always convex, since $A^\dagger > 0$. If $B^{\dagger 2} - A^\dagger C^\dagger > 0$, then an asymptotic $100(1-\alpha)$ percent confidence interval for Δ is given by $[\Delta_l(\mathcal{T}), \Delta_u(\mathcal{T})]$, where $\Delta_l(\mathcal{T}) = \max\{(B^\dagger - \sqrt{B^{\dagger 2} - A^\dagger C^\dagger})/A^\dagger, -1\}$ and $\Delta_u(\mathcal{T}) = \min\{(B^\dagger + \sqrt{B^{\dagger 2} - A^\dagger C^\dagger})/A^\dagger, 1\}$ are the two distinct roots in the interval $(-1, 1)$ such that the equality in (2.21) holds. Following similar arguments to those presented in Section 2.1, we obtain the following three estimators.

First, consider the use of $\hat{\mathcal{T}}_1 = \hat{\pi}_1 + \hat{\pi}_0$ to estimate $\mathcal{T}$. We obtain an asymptotic $100(1-\alpha)$ percent confidence interval for Δ of the form

$$[\Delta_l(\hat{\mathcal{T}}_1), \Delta_u(\hat{\mathcal{T}}_1)]. \tag{2.22}$$

Note that when either of the $\hat{\pi}_i = 0$ or 1, we propose using $\hat{\pi}_i^*(1 - \hat{\pi}_i^*)f(\mathbf{m}_i, \hat{\rho}_i)/m_{i.}$ to estimate $\mathrm{Var}(\hat{\pi}_i)$ in (2.20) and $\hat{\mathcal{T}}_2 = \hat{\pi}_1^* + \hat{\pi}_0^*$ to estimate $\mathcal{T}$ in (2.22), where $\hat{\pi}_i^* = (X_{i..} + 0.5)/(m_{i.} + 1)$.

Following Beal (1987), we may also consider using $\hat{\mathcal{T}}_2$ to estimate $\mathcal{T}$ regardless of whether $\hat{\pi}_i = 0$ or 1. Thus, we obtain an asymptotic $100(1-\alpha)$ percent confidence interval for Δ given by

$$[\Delta_l(\hat{\mathcal{T}}_2), \Delta_u(\hat{\mathcal{T}}_2)]. \tag{2.23}$$

Finally, following Newcombe (1998), we may extend (2.7) to account for the intraclass correlation between responses within clusters. We obtain an asymptotic $100(1-\alpha)$ percent confidence interval for Δ given by

$$\begin{aligned}[&\max\{\hat{\Delta} - Z_{\alpha/2}\sqrt{f(\mathbf{m}_1, \hat{\rho}_1)l_1(1 - l_1)/m_{1.} + f(\mathbf{m}_0, \hat{\rho}_0)u_0(1 - u_0)/m_{0.}}, -1\},\\ &\min\{\hat{\Delta} + Z_{\alpha/2}\sqrt{f(\mathbf{m}_1, \hat{\rho}_1)u_1(1 - \mu_1)/m_{1.} + f(\mathbf{m}_0, \hat{\rho}_0)l_0(1 - l_0)/m_{0.}}, 1\}],\end{aligned} \tag{2.24}$$

where l_i and u_i are the smaller and larger roots of $\pi_i : |\hat{\pi}_i - \pi_i| = Z_{\alpha/2}\sqrt{f(\mathbf{m}_i, \hat{\rho}_i)\pi_i(1 - \pi_i)/m_{i.}}$, for $i = 1, 0$. Note that when $m_{ij} = 1$ for all i and j, all interval estimators (2.20) and (2.22)–(2.24) under cluster sampling reduce to (2.3) and (2.5)–(2.7) under independent binomial sampling. Lui (2001a) evaluates and compares the performance of estimators (2.20), (2.22), and (2.23) in the context of repeated measurements per subject, focusing on situations in which the cluster size is small (e.g., households), and finds that (2.22) and (2.23) generally outperform (2.20).

Example 2.5 Consider the data in Table 1.1 for an educational intervention program with emphasis on behavior change with regard to solar protection (Mayer *et al.*, 1997). We assign 29 classes to each of the intervention and placebo groups. Because classes are randomly assigned to one of these two groups, we may assume that the intraclass correlations of the intervention and placebo groups are equal (i.e., $\rho_1 = \rho_0 = \rho_c$). We use $\hat{\rho}_c = (m_{1.}\hat{\rho}_1 + m_{0.}\hat{\rho}_0)/(m_{1.} + m_{0.})$ to estimate this common intraclass correlation, and obtain $\hat{\rho}_c = 0.30$. Let π_1 and π_0 denote the proportion of children with an inadequate level of solar protection in the intervention and control groups, respectively. Applying (2.20) and (2.22)–(2.24), we obtain 95% confidence intervals for Δ of $[-0.405, 0.013]$, $[-0.392, 0.017]$, $[-0.392, 0.017]$, and $[-0.385, 0.016]$. Because both $\hat{\pi}_1 = 0.422$ and $\hat{\pi}_0 = 0.618$ are not close to the boundary of 0, the interval estimators considered here are all adequate for use (Lui, 2001a).

2.4 PAIRED-SAMPLE DATA

To increase the efficiency of a clinical trial, we may consider matching subjects with respect to some strong nuisance confounders to form paired-sample data. For each given matched pair, we randomly assign one subject to receive the experimental treatment under investigation and the other to receive the standard treatment or placebo. We then want to compare the probability of responses between the two treatments. Similarly, in surveys we may follow a group of sampled subjects and record their responses on two occasions. We then want to investigate whether there is a change in response rates between these two occasions. For example, consider a example (Agresti, 1990, p. 350), in which a random sample of 1600 voting-age Americans is taken. Among these, 944 people originally indicate approval of the President's performance in office, but only 880 people indicate approval a month later. We want to study whether there is a significant change in approval rates between the two surveys. Because the responses between the two surveys are taken from the same group of subjects, the responses on the same subject are likely correlated. Interval estimators of the RD presented for independent binomial sampling are inappropriate for use in this situation. For clarity, we use the following table to summarize the possible outcomes:

		Second survey		
		Yes	No	
First survey	Yes	π_{11}	π_{10}	$\pi_{1.}$
	No	π_{01}	π_{00}	$\pi_{0.}$
		$\pi_{.1}$	$\pi_{.0}$	

where $0 < \pi_{ij} < 1$ ($i = 0, 1$ and $j = 0, 1$) are the cell probabilities, $\pi_{i.} = \pi_{i1} + \pi_{i0}$, and $\pi_{.j} = \pi_{1j} + \pi_{0j}$. We define the RD in paired-sample data as $\Delta = \pi_{1.} - \pi_{.1} = \pi_{10} - \pi_{01}$. The range of Δ is $-1 < \Delta < 1$. Let $\mathcal{T} = \pi_{10} + \pi_{01}$ represent the probability of discordance between responses within a given pair. Thus, we have $\pi_{10} = (\mathcal{T} + \Delta)/2$ and $\pi_{01} = (\mathcal{T} - \Delta)/2$.

Suppose that we have n matched pairs. Let N_{ij} denote the observed frequency of pairs falling in the cell with probability π_{ij}. The random vector $\mathbf{N}' = (N_{11}, N_{10}, N_{01}, N_{00})$ then follows the multinomial distribution with parameter n and $\boldsymbol{\pi}' = (\pi_{11}, \pi_{10}, \pi_{01}, \pi_{00})$:

$$\mathbf{f_N}(\mathbf{n}|n, \boldsymbol{\pi}) = \frac{n!}{n_{11}!n_{10}!n_{01}!n_{00}} \pi_{11}^{n_{11}} \pi_{10}^{n_{10}} \pi_{01}^{n_{01}} \pi_{00}^{n_{00}}, \tag{2.25}$$

where $\mathbf{n} = (n_{11}, n_{10}, n_{01}, n_{00})'$, $N_{ij} \geq 0$, $\sum_i \sum_j N_{ij} = n$, and $\sum_i \sum_j \pi_{ij} = 1$.

Under (2.25), the MLE $\hat{\pi}_{ij}$ of π_{ij} is N_{ij}/n. Define $\hat{\Delta} = \hat{\pi}_{10} - \hat{\pi}_{01}$. Note that the expectation $\mathrm{E}(\hat{\Delta}) = \Delta$. Note further that we can easily show that the variance $\mathrm{Var}(\hat{\Delta}) = [\pi_{10} + \pi_{01} - (\pi_{10} - \pi_{01})^2]/n$ (**Exercise 2.7**). These lead us to obtain an asymptotic $100(1 - \alpha)$ percent confidence interval for Δ of

$$[\max\{-1, \hat{\Delta} - Z_{\alpha/2}\sqrt{\widehat{\mathrm{Var}}(\hat{\Delta})}\}, \quad \min\{1, \hat{\Delta} + Z_{\alpha/2}\sqrt{\widehat{\mathrm{Var}}(\hat{\Delta})}\}]. \tag{2.26}$$

where $\widehat{\mathrm{Var}}(\hat{\Delta}) = [\hat{\pi}_{10} + \hat{\pi}_{01} - (\hat{\pi}_{10} - \hat{\pi}_{01})^2]/n$. Except for the minor adjustment to ensure the resulting confidence limits fall in the range $(-1, 1)$ of Δ, the above interval estimator is, in fact, the most frequently used interval estimator of Δ for paired-sample data and appears in many textbooks (Fleiss, 1981; Agresti, 1990; Dixon and Massey, 1969; Selvin, 1996). Note that interval estimator (2.26) depends on only the marginal cell frequencies (N_{10}, N_{01}), which have marginal probability mass function

$$\begin{aligned} f_{N_{10},N_{01}}(n_{10}, n_{01}|n, \Delta, \mathcal{T}) &= \frac{n!}{n_{10}!n_{01}!(n - n_{10} - n_{01})!} \\ &\times [(\Delta + \mathcal{T})/2]^{n_{10}}[(\mathcal{T} - \Delta)/2]^{n_{01}}(1 - \mathcal{T})^{(n - n_{10} - n_{01})}. \end{aligned} \tag{2.27}$$

Given a sample vector (n_{10}, n_{01}), the log-likelihood of (2.27) is then

$$\begin{aligned} \log\{L(\Delta, \mathcal{T}|n_{10}, n_{01})\} &= C + n_{10}\log(\Delta + \mathcal{T}) + n_{01}\log(\mathcal{T} - \Delta) \\ &+ (n - n_{10} - n_{01})\log(1 - \mathcal{T}), \end{aligned} \tag{2.28}$$

where C is a constant that does not depend on parameters Δ and $\mathcal{T}$. Under the trinomial distribution (2.27), we can also show that the MLEs of Δ and $\mathcal{T}$ are given by $\hat{\Delta} = \hat{\pi}_{10} - \hat{\pi}_{01}$ and $\hat{\mathcal{T}} = \hat{\pi}_{10} + \hat{\pi}_{01}$, respectively (**Exercise 2.8**). We can further derive interval estimator (2.26) on the basis of the asymptotic properties of the MLE $\hat{\Delta}$ (**Exercise 2.9**).

Consider testing $H_0 : \Delta = \Delta_0$ versus $H_a : \Delta \neq \Delta_0$. When one applies the asymptotic likelihood ratio test on the basis of the log-likelihood (2.28), the acceptance region (Casella and Berger, 1990) consists of all sample points (n_{10}, n_{01}) such that

$$\begin{aligned}2[n_{10}\log(\hat{\Delta}+\hat{T})+n_{01}\log(\hat{T}-\hat{\Delta})+(n-n_{10}-n_{01})\log(1-\hat{T})\\ -(n_{10}\log(\Delta_0+\hat{T}(\Delta_0))+n_{01}\log(\hat{T}(\Delta_0)-\Delta_0)\\ +(n-n_{10}-n_{01})\log(1-\hat{T}(\Delta_0)))]\le\chi^2_\alpha,\end{aligned}$$

where $\hat{T}(\Delta_0)$ denotes the restricted MLE of T, given a fixed Δ_0, and χ^2_α is the upper 100αth percentile of the central χ^2 distribution with one degree of freedom. Therefore, by inverting the acceptance region (Casella and Berger, 1990; see also Appendix), we obtain an asymptotic $100(1-\alpha)$ percent confidence interval for Δ given by

$$[\Delta_l^*, \Delta_u^*], \tag{2.29}$$

where $-1<\Delta_l^*<\Delta_u^*<1$ are the smaller and the larger roots of Δ_0 such that

$$\begin{aligned}2[n_{10}\log(\hat{\Delta}+\hat{T})+n_{01}\log(\hat{T}-\hat{\Delta})+(n-n_{10}-n_{01})\log(1-\hat{T})\\ -(n_{10}\log(\Delta_0+\hat{T}(\Delta_0))+n_{01}\log(\hat{T}(\Delta_0)-\Delta_0)\\ +(n-n_{10}-n_{01})\log(1-\hat{T}(\Delta_0)))]=\chi^2_\alpha.\end{aligned} \tag{2.30}$$

Details of how to find the restricted MLE $\hat{T}(\Delta_0)$, given $\Delta=\Delta_0$, appear in **Exercise 2.11**.

Note that we can rewrite $\mathrm{Var}(\hat{\Delta})$ as $(T-\Delta^2)/n$. Thus, if n is large, the probability $P((\hat{\Delta}-\Delta)^2/(T-\Delta^2)/n)\le Z^2_{\alpha/2})\doteq 1-\alpha$. Because $\hat{T}$ is a consistent estimator of T, the above probability still holds when we substitute $\hat{T}$ for T (Casella and Berger, 1990). These results lead to the following quadratic equation:

$$A^\ddagger\Delta^2-2B^\ddagger\Delta+C^\ddagger\le 0, \tag{2.31}$$

where $A^\ddagger=(1+Z^2_{\alpha/2}/n)$, $B^\ddagger=\hat{\Delta}$, and $C^\ddagger=\hat{\Delta}^2-Z^2_{\alpha/2}\hat{T}/n$. An asymptotic $100(1-\alpha)$ percent confidence interval for Δ is then (Lui, 1998; May and Johnson, 1997)

$$[\max\{(B^\ddagger-\sqrt{B^{\ddagger 2}-A^\ddagger C^\ddagger})/A^\ddagger,-1\},\min\{(B^\ddagger+\sqrt{B^{\ddagger 2}-A^\ddagger C^\ddagger})/A^\ddagger,1\}]. \tag{2.32}$$

Note that because $A^\ddagger>0$, (2.31) is convex. Furthermore, if both n_{10} and n_{01} were positive, then the condition that $B^{\ddagger 2}-A^\ddagger C^\ddagger>0$ would be true and hence the confidence limits (2.32) would exist.

In order to attempt to improve the normal approximation of $\hat{\Delta}$, we may consider use of the $\tanh^{-1}(x)=\frac{1}{2}\log((1+x)/(1-x))$ transformation (Edwardes, 1995). Using the delta method, we can easily show that the estimated asymptotic variance $\widehat{\text{Var}}(\tanh^{-1}(\hat{\Delta})) = \widehat{\text{Var}}(\hat{\Delta})/(1-\hat{\Delta}^2)^2$ (**Exercise 2.12**). Therefore, an asymptotic $100(1-\alpha)$ percent confidence interval for Δ is given by

$$[\tanh(\tanh^{-1}(\hat{\Delta}) - Z_{\alpha/2}\sqrt{\widehat{\text{Var}}(\hat{\Delta})}/(1-\hat{\Delta}^2)),$$
$$\tanh(\tanh^{-1}(\hat{\Delta}) + Z_{\alpha/2}\sqrt{\widehat{\text{Var}}(\hat{\Delta})}/(1-\hat{\Delta}^2))]. \quad (2.33)$$

Note that when $n_{10} = 0$, $n_{01} = 0$, or $n - n_{10} - n_{01} = 0$, the respective estimates $\hat{\pi}_{10}$, $\hat{\pi}_{01}$, or $1 - \hat{\mathcal{T}}$ equal 0, on the boundary of the range for the corresponding parameters. Therefore, all the interval estimators (2.26), (2.29), (2.32), and (2.33) are inappropriate for use in these cases. Thus, whenever any of the above cell frequencies equals 0, we apply the adjustment procedure for sparse data by adding 0.50 to each cell. Lui (1998) notes that when n is small (less than 30), $\mathcal{T} \geq 0.30$, and $\Delta \leq 0.20$, applying interval estimator (2.26) is likely to produce a confidence interval with coverage probability less than the desired confidence level, but with average length longer than those of (2.29) and (2.32). Lui (2002) further notes that except for the few extreme cases in which π_1 is large (0.50 or greater) and π_0 is small (0.025 or less) (or π_1 is small (0.025 or less) and π_0 is large (0.5 or greater)), interval estimator (2.32) is generally preferable to (2.33). However, when n is small (less than 30) in these extreme cases, interval estimator (2.33) can actually perform quite well without essentially losing efficiency as compared with (2.29) and hence (2.33) is recommended for use on account of its simplicity. When n is large, all interval estimators considered here are essentially equivalent with respect to coverage probability and average length. To avoid use of an iterative numerical procedure in application of (2.29) in this case, interval estimators (2.26), (2.32), and (2.33) can be employed.

Example 2.6 Consider the example of a random sample of 1600 voting-age Americans (Agresti, 1990, p. 350). There are (n_{10} =)150 people who originally indicate approval of the President's performance in office, but indicate disapproval in the second survey, while there are only (n_{01} =)86 people who originally indicate disapproval but indicate approval in the second survey. Given these data, we obtain 95% confidence intervals for $\Delta(= \pi_{10} - \pi_{01})$ of [0.0213, 0.0587], [0.0214, 0.0589], [0.0212, 0.0586], and [0.0213, 0.0587] using (2.26), (2.29), (2.32), and (2.33), respectively. Because the sample size in this example is large, all interval estimators considered here are adequate for use. Since all these resulting confidence intervals exclude 0, there is significant evidence at the 5% level that the approval rate in the first survey is higher than that in the second survey.

Example 2.7 Consider the numerical example in Rosner, (1990, pp. 342–343), in which we compare two treatments for a rare form of cancer. Within each pair, we randomly assign patients to receive either chemotherapy or surgery, and determine vital status, survival or death, at the end of a 5-year follow-up. There are (n =)621 pairs of patients matched with respect to age, sex, and clinical condition. We obtain $n_{10} = 16$ (the number of pairs in which the patient receiving chemotherapy survives, but the patient receiving surgery dies), and $n_{01} = 5$ (the number of pairs in which the patient receiving surgery survives, but the patient receiving chemotherapy dies). Given these data, we obtain 95% confidence intervals of [0.0033, 0.0321], [0.0037, 0.0335], [0.0033, 0.0320], and [0.0033, 0.0321] using estimators (2.26), (2.29), (2.32), and (2.33), respectively. Because all these confidence intervals exclude 0, using any of these interval estimators indicates significant evidence at the 5% level that chemotherapy is preferable to surgery with respect to vital status at 5-year follow-up. However, this improvement in survival rate over the follow-up period is small.

Using the Dirichlet-multinomial model, Lui (2001b) has extended the discussion of the estimation of the RD for paired-sample data to the situation where pairs of observations are collected under cluster sampling. Lui (2000) also discusses estimation of the RD with regard to attack rates between successive infections for an incomplete 2×2 table.

2.5 INDEPENDENT NEGATIVE BINOMIAL SAMPLING (INVERSE SAMPLING)

When the underlying disease is rare, the number of cases in a sample can be small or even 0 under binomial sampling. To ensure that we obtain an appropriate number of cases in our sample and to control the relative error of estimation (Cochran, 1977; see also **Exercise 1.11**), we may employ inverse sampling (Haldane, 1945) or negative binomial sampling. Therefore, we now discuss estimation of the RD under this sampling distribution.

Suppose that we employ independent inverse sampling, in which we continue sampling subjects until we obtain a predetermined number $x_i > 0$ of cases from the exposed ($i = 1$) and the unexposed ($i = 0$) groups, respectively. Let Y_i denote the number of non-cases collected before obtaining exactly the desired number, x_i, of cases from group i. The random variable Y_i then follows the negative binomial distribution (1.13) with parameters x_i and π_i. Under independent negative binomial sampling, we can easily show that the MLEs of $\mathcal{T}$ and Δ are $\hat{\mathcal{T}} = \hat{\pi}_1 + \hat{\pi}_0$ and $\hat{\Delta} = \hat{\pi}_1 - \hat{\pi}_0$, respectively, where $\hat{\pi}_i = x_i/N_i$, $N_i = x_i + Y_i$ (**Exercise 2.13**) with the estimated asymptotic variance of $\hat{\Delta}$ equal to (**Exercise 2.14**)

$$\widehat{\mathrm{Var}}(\hat{\Delta}) = \hat{\pi}_1^2(1 - \hat{\pi}_1)/x_1 + \hat{\pi}_0^2(1 - \hat{\pi}_0)/x_0. \tag{2.34}$$

Thus, an asymptotic $100(1 - \alpha)$ percent confidence interval for Δ is given by

$$[\max\{-1, \hat{\Delta} - Z_{\alpha/2}\sqrt{\widehat{\mathrm{Var}}(\hat{\Delta})}\}, \min\{1, \hat{\Delta} + Z_{\alpha/2}\sqrt{\widehat{\mathrm{Var}}(\hat{\Delta})}\}]. \tag{2.35}$$

Note that $\hat{\pi}_i$ is a biased estimator of π_i under (1.13), and hence $\hat{\pi}_1 - \hat{\pi}_0$ is a biased estimator of Δ as well. On the other hand, it is well known that the UMVUE of π_i is $\hat{\pi}_i^{(u)} = (x_i - 1)/(x_i + Y_i - 1)$ when $x_i > 1$ (**Exercise 1.12**; Haldane, 1945; Mikulski and Smith, 1976). Thus, the UMVUE of Δ is $\hat{\Delta}^{(u)} = \hat{\pi}_1^{(u)} - \hat{\pi}_0^{(u)}$. Best (1974) derives the variance of $\hat{\pi}_i^{(u)}$ in closed form:

$$\mathrm{Var}(\hat{\pi}_i^{(u)}) = (x_i - 1)(1 - \pi_i)\left[\sum_{k=2}^{x_i-1}(-\pi_i/(1-\pi_i))^k/(x_i - k)\right.$$
$$\left. - (-\pi_i/(1-\pi_i))^{x_i}\log(\pi_i)\right] - \pi_i^2. \qquad (2.36)$$

Thus, the variance $\mathrm{Var}(\hat{\Delta}^{(u)})$ is simply given by $\sum_{i=0}^{1}\mathrm{Var}(\hat{\pi}_i^{(u)})$. Furthermore, since

$$\widehat{\mathrm{Var}}(\hat{\Delta}^{(u)}) = \hat{\pi}_1^{(u)}(1 - \hat{\pi}_1^{(u)})/(x_1 + Y_1 - 2) + \hat{\pi}_0^{(u)}(1 - \hat{\pi}_0^{(u)})/(x_0 + Y_0 - 2)$$

for $x_i > 2$ is an unbiased estimator of the variance $\mathrm{Var}(\hat{\Delta}^{(u)})$ (Finney, 1949; **Exercise 1.13**), we obtain an asymptotic $100(1-\alpha)$ percent confidence interval for Δ of

$$[\max\{\hat{\Delta}^{(u)} - Z_{\alpha/2}\sqrt{\widehat{\mathrm{Var}}(\hat{\Delta}^{(u)})}, -1\}, \min\{\hat{\Delta}^{(u)} + Z_{\alpha/2}\sqrt{\widehat{\mathrm{Var}}(\hat{\Delta}^{(u)})}, -1\}]. \quad (2.37)$$

Consider testing $H_0 : \Delta = \Delta_0$ versus $H_a : \Delta \neq \Delta_0$. When applying the asymptotic likelihood ratio test, the acceptance region consists of all (Y_1, Y_0) such that

$$2[x_1\log(\hat{\Delta} + \hat{T}) + y_1\log(2 - \hat{\Delta} - \hat{T}) + x_0\log(\hat{T} - \hat{\Delta})$$
$$+ y_0\log(2 - \hat{T} + \hat{\Delta}) - x_1\log(\Delta_0 + \hat{T}(\Delta_0)) - y_1\log(2 - \Delta_0 - \hat{T}(\Delta_0))$$
$$- x_0\log(\hat{T}(\Delta_0) - \Delta_0) - y_0\log(2 - \hat{T}(\Delta_0) + \Delta_0)] \leq \chi^2_\alpha, \qquad (2.38)$$

where $\hat{T}(\Delta_0)$ denotes the restricted MLE of T, given a fixed Δ_0, and χ^2_α is the upper 100αth percentile of the central χ^2 distribution with one degree of freedom. Therefore, we can obtain an asymptotic $100(1-\alpha)$ percent confidence interval for Δ by simply inverting the acceptance region (Casella and Berger, 1990; Cox and Oakes, 1984; see also Appendix) to give

$$[\Delta_l^{**}, \Delta_u^{**}], \qquad (2.39)$$

where $-1 < \Delta_l^{**} < \Delta_u^{**} < 1$ are the smaller and the larger roots of Δ_0 such that equality holds in (2.38). Details of the doubly-iterative procedure for finding these two roots of Δ_0 can be found in the Appendix of Lui (1999). Note that we can also derive an asymptotic likelihood ratio test-based confidence interval under binomial sampling as for deriving interval estimator (2.39). In fact, when $X_i = x_i$ and $n_i - X_i = Y_i$, the asymptotic likelihood ratio test-based confidence intervals under these two different samplings are identical. Note also that the likelihood ratio test-based confidence interval (2.39) is generally preferable to Wald's confidence interval (2.35).

Lui (1999) evaluates and compares the performance of the 95% confidence intervals (2.35), (2.37), and (2.39) in a variety of situations. He finds that even when both x_i are as small as 5, the coverage probability of all these estimators can be either larger than or approximately equal to the desired 95% confidence level. When both $x_i \geq 50$, estimators (2.35), (2.37), and (2.39) are essentially equivalent with respect to both coverage probability and expected length. Furthermore, when $\pi_1 \leq 0.10$, Lui notes that the expected length of (2.37) seems to be the shortest, while maintaining a coverage probability equal to or greater than the desired 95% confidence level. Because applying interval estimator (2.37) does not involve a numerical procedure, we may wish to use this estimator in these situations.

Note that the probability that the MLE $\hat{\pi}_i$ is on the boundary as a result of either Y_i being equal to 0 can be shown to be $\pi_1^{x_1} + \pi_0^{x_0} - \pi_1^{x_1}\pi_0^{x_0}$ (**Exercise 2.15**). Because the underlying proportions π_i of cases are frequently small in practice, this probability should be negligible in most practical instances. On the other hand, if this should occur, we may apply the commonly used adjustment of adding 0.50 to each of Y_i whenever either of Y_i equals 0 to avoid this practical concern in estimation of variances $\text{Var}(\hat{\Delta})$ or $\text{Var}(\hat{\Delta}^{(u)})$.

Example 2.8 Suppose that we wish to provide an interval estimate of the prevalence RD between two populations. Suppose further that from each of these two populations we decide to continue independently sampling subjects until we obtain exactly ($x_1 = x_0 =$)30 cases. Suppose we obtain $y_1 = 120$ and $y_0 = 270$. On the basis of these data, the MLE $\hat{\Delta}$ and the UMVUE $\hat{\Delta}^{(u)}$ are then 0.075 and 0.073, respectively. Estimators (2.35), (2.37), and (2.39) give [0.028, 0.172], [0.026, 0.170], and [0.031, 0.176], respectively. Since the predetermined numbers of index cases are reasonably large, these interval estimates are all similar to one another.

Note that the total number n_i of subjects under binomial sampling is fixed, but the number of cases x_i is random and can be 0 with a positive probability. By contrast, the total number of subjects N_i under inverse sampling is random, but the number of cases x_i is fixed and is determined by the investigator. We can show that $E(N_i) = x_i/\pi_i$ (**Exercise 2.16**). When the underlying disease is not rare, we may wish to use binomial sampling due to its simplicity and many well-understood statistical properties. When the underlying disease is rare and the data arrive sequentially, it may be natural to consider use of inverse sampling so that one can ensure that the relative error is less than or equal to a given desired precision. On the other hand, if the underlying disease were chronic, it would be practically impossible to employ inverse sampling for cohort studies. This is because it is not appropriate to wait for a long time to determine the disease status of each sampled subject before we decide if we should sample the next subject. When the underlying disease is acute, or when we use a historical cohort design or take a survey, in which the disease status for each sampled subject can be quickly determined, inverse sampling can be a useful alternative sampling design to binomial sampling.

2.6 INDEPENDENT POISSON SAMPLING

When employing a cohort study design to investigate the effect of a risk factor on the incidence rate of a chronic disease, such as cancer (Rosner, 2000) or cardiovascular disease (Doll and Hill, 1966), we often need to follow up a large number of subjects for a long period of time. Because some of these subjects are likely to drop out of the study during the lengthy follow-up period, the number of person-years at risk will likely vary between comparison groups. In this situation the Poisson model, which can easily account for various lengths in person-years at risk between groups, is often assumed (Breslow, 1984; Newman, 2001). In fact, when the underlying disease is rare and the occurrence of a case is random, one can show mathematically that the incident number of cases in a fixed time interval follows a Poisson process (Bailey, 1964).

Suppose that we want to compare the incidence rates λ_i of two comparison groups distinguished by exposure ($i = 1$) or non-exposure ($i = 0$) to a risk factor. Suppose further that we follow n_i subjects and obtain X_i cases from group i. We assume that X_i follows the Poisson distribution with parameter $n_i^*\lambda_i$, where n_i^* is the number of person-years at risk over these n_i subjects. The likelihood function is then given by

$$\prod_{i=0}^{1}(n_i^*\lambda_i)^{X_i}\exp(-n_i^*\lambda_i)/X_i!, \tag{2.40}$$

where $X_i = 0, 1, \ldots$. The incidence RD, denoted by Δ^*, is simply equal to $\lambda_1 - \lambda_0$. Define $T^* = \lambda_1 + \lambda_0$. We can easily show that the MLEs of Δ^* and T^* are given by $\hat{\Delta}^* = \hat{\lambda}_1 - \hat{\lambda}_0$ and $\hat{T}^* = \hat{\lambda}_1 + \hat{\lambda}_0$, respectively, where $\hat{\lambda}_i = X_i/n_i^*$ (**Exercise 2.22**). Furthermore, the variance of $\hat{\Delta}^*$ is $\mathrm{Var}(\hat{\Delta}^*) = (T^* + \Delta^*)/(2n_1^*) + (T^* - \Delta^*)/(2n_0^*)$. Thus, we obtain an asymptotic $100(1-\alpha)$ percent confidence interval for Δ^*, using Wald's statistic, of

$$\begin{aligned}[\hat{\Delta}^* - Z_{\alpha/2}\sqrt{(\hat{T}^* + \hat{\Delta}^*)/(2n_1^*) + (\hat{T}^* - \hat{\Delta}^*)/(2n_0^*)},\\ \hat{\Delta}^* + Z_{\alpha/2}\sqrt{(\hat{T}^* + \hat{\Delta}^*)/(2n_1^*) + (\hat{T}^* - \hat{\Delta}^*)/(2n_0^*)}].\end{aligned} \tag{2.41}$$

Because the sampling distribution of $\hat{\Delta}^*$ can be skewed, interval estimator (2.41) may not perform well, especially when the underlying disease is rare. When both n_i^* are large, we have the probability $P(((\hat{\Delta}^* - \Delta^*)/\sqrt{\mathrm{Var}(\hat{\Delta}^*)})^2 \le Z_{\alpha/2}^2) \doteq 1-\alpha$. This leads us to consider the following quadratic equation in Δ^* (**Exercise 2.23**):

$$\mathfrak{A}(\Delta^*)^2 - 2\mathfrak{B}\Delta^* + \mathfrak{C} \le 0, \tag{2.42}$$

where $\mathfrak{A} = 1$, $\mathfrak{B} = \hat{\Delta}^* + (1/n_1^* - 1/n_0^*)Z_{\alpha/2}^2/4$, and $\mathfrak{C} = (\hat{\Delta}^*)^2 - (1/n_1^* + 1/n_0^*)$ $\hat{T}^* Z_{\alpha/2}^2/2$. If either x_i is positive, we can show that the inequality $\mathfrak{B}^2 - \mathfrak{A}\mathfrak{C} >$ 0 holds and hence the two distinct roots of (2.42) exist. An approximately $100(1-\alpha)$ percent confidence interval for Δ^* is then given by

$$[(\mathfrak{B} - \sqrt{\mathfrak{B}^2 - \mathfrak{A}\mathfrak{C}})/\mathfrak{A}, (\mathfrak{B} + \sqrt{\mathfrak{B}^2 - \mathfrak{A}\mathfrak{C}})/\mathfrak{A}]. \quad (2.43)$$

Note that when x_1 or x_0 equals 0, using the variance estimate $\widehat{\text{Var}}(\hat{\Delta}^*)(= \hat{\lambda}_1/n_1^* + \hat{\lambda}_0/n_0^*)$ tends to underestimate the true variance $\text{Var}(\hat{\Delta}^*)$. To alleviate this concern, we may employ the *ad hoc* adjustment procedure for sparse data using $(x_i + 0.50)/(n_i^* + 0.50)$ to estimate λ_i for both i whenever x_1 or x_0 equals 0. Thus, with use of this adjustment, the confidence limits (2.43) always exist.

Based on Monte Carlo simulation, Lui and Lin (2003) evaluate and compare the performance of (2.41) with (2.43) as well as the other two test-based interval estimators involving iterative numerical procedures. Lui and Lin (2003) find that when the underlying disease is rare, the coverage probability of (2.41) tends to be smaller than the desired confidence level. Interval estimator (2.43) can generally improve the coverage probability of (2.41) and perform reasonable well in a variety of situations. Lui and Lin further observe that the two test-based interval estimators (which are not presented here to save space) may even be slightly preferable to (2.43). When the underlying disease in the non-exposure group is not rare ($\lambda_0 \geq 0.10$) and the number of person-years at risk is reasonably large, however, all interval estimators are essentially equivalent. In this case, we may use (2.41) or (2.43) for simplicity. See Lui and Lin (2003) for further details.

Example 2.9 Consider the data in Table 2.4 collected from the Nurses' Health Study to assess the effect of current use of estrogen replacement therapy and the risk of breast cancer (Colditz *et al.*, 1990). For illustration purposes, let us consider only the data for women aged 50–54 years (Rosner, 2000, p. 695). There were 51 cases amounting 24 948 person-years at risk in the group of current users

Table 2.3 The number of deaths/the number of cases with breast cancer at stages I, II, and III between the groups with low and high levels of estrogen receptor in 192 women over a maximum 5 years of follow-up.

		Disease stage		
		I	II	III
Receptor level	Low	2/12	9/22	12/14
	High	5/55	17/74	9/15

Source: Newman (2001).

Table 2.4 The number of cases with breast cancer/the number of person-years at risk between current and never users of hormone in postmenopausal women of ages 39–64.

Age	Current users	Never users
39–44	12/10199	5/4722
45–49	22/14044	26/20812
50–54	51/24948	129/71746
55–59	72/21576	159/73413
60–64	23/4876	35/15773

Sources: These data are abstracted from a data set originally reported by Colditz *et al.* (1990) and quoted by Rosner (2000).

$(i = 1)$ and 129 cases amounting 71746 person-years at risk in the group of non-users $(i = 0)$. The MLE $\hat{\Delta}^*(= \hat{\lambda}_1 - \hat{\lambda}_0)$ is 0.00025. Applying (2.41) and (2.43), the 95% confidence intervals for Δ^* are given by [−0.00039, 0.00089] and [−0.00037, 0.00091], respectively. Since both interval estimates include 0, there is no significant evidence at the 5% level that current hormone use in postmenopausal women aged 50–54 increases the risk of breast cancer.

2.7 STRATIFIED POISSON SAMPLING

When comparing incidence rates between exposed $(i = 1)$ and non-exposed $(i = 0)$ groups, we may often need to adjust for the effect of confounders to avoid inferential bias. For example, when assessing the effect of estrogen replacement therapy on the risk of breast cancer in Table 2.4, we adjust for the effect of age, believed to be associated with hormone use and breast cancer. Here, we focus discussion on using stratified analysis to control the confounders. From each stratum $j (j = 1, 2, \ldots, S)$, suppose that we obtain X_{ij} cases amounting to n^*_{ij} person-years at risk in group $i (i = 1, 0)$. We assume that the random variables X_{ij} independently follow the Poisson distribution with parameters $\lambda_{ij} n^*_{ij}$, where λ_{ij} is the disease incidence rate. Breslow and Day (1987) provide arguments to justify the fact that the random variables X_{ij} can still be regarded as independent even when the same person may contribute observation time to several contiguous age categories. On the basis of the above model assumptions, the likelihood is

$$\prod_{i=0}^{1} \prod_{j=1}^{S} (\lambda_{ij} n^*_{ij})^{X_{ij}} \exp(-(\lambda_{ij} n^*_{ij}))/X_{ij}!. \tag{2.44}$$

The incidence RD in stratum j is then simply equal to $\Delta^*_j = \lambda_{1j} - \lambda_{0j}$. We can easily show that the MLE of Δ^*_j is $\hat{\Delta}^*_j$, where $\hat{\Delta}^*_j = \hat{\lambda}_{1j} - \hat{\lambda}_{0j}$ and $\hat{\lambda}_{ij} = X_{ij}/n^*_{ij}$. In the following discussion, we assume that $\Delta_1^* = \Delta_2^* = \ldots = \Delta_S^*$ and denote this common incidence RD by Δ_c^*. We are interested in providing an interval estimator of Δ_c^*.

First, consider the most commonly used interval estimator based on the WLS point estimator $\hat{\Delta}_{\text{WLS}}^* = \sum \hat{\mathcal{W}}_j \hat{\Delta}_j^* / \sum \hat{\mathcal{W}}_j$, where $\hat{\mathcal{W}}_j = \widehat{\text{Var}}(\hat{\Delta}_j^*)^{-1} = (\hat{\lambda}_{1j}/n_{1j}^* + \hat{\lambda}_{0j}/n_{0j}^*)^{-1}$. Thus, we obtain an approximately $100(1-\alpha)$ percent confidence interval for Δ_c^* given by

$$\left[\hat{\Delta}_{\text{WLS}}^* - Z_{\alpha/2}\Big/\sqrt{\sum \hat{\mathcal{W}}_j},\quad \hat{\Delta}_{\text{WLS}}^* + Z_{\alpha/2}\Big/\sqrt{\sum \hat{\mathcal{W}}_j}\right]. \tag{2.45}$$

When the number of person-years at risk (or when the expected number of cases) is small in each stratum, the WLS interval estimator (2.45) will likely not perform well. Thus, we consider interval estimator based on the Mantel–Haenszel estimator (Greenland and Robins, 1985)

$$\hat{\Delta}_{\text{MH}}^* = \left(\sum_j n_{1j}^* n_{0j}^* \hat{\Delta}_j^* / n_{.j}^*\right) \Big/ \left(\sum_j n_{0j}^* n_{1j}^* / n_{.j}^*\right). \tag{2.46}$$

Note that $\hat{\Delta}_{\text{MH}}^*$ is actually identical to the WLS estimator $\hat{\Delta}_{\text{WLS}}^*$ with weights inversely proportional to $\text{Var}(\hat{\Delta}_j^*)$ under the null condition that $\Delta_j^* = 0$ for all j. Furthermore, one can easily show that the variance of $\hat{\Delta}_{\text{MH}}^*$ (**Exercise 2.25**) is given by

$$\text{Var}(\hat{\Delta}_{\text{MH}}^*) = \sum_j \left[\left(\frac{n_{0j}^*}{n_{.j}^*}\right)^2 n_{1j}^* \lambda_{1j} + \left(\frac{n_{1j}^*}{n_{.j}^*}\right)^2 n_{0j}^* \lambda_{0j}\right] \Big/ \left(\sum_j n_{0j}^* n_{1j}^* / n_{.j}^*\right)^2, \tag{2.47}$$

which is a function of unknown parameters λ_{ij}. We can simply substitute the MLEs $\hat{\lambda}_{ij}$ for λ_{ij} in (2.47) to obtain the estimated variance

$$\widehat{\text{Var}}(\hat{\Delta}_{\text{MH}}^*) = \sum_j \left[\left(\frac{n_{0j}^*}{n_{.j}^*}\right)^2 X_{1j} + \left(\frac{n_{1j}^*}{n_{.j}^*}\right)^2 X_{0j}\right] \Big/ \left(\sum_j n_{0j}^* n_{1j}^* / n_{.j}^*\right)^2. \tag{2.48}$$

Thus, an asymptotic $100(1-\alpha)$ percent confidence interval for Δ_c^* based on $\hat{\Delta}_{\text{MH}}^*$ is given by

$$[\hat{\Delta}_{\text{MH}}^* - Z_{\alpha/2}\sqrt{\widehat{\text{Var}}(\hat{\Delta}_{\text{MH}}^*)},\quad \hat{\Delta}_{\text{MH}}^* + Z_{\alpha/2}\sqrt{\widehat{\text{Var}}(\hat{\Delta}_{\text{MH}}^*)}]. \tag{2.49}$$

To improve the normal approximation of $\hat{\Delta}_{\text{MH}}^*$, following Edwardes (1995), we may also consider using the $\tanh^{-1}(x)$ transformation. Using the delta method, we can show that the asymptotic variance $\widehat{\text{Var}}(\tanh^{-1}(\hat{\Delta}^*_{MH})) = \widehat{\text{Var}}(\hat{\Delta}^*_{MH})/(1 - (\hat{\Delta}^*_{MH})^2)^2$ (**Exercise 2.12**). Therefore, we obtain an asymptotic $100(1-\alpha)$ percent confidence intervals for Δ^*_c given by

$$[\tanh\{\tanh^{-1}(\hat{\Delta}^*_{MH}) - Z_{\alpha/2}\sqrt{\widehat{\text{Var}}(\hat{\Delta}^*_{MH})/(1 - (\hat{\Delta}^*_{MH})^2)}\},$$

$$\tanh\{\tanh^{-1}(\hat{\Delta}^*_{MH}) + Z_{\alpha/2}\sqrt{\widehat{\text{Var}}(\hat{\Delta}^*_{MH})/(1 - (\hat{\Delta}^*_{MH})^2)}\}]. \quad (2.50)$$

Based on Monte Carlo simulation, Lui (2003) evaluates and compares the performance of several interval estimators of Δ^*_c, including (2.45), (2.49), and (2.50). Because the number of person-years at risk is usually quite large in cohort studies, if the underlying disease were not rare, all these three interval estimators would be appropriate for use. However, when the number of person-years at risk is moderate, the coverage probability of (2.45) can be much less than the desired confidence level, especially when the number S of strata is large and the underlying disease is rare. By contrast, interval estimators (2.49) and (2.50) can consistently perform well. Thus, in these cases we may wish to use the latter. To test the homogeneity of the incidence RD across strata, given a reasonably large number of person-years at risk in each stratum, we may apply the WLS test statistic $\sum \hat{W}_j(\hat{\Delta}^*_j - \hat{\Delta}^*_{WLS})^2$, where $\hat{W}_j = \widehat{\text{Var}}(\hat{\Delta}^*_j)^{-1}$. When $\sum_j \hat{W}_j(\hat{\Delta}^*_j - \hat{\Delta}^*_{WLS})^2 > \chi^2_{S-1,\alpha}$, we reject H_0 : $\Delta^*_1 = \Delta^*_2 = \ldots = \Delta^*_S$ at level α. If we should reject this H_0 at a small given level α, we may not want to provide a summary estimator for Δ^*_c to avoid overlooking a possibly important association between the exposure and the disease at certain stratum levels.

Example 2.10 Consider the data (Table 2.4) discussed in Example 2.9. Suppose that we wish to provide a summary estimate of the effect due to estrogen replacement therapy on the risk of breast cancer in postmenopausal women, while controlling the confounding effect of age by means of stratified analysis. Applying the WLS statistic to test the homogeneity of RD, we obtain a p-value of 0.124. Thus, there is only weak evidence that RD varies between strata. Assuming constant RD across age categories, we obtain the point estimates $\hat{\Delta}^*_{WLS} = 0.00054$ and $\hat{\Delta}^*_{MH} = 0.00072$. Applying interval estimators (2.45), (2.49), and (2.50), we obtain 95% confidence intervals for Δ^*_c given by [0.00014, 0.00093], [0.00030, 0.00114], and [0.00030, 0.00114], respectively. We can see that the resulting WLS interval estimate (2.45) tends to shift to the left as compared with the other two estimates. Although the number of person-years at risk in Table 2.4 is quite large, the underlying disease rate is (as given by the above two point estimates) really small. In this case, we may want to apply the WLS interval estimator (2.45) with caution. Because all the above lower limits are above 0, there is significant evidence at the 5% level that use of estrogen replacement therapy can increase the risk of breast cancer. Based on the above resulting interval estimates using

(2.49) and (2.50), we have 95% confidence that the increase in the number of cases of breast cancer due to estrogen use should be between 30 and 114 per 100000 person-years.

EXERCISES

2.1. Under independent binomial sampling, show that the difference $\hat{\Delta}$ between the two sample proportions in (2.1) is an unbiased estimator of RD with variance $\mathrm{Var}(\hat{\Delta})$ given by $\pi_1(1-\pi_1)/n_1 + \pi_0(1-\pi_0)/n_0$.

2.2. Show that $\hat{\pi}_i(1-\hat{\pi}_i)/(n_i-1)$ is an unbiased estimator of $\pi_i(1-\pi_i)/n_i$, where $\hat{\pi}_i = X_i/n_i$, and X_i follows the binomial distribution $\binom{n_i}{x_i}\pi_i^{x_i}(1-\pi_i)^{n_i-x_i}$, $x_i = 0, 1, \ldots, n_i$.

2.3. From the results of **Exercise 2.1**, show that the inequality $((\hat{\Delta} - \Delta)/\sqrt{\mathrm{Var}(\hat{\Delta})})^2 \leq Z^2_{\alpha/2}$ can be rewritten as $A\Delta^2 - 2B\Delta + C \leq 0$, where the coefficients A, B, and C are defined in (2.4).

2.4. When the weight is proportional to the sample size n_i from population $i(i = 1, 0)$, conditional upon $\Delta = \pi_1 - \pi_0$, show that $\hat{\pi}_{1|\Delta} = \bar{\pi} + \Delta n_0/n_.$ and $\hat{\pi}_{0|\Delta} = \bar{\pi} - \Delta n_1/n$, where $\bar{\pi} = (n_1\hat{\pi}_1 + n_0\hat{\pi}_0)/n_.$ and $n_. = n_1 + n_0$, are the weighted least-squares estimators of π_1 and π_0, respectively (Wallenstein, 1997).

2.5. On the basis of the weighted least-squares estimators in **Exercise 2.4**, following similar arguments to those for deriving (2.4), show that we can derive the quadratic equation $\mathcal{A}\Delta^2 - 2\mathcal{B}\Delta + \mathcal{C} \leq 0$, where $\mathcal{A} = 1 + [(1/n_1)(n_0/n_.)^2 + (1/n_0)(n_1/n_.)^2]Z^2_{\alpha/2}$, $\mathcal{B} = \hat{\pi}_1 - \hat{\pi}_0 + (1-2\bar{\pi})[(1/n_1)(n_0/n_.) - (1/n_0)(n_1/n_.)]$ $Z^2_{\alpha/2}/2$, and $\mathcal{C} = (\hat{\pi}_1 - \hat{\pi}_0)^2 - \bar{\pi}(1-\bar{\pi})[1/n_1 + 1/n_0]Z^2_{\alpha/2}$. If $\mathcal{B}^2 - \mathcal{AC} > 0$, an approximately $100(1-\alpha)$ percent confidence interval for Δ would be given by $[\max\{(\mathcal{B} - \sqrt{\mathcal{B}^2 - \mathcal{AC}})/\mathcal{A}, -1\}, \min\{(\mathcal{B} + \sqrt{\mathcal{B}^2 - \mathcal{AC}})/\mathcal{A}, 1\}]$. As noted by Wallenstein (1997), this interval estimator is valid only when the estimates $\hat{\pi}_{1|\Delta} = \bar{\pi} + \Delta n_0/n_.$ and $\hat{\pi}_{0|\Delta} = \bar{\pi} - \Delta n_1/n_.$ fall between 0 and 1 when Δ is replaced by either of these limits. A tedious *ad hoc* procedure to solve this problem when either of these estimates $\hat{\pi}_{1|\Delta}$ or $\hat{\pi}_{0|\Delta}$ does not fall in (0, 1) can be found in Wallenstein's paper.

2.6. If the random variables $X_s (s = 1, 2, \ldots, S)$ are independently distributed with a common mean μ and known variance σ_s^2, then the weighted average estimator $\sum_s W_s^* X_s$, where W_s^* is proportional to $1/\sigma_s^2$, has the minimum variance among all possible linear combinations $\sum_s W_s X_s$. Note that because $\sum_s W_s = 1$, $\sum_s W_s X_s$ is an unbiased estimator of μ.

2.7. Under the multinomial distribution (2.25), show that the expectation $E(\hat{\pi}_{10} - \hat{\pi}_{01}) = \Delta$, where $\hat{\pi}_{ij} = N_{ij}/n$, and the variance $\mathrm{Var}(\hat{\pi}_{10} - \hat{\pi}_{01}) = [\pi_{10} + \pi_{01} - (\pi_{10} - \pi_{01})^2]/n$.

2.8. Under the trinomial distribution (2.27), show that the MLEs of Δ and $\mathcal{T}$ are $\hat{\Delta} = \hat{\pi}_{10} - \hat{\pi}_{01}$ and $\hat{\mathcal{T}} = \hat{\pi}_{10} + \hat{\pi}_{01}$, respectively. (Hint: solve the equations $\partial \log(L)/\partial \Delta = n_{10}/(\Delta + \mathcal{T}) - n_{01}/(\mathcal{T} - \Delta) = 0$ and $\partial \log(L)/\partial \mathcal{T} = n_{10}/(\Delta + \mathcal{T}) + n_{01}/(\mathcal{T} - \Delta) - (n - n_{10} - n_{01})/(1 - \mathcal{T}) = 0$.)

2.9. Under distribution (2.27), show that the asymptotic variance of $\hat{\Delta}(= \hat{\pi}_{10} - \hat{\pi}_{01})$ obtained from the inverse of the Fisher information matrix is equal to $[\pi_{10} + \pi_{01} - (\pi_{10} - \pi_{01})^2]/n$.

2.10. Show that for the case where $n_{1s} = n_{0s} = 1$, the estimated variance $\widehat{\text{Var}}(\hat{\Delta}_{MH})$ (2.12) reduces to $[\hat{\pi}_{10} + \hat{\pi}_{01} - (\hat{\pi}_{10} - \hat{\pi}_{01})^2]/n$ when the number S of strata is n.

2.11. Discuss how to find the restricted MLE of $\mathcal{T}$, given $\Delta = \Delta_0$ under the trinomial distribution (2.27). (Hint: Given a fixed $\Delta = \Delta_0$, where $-1 < \Delta_0 < 1$, as $\mathcal{T}$ increases from $\max\{\Delta_0, -\Delta_0\}$ to 1, the value of $\partial \log(L)/\partial \mathcal{T}$ monotonically decreases from ∞ to $-\infty$. Furthermore, $\partial \log(L)/\partial \mathcal{T}$ is a continuous function of $\mathcal{T}$ from $\max\{\Delta_0, -\Delta_0\}$ to 1. Based on these results, the restricted MLE $\hat{\mathcal{T}}(\Delta_0)$ of $\mathcal{T}$ is simply the unique root, for $\mathcal{T}$ falling in the range $\max\{\Delta_0, -\Delta_0\} < \mathcal{T} < 1$, of the equation $\partial \log(L)/\partial \mathcal{T} = 0$ with Δ replaced by Δ_0. For a fixed Δ_0, we can obtain the restricted MLE $\hat{\mathcal{T}}(\Delta_0)$ by means of trial and error.)

2.12. Under the trinomial distribution (2.27), show that the estimated asymptotic variance of $\tanh^{-1}(\hat{\Delta})$ is given by $\widehat{\text{Var}}(\tanh^{-1}(\hat{\Delta})) = \widehat{\text{Var}}(\hat{\Delta})/(1 - \hat{\Delta}^2)^2$.

2.13. Show that when the random variables $Y_i (i = 1, 0)$ independently follow the negative binomial distribution (1.13) with parameters x_i and π_i, the MLEs of $\mathcal{T} = \pi_1 + \pi_0$ and $\Delta = \pi_1 - \pi_0$ are given by $\hat{\mathcal{T}} = \hat{\pi}_1 + \hat{\pi}_0$ and $\hat{\Delta} = \hat{\pi}_1 - \hat{\pi}_0$, respectively, where $\hat{\pi}_i = x_i/(x_i + Y_i)$.

2.14. Under independent negative binomial sampling, show that the asymptotic variance of the MLE $\hat{\Delta} = \hat{\pi}_1 - \hat{\pi}_0$ is $\pi_1^2(1 - \pi_1)/x_1 + \pi_0^2(1 - \pi_0)/x_0$.

2.15. Show that the probability that either Y_1 or Y_0 equals 0 is $\pi_1^{x_1} + \pi_0^{x_0} - \pi_1^{x_1}\pi_0^{x_0}$, where the Y_i independently follow the negative binomial distribution (1.13) with parameters x_i and π_i.

2.16. Under negative binomial sampling (1.13) with parameters x_i and π_i, show that the expectation $E(N_i)$, where $N_i = x_i + Y_i$ is the total number of sampled subjects, is equal to x_i/π_i.

2.17. Suppose that we compare two treatments under independent binomial sampling. Suppose further that we obtain 15 subjects with positive responses out of 100 subjects assigned to the experimental treatment ($i = 1$), and only 5 out of 80 subjects assigned to the standard treatment ($i = 0$). Let π_i denote the proportion of positive responses for treatment $i (i = 1, 0)$. What are the 95% confidence intervals for $\Delta(= \pi_1 - \pi_0)$ when we use (2.3) and (2.5)–(2.7)?

2.18. Consider the data consisting of 192 cases with breast cancer at stage I, II, or III, collected by the Northern Alberta Breast Cancer Registry (Newman, 2001). One important predictor of survival for beast cancer is the amount of estrogen receptor that is present in breast tissue. For illustration purposes, the estrogen receptor variable has been dichotomized into low ($i = 1$) and high ($i = 0$) levels using a conventional cutoff value. As reported elsewhere (Newman, 2001, p. 99), 23 out of 48 cases in the low-level group died of breast cancer over a maximum 5 years of follow-up, while only 31 out of 144 cases died in the of high-level group. (a) What is the point estimate $\hat{\Delta}$ (2.1)? (b) What are the 95% confidence intervals for the RD ($= \pi_1 - \pi_0$, where π_i denotes the mortality risk for group i, $i = 1, 0$) when we use (2.3) and (2.5)–(2.7)?

2.19. Consider the data consisting of 192 females with breast cancer in Table 2.3 (Newman, 2001, p. 126). The data are stratified by various stages of breast cancer. Assume that the underlying RD $\Delta_s (= \pi_{1s} - \pi_{0s}, s = 1, 2, 3)$ of death rates between the two groups with low and high levels of estrogen receptor is constant across different stages and equals Δ_c.
(a) What are the point estimates of Δ_c using (2.9) and (2.11)?
(b) What are the 95% confidence intervals for Δ_c using (2.10) and (2.13)?
(c) If we apply (2.14)–(2.16) to examine the assumption that the above assumption Δ_s is constant, what are the corresponding p-values?

2.20. Dixon and Massey (1969, p. 250) present data on 105 individuals who are asked a question on public affairs both before and after a propaganda lecture; a yes/no answer is required. Fifteen individuals answer no before and yes after the propaganda, while nine answer yes before and no after the propaganda. What are the 95% confidence intervals for the difference in the 'no' rates before and after propaganda using (2.26), (2.29), (2.32), and (2.33)?

2.21. Suppose that we decide to continue independently sampling subjects until we obtain exactly ($x_1 = x_0 =$)20 cases from each of the two comparison populations ($i = 1, 0$), respectively. Suppose further that we obtain $y_1 = 60$ and $y_0 = 180$.
(a) Calculate the MLE $\hat{\Delta}$ and the UMVUE $\hat{\Delta}^{(u)}$ of $\Delta (= \pi_1 - \pi_0)$.
(b) Using estimators (2.35), (2.37), and (2.39), what are the 95% confidence intervals for Δ?

2.22. Assume that the numbers $X_i (i = 1, 0)$ of cases for group i independently follow a Poisson distribution $(n_i^* \lambda_i)^{X_i} \exp(-n_i^* \lambda_i)/X_i!$, where $X_i = 0, 1, \ldots$ and n_i^* is the number of person-years at risk.
(a) Show that the MLEs of $\Delta^* (= \lambda_1 - \lambda_0)$ and $T^* (= \lambda_1 + \lambda_0)$ are given by $\hat{\Delta}^* = \hat{\lambda}_1 - \hat{\lambda}_0$ and $\hat{T}^* = \hat{\lambda}_1 + \hat{\lambda}_0$, respectively, where $\hat{\lambda}_i = X_i/n_i^*$.
(b) Show that $\hat{\Delta}^*$ is an unbiased estimator of $\lambda_1 - \lambda_0$.
(c) Show that the variance of $\hat{\Delta}^*$ is $\mathrm{Var}(\hat{\Delta}^*) = (T^* + \Delta^*)/(2n_1^*) + (T^* - \Delta^*)/(2n_0^*)$.

2.23. Show that from $P(((\hat{\Delta}^* - \Delta^*)/\sqrt{\text{Var}(\hat{\Delta}^*)})^2 \leq Z^2_{\alpha/2}) \doteq 1 - \alpha$, we can obtain an asymptotic $100(1-\alpha)$ percent confidence interval for Δ^* given by the two distinct roots of the following quadratic equation in Δ^*: $\mathfrak{A}(\Delta^*)^2 - 2\mathfrak{B}\Delta^* + \mathfrak{C} = 0$, where $\mathfrak{A} = 1$, $\mathfrak{B} = \hat{\Delta}^* + (1/n_1^* - 1/n_2^*)Z^2_{\alpha/2}/4$, and $\mathfrak{C} = (\hat{\Delta}^*)^2 - (1/n_1^* + 1/n_2^*)\hat{T}^* Z^2_{\alpha/2}/2$, if $\mathfrak{B}^2 - \mathfrak{A}\mathfrak{C} > 0$.

2.24. Consider only those women aged 39–44 years in the study of breast cancer in postmenopausal women (Table 2.4). There are 5 cases over 4722 person-years in the group who have never used estrogen replacement therapy ($i = 0$) and 12 cases over 10199 person-years among current users ($i = 1$).
(a) What is the MLE $\hat{\lambda}_1 - \hat{\lambda}_0$?
(b) What is the 95% confidence interval for Δ^* when we apply $\hat{\lambda}_1 - \hat{\lambda}_0 \pm Z_{\alpha/2}\sqrt{\hat{\lambda}_1/n_1^* + \hat{\lambda}_0/n_0^*}$?
(c) What is the 95% confidence interval for $\lambda_1 - \lambda_0$ using interval estimator (2.43)?

2.25. Show that the variance of $\hat{\Delta}^*_{MH}$ (2.46) under stratified Poisson sampling is given by

$$\text{Var}(\hat{\Delta}^*_{MH}) = \sum_j \left[\left(\frac{n_{0j}^*}{n_{.j}^*}\right)^2 n_{1j}^*\lambda_{1j} + \left(\frac{n_{1j}^*}{n_{.j}^*}\right)^2 n_{0j}^*\lambda_{0j} \right] \Big/ \left(\sum_j n_{0j}^* n_{1j}^*/n_{.j}^*\right)^2.$$

2.26. Suppose that we concentrate our attention on studying the effect of hormone use on the risk of breast cancer for women aged between 39 and 54 years in Table 2.4.
(a) What is the p-value of the test the homogeneity of the RD across these three different age categories?
(b) What are the point estimates using $\hat{\Delta}^*_{WLS}$ and $\hat{\Delta}^*_{MH}$ of the underlying common Δ^*_c?
(c) What are the 95% confidence intervals for Δ^*_c using interval estimators (2.45), (2.49), and (2.50)?

REFERENCES

Agresti, A. (1990) *Categorical Data Analysis*. Wiley, New York.
Aspirin Myocardial Infarction Study Research Group (1980) A randomized controlled trial of aspirin in persons recovered from myocardial infarction. *Journal of the American Medical Association*, **243**, 661–669.
Bailey, N. T. J. (1964) *The Elements of Stochastic Processes with Application to the Natural Sciences*. Wiley, New York.
Beal, S. L. (1987) Asymptotic confidence intervals for the difference between two binomial parameters for use with small samples. *Biometrics*, **43**, 941–950.
Berkson, J. (1958) Smoking and lung cancer: some observations on two recent reports. *Journal of the American Statistical Association*, **53**, 28–38.

Best, D. J. (1974) The variance of the inverse binomial estimator. *Biometrika*, **67**, 385–386.

Breddin, K., Loew, D., Lechner, K., Uberla, K. and Walter, E. (1979) Secondary prevention of myocardial infarction. Comparison of acetylsalicylic acid, phenprocoumon and placebo. A multicenter two year-prospective study. *Thrombosis and Haemostasis*, **40**, 225–236.

Breslow, N. E. (1984) Elementary methods of cohort analysis. *International Journal of Epidemiology*, **13**, 112–115.

Breslow, N. E. and Day, N. E. (1987) *Statistical Methods in Cancer Research, Vol II: The Design and Analysis of Cohort Studies*. International Agency for Research on Cancer, Lyon, France.

Canner, P. L. (1987) An overview of six clinical trials of aspirin in coronary heart disease. *Statistics in Medicine*, **6**, 255–263.

Casella, G. and Berger, R. L. (1990) *Statistical Inference*. Duxbury, Belmont, CA.

Cochran, W. G. (1977) *Sampling Techniques*, 3rd edition. Wiley, New York.

Coe, P. R. and Tamhane, A. C. (1993) Small sample confidence intervals for the difference, ratio and odds ratio of 2 success probabilities. *Communications in Statistics – Simulation and Computation*, **22**, 925–938.

Colditz, G. A., Stampfer, M. J., Willett, W. C., Hennekens, C. H., Rosner, B. and Speizer, F. E. (1990) Prospective study of estrogen replacement therapy and risk of breast cancer in post-menopausal women. *Journal of the American Medical Association*, **264**, 2648–2653.

Cooper, M. R., Dear, K. B. G., McIntyre, O. R., Ozer, H., Ellerton, J., Cannellos, G., Bernhardt, B., Duggan, B., Faragher, D. and Schiffer, C. (1993) A randomized clinical trial comparing melphalan/prednisone with or without interferon α-2b in newly diagnosed patients with multiple myeloma: A Cancer and Leukemia Group B study. *Journal of Clinical Oncology*, **11**, 155–160.

Cornfield, J. (1978) Randomization by group: a formal analysis. *American Journal of Epidemiology*, **108**, 100–102.

Coronary Drug Project Research Group. (1976) Aspirin in coronary heart disease. *Journal of Chronic Diseases*, **29**, 625–642.

Cox, D. R. and Oakes, D. (1984) *Analysis of Survival Data*. Chapman & Hall, London.

Dixon, W. J. and Massey, F. J. Jr. (1969) *Introduction to Statistical Analysis*, 3rd edition. McGraw-Hill, New York.

Doll, R. and Hill, A. B. (1966) Mortality of British doctors in relation to smoking: observations on coronary thrombosis. *National Cancer Institute Monograph*, **19**, 205–268.

Donner, A., Birkett, N. and Buck, C. (1981) Randomization by cluster sample size requirements and analysis. *American Journal of Epidemiology*, **14**, 906–914.

Edwardes, M. D. (1995) A confidence interval for $\Pr(X < Y) - \Pr(X > Y)$ estimated from simple cluster samples. *Biometrics*, **51**, 571–578.

Elston, R. C. (1977) Response to query: estimating 'inheritability' of a dichotomous trait. *Biometrics*, **33**, 232–233.

Elwood, P. C. and Sweetnam, P. M. (1979) Aspirin and secondary mortality after myocardial infarction. *Lancet*, **2**, 1313–1315.

Elwood, P. C., Cochrane, A. L., Burr, M. L., Sweetnam, P. M., Williams, G., Welsby, E., Hughes, S. J. and Renton, R. (1974) A randomized controlled trial of acetyl salicylic acid in the secondary prevention of mortality from myocardial infarction. *British Medical Journal*, **1**, 436–440.

Finney, D. J. (1949) On a method of estimating frequencies. *Biometrika*, **36**, 233–234.

Fisher, R. A. (1928) On a distribution yielding the error functions of several well known statistics. In J. C. Fields (ed.), *Proceedings of the International Mathematical Congress*, Vol. 2. University of Toronto Press, Toronto, pp. 805–813.

Fleiss, J. L. (1981) *Statistical Methods for Rates and Proportions*, 2nd edition. Wiley, New York.

Fleiss, J. L. (1986) *The Design and Analysis of Clinical Experiments*. Wiley, New York.

Greenland, S. and Robins, J. M. (1985) Estimation of a common effect parameter from sparse follow-up data. *Biometrics*, **41**, 55–68.
Haldane, J. B. S. (1945) On a method of estimating frequencies. *Biometrika*, **33**, 222–225.
Hauck, W. W. and Anderson, S. (1986) A comparison of large-sample confidence interval methods for the difference of two binomial probabilities. *American Statistician*, **40**, 318–322.
Herrera, M. G., Nestel, P., El Amin, A., Fawzi, W. W., Mohamed, K. A. and Weld, L. (1992) Vitamin A supplementation and child survival. *Lancet*, **340**, 267–271.
Johnson, N. L. and Kotz, S. (1970) *Distributions in Statistics: Continuous Univariate Distributions 2*. Wiley, New York.
Klar, N. and Donner, A. (2001) Current and future challenges in the design and analysis of cluster randomization trials. *Statistics in Medicine*, **20**, 3729–3740.
Lipsitz, S. R., Dear, K. B. G., Laird, N. M. and Molenberghs, G. (1998) Tests for homogeneity of the risk difference when data are sparse. *Biometrics*, **54**, 148–160.
Lui, K.-J. (1998) Confidence intervals for differences in correlated binary proportions. *Statistics in Medicine*, **17**, 2017–2021.
Lui, K.-J. (1999) Interval estimation of simple difference under independent negative binomial sampling. *Biometrical Journal*, **41**, 83–92.
Lui, K.-J. (2000) Confidence intervals of the simple difference between the proportions of a primary infection and a secondary infection, given the primary infection. *Biometrical Journal*, **42**, 59–69.
Lui, K.-J. (2001a) Interval estimation of simple difference in dichotomous data with repeated measurements. *Biometrical Journal*, **43**, 845–861.
Lui, K.-J. (2001b) A note on interval estimation of the simple difference in data with correlated matched pairs. *Biometrical Journal*, **43**, 235–247.
Lui, K.-J. (2002) Notes on estimation of the general odds ratio and the general risk difference for paired-sample data. *Biometrical Journal*, **44**, 957–968.
Lui, K.-J. (2003) Notes on interval estimation of a common rate difference under stratified Poisson sampling. Department of Mathematics and Statistics, San Diego State University.
Lui, K.-J. and Kelly, C. (2000) A revisit on tests for homogeneity of the risk difference. *Biometrics*, **56**, 309–315.
Lui, K.-J. and Lin, C. D. (2003) Four confidence intervals of rate difference under Poisson distribution. *Journal of Probability and Statistical Science*. To appear.
Lui, K.-J., Cumberland, W. G., and Kuo, L. (1996) An interval estimate for the intraclass correlation in beta-binomial sampling. *Biometrics*, **52**, 412–425.
Lui, K.-J., Mayer, J. A. and Eckhardt, L. (2000) Confidence intervals for the risk ratio under cluster sampling based on the beta-binomial model. *Statistics in Medicine*, **19**, 2933–2942.
May, W. L. and Johnson, W. D. (1997) Confidence intervals for differences in correlated binary proportions. *Statistics in Medicine*, **16**, 2127–2136.
Mayer, J., Slymen, D. J., Eckhardt, L., *et al.* (1997) Reducing ultraviolet radiation exposure in children. *Preventive Medicine*, **26**, 516–522.
Mee, R. W. (1984) Confidence bounds for the difference between two probabilities. *Biometrics*, **40**, 1175–1176.
Miettinen, O. S. (1985) *Theoretical Epidemiology: Principles of Occurrence Research in Medicine*. Wiley, New York.
Mikulski, P. W. and Smith, P. J. (1976) A variance bound for unbiased estimation in inverse sampling. *Biometrika*, **63**, 216–217.
Newcombe, R. G. (1998) Interval estimation for the difference between independent proportions: comparison of eleven methods. *Statistics in Medicine*, **17**, 873–890.
Newman, S. C. (2001) *Biostatistical Methods in Epidemiology*. Wiley, New York.
Persantine-Aspirin Reinfarction Study Research Group (1980) Persantine and aspirin in coronary heart disease. *Circulation*, **62**, 449–461.

Rosner, B. (1990) *Fundamentals of Biostatistics*. PWS-Kent, Boston.

Rosner, B. (2000) *Fundamentals of Biostatistics*, 5th edition. Duxbury, Pacific Grove, CA.

Santner, T. J. and Snell, M. K. (1980) Small-sample confidence intervals for $p_1 - p_2$ and p_1/p_2 in 2×2 contingency tables. *Journal of the American Statistical Association*, **73**, 386–394.

Santner, T. J. and Yamagani, S. (1993) Invariant small sample confidence intervals for the difference of two success probabilities. *Communications in Statistics – Simulation and Computation*, **22**, 33–59.

Sato, T. (1989) On the variance estimator for the Mantel–Haenszel risk difference. *Biometrics*, **45**, 1323–1324.

Selvin, S. (1996). *Statistical Analysis of Epidemiologic Data*. Oxford University Press, New York.

Sharief, M., Viteri, C., Ben-Menachem, E., Weber, M., Reife, R., Pledger, G. and Karim, R. (1996) Double-blind, placebo-controlled study of topiramate in patients with refractory partial epilepsy. *Epilepsy Research*, **25**, 217–224.

Wallenstein, S. (1997) A non-iterative accurate asymptotic confidence interval for the difference between two proportions. *Statistics in Medicine*, **16**, 1329–1336.

3

Relative Difference

To measure the excess effect due to a risk factor on the probability of having a given disease or to quantify the efficacy of a treatment to reduce the risk of developing an undesirable outcome, Sheps (1958, 1959) proposes use of the relative difference (Fleiss, 1981). To clarify the practical meaning of this index, we use the following two examples given by Sheps (1958, 1959). In the first example, Sheps considers a study of the association between smoking and the rate of mortality from lung cancer. Sheps maintains that the excess risk of death associated with smoking can only affect those individuals who would not have died from lung cancer if they had not smoked. Thus, it is reasonable to postulate that $\pi_1 > \pi_0$ and $\pi_1 = \pi_0 + \delta(1 - \pi_0)$, where π_1 and π_0 denote the lung cancer mortality rates in the smoking and non-smoking populations, respectively, and the parameter δ denotes the proportion of subjects who die from lung cancer among those who would otherwise have escaped death from lung cancer if they had not smoked. Thus, the relative difference $\delta = (\pi_1 - \pi_0)/(1 - \pi_0)$ represents the additional risk of dying from lung cancer attributed to smoking. In the second example, Sheps (1958, 1959) considers a vaccine trial for poliomyelitis. Let π_1 and π_0 denote the proportions of subjects who are free of poliomyelitis in the vaccinated and unvaccinated groups. Because the vaccine is expected to protect a fraction δ of those who would have developed poliomyelitis if they had not been vaccinated, we may assume that $\pi_1 > \pi_0$ and $\pi_1 = \pi_0 + \delta(1 - \pi_0)$. Thus, the relative difference δ represents the protection effect due to vaccinating those subjects who would otherwise have had poliomyelitis. In this example, note that the relative difference δ is actually equal to $1 - \phi$, where $\phi = (1 - \pi_1)/(1 - \pi_0)$ represents the relative risk of poliomyelitis between the vaccinated and the unvaccinated groups. The relative difference in this context is also called the relative risk reduction (Laupacis *et al.*, 1988; Hutton, 2000). Note also that the range for both δ and ϕ considered here is [0, 1]. When there is no association between the exposure and the outcome, $\delta = 0$. When the probability π_0 is extremely small ($\doteq 0$), the relative difference δ approximately equals the risk difference Δ discussed in Chapter 2.

Statistical Estimation of Epidemiological Risk K-J. Lui
 ISBN: 0-470-85071-X (HB)

We generally use the relative difference in cohort studies or randomized clinical trials. We first discuss estimation of the relative difference when sampling subjects independently from two comparison populations. We then discuss estimation of the relative difference when using stratified analysis to control confounders or when analyzing data collected from a multicenter study design. We further discuss estimation of the relative difference under independent cluster sampling. This is useful for cluster randomization trials. We also discuss estimation of the relative difference for paired-sample data in which a matched-pair design is employed to increase the efficiency of estimation in clinical trials. Finally, to ensure that we can collect an adequate number of cases in the sample when the underlying disease is rare, we consider estimating the relative difference under independent inverse sampling.

3.1 INDEPENDENT BINOMIAL SAMPLING

Suppose that we independently take a random sample of n_i subjects from the population with exposure ($i = 1$) and from the population with non-exposure ($i = 0$) to a risk factor, respectively. Suppose further that the risk factor under investigation tends to increase the probability π_i of possessing the underlying disease of interest so that $\pi_1 > \pi_0$. Suppose we obtain X_i cases. The random variable X_i then follows the binomial distribution (1.1) with parameters n_i and π_i.

First, note that the maximum likelihood estimator of π_i under independent binomial sampling is simply $\hat{\pi}_i = X_i/n_i$. Because the MLE is invariant with respect to functional transformation (Casella and Berger, 1990), the MLE of the relative difference δ for $\hat{\pi}_1 > \hat{\pi}_0$ is simply (Sheps, 1959)

$$\hat{\delta} = (\hat{\pi}_1 - \hat{\pi}_0)/(1 - \hat{\pi}_0). \tag{3.1}$$

Furthermore, using the delta method (Bishop *et al.*, 1975), we obtain that the estimated asymptotic variance of $\hat{\delta}$ is equal to (**Exercise 3.1**)

$$\widehat{\text{Var}}(\hat{\delta}) = \hat{\phi}^2\{\hat{\pi}_1/[n_1(1 - \hat{\pi}_1)] + \hat{\pi}_0/[n_0(1 - \hat{\pi}_0)]\}, \tag{3.2}$$

where $\hat{\phi} = (1 - \hat{\pi}_1)/(1 - \hat{\pi}_0)$. On the basis of (3.1) and (3.2), an asymptotic $100(1 - \alpha)$ percent confidence interval for the relative difference δ is given by

$$[\max\{\hat{\delta} - Z_{\alpha/2}\sqrt{\widehat{\text{Var}}(\hat{\delta})}, 0\}, \min\{\hat{\delta} + Z_{\alpha/2}\sqrt{\widehat{\text{Var}}(\hat{\delta})}, 1\}] \tag{3.3}$$

where Z_α is the upper 100αth percentile of the standard normal distribution. When both the sample size n_i and the underlying disease rate π_i are small, the sampling distribution of $\hat{\delta}$ may be skewed and hence the interval estimator (3.3) is unlikely to perform well. To improve the performance of (3.3), Walter (1975) suggests using a logarithmic transformation $\log(1 - x)$. Using the delta method again, we can easily show that an estimated asymptotic variance of $\log(\hat{\phi})$ is given by (**Exercise 3.2**)

$$\widehat{\text{Var}}(\log(\hat{\phi})) = \hat{\pi}_1/[n_1(1-\hat{\pi}_1)] + \hat{\pi}_0/[n_0(1-\hat{\pi}_0)]. \tag{3.4}$$

This leads to an asymptotic $100(1-\alpha)$ percent confidence interval for δ given by

$$[1-\min\{\hat{\phi}\exp(Z_{\alpha/2}\sqrt{\widehat{\text{Var}}(\log(\hat{\phi}))}), 1\}, 1-\hat{\phi}\exp(-Z_{\alpha/2}\sqrt{\widehat{\text{Var}}(\log(\hat{\phi}))})]. \tag{3.5}$$

Using a principle analogous to that of Fieller's theorem (Casella and Berger, 1990), we define $Z = (1-\hat{\pi}_1) - \phi(1-\hat{\pi}_0)$. We can easily see that the expectation $E(Z) = 0$. When both n_i are large, we may claim that the probability $P((Z/\sqrt{\text{Var}(Z)})^2 \leq Z^2_{\alpha/2}) \doteq 1-\alpha$. This leads us to consider the following quadratic equation in ϕ (**Exercise 3.3**):

$$A\phi^2 - 2B\phi + C \leq 0, \tag{3.6}$$

where $A = (1-\hat{\pi}_0)^2 - Z^2_{\alpha/2}\hat{\pi}_0(1-\hat{\pi}_0)/n_0$, $B = (1-\hat{\pi}_1)(1-\hat{\pi}_0)$, and $C = (1-\hat{\pi}_1)^2 - Z^2_{\alpha/2}\hat{\pi}_1(1-\hat{\pi}_1)/n_1$. If $A > 0$ and $B^2 - AC > 0$, then an asymptotic $100(1-\alpha)$ percent confidence interval for δ will be given by

$$[1-\min\{(B+\sqrt{B^2-AC})/A, 1\}, 1-\max\{(B-\sqrt{B^2-AC})/A, 0\}]. \tag{3.7}$$

By definition, since $\delta > 0$, if the resulting estimate $\hat{\delta}$ (3.1) is negative, we would set the point estimate of δ equal to 0. When either of the $\hat{\pi}_i$ equals 0 or 1 in (3.3) and (3.5), using the estimates $\widehat{\text{Var}}(\hat{\delta})$ (3.2) and $\widehat{\text{Var}}(\log(\hat{\delta}))$ (3.4) is certainly inappropriate, as is use of (3.7). In these cases, we recommend the *ad hoc* adjustment procedure for sparse data of using $(X_i + 0.5)/(n_i + 1)$ to estimate π_i. Note that when deriving asymptotic variances (3.2) and (3.4), we do not account for the range of δ or ϕ. Therefore, if our goal is to produce an interval estimator of δ, we will recommend using $\hat{\delta}$ in (3.3) rather than 0 if $\hat{\delta} < 0$. Similarly, we will recommend using $\hat{\phi}$ in (3.5) rather than 1 if $\hat{\phi} > 1$. After obtaining the resulting interval estimate, however, we will adjust these limits to ensure that they fall in the range [0, 1] for δ. Similar principles to these are used throughout this chapter.

Example 3.1 To illustrate the use of interval estimators (3.3), (3.5), and (3.7), consider the data obtained from the Framingham epidemiologic study of heart disease (Dawber *et al.*, 1957; Sheps, 1959). During a 4-year period, there were 20 new cases of arteriosclerotic heart disease (ASHD) among 176 patients in the two highest weight categories, and 32 new cases among 717 patients in the two lowest weight categories. We assume that obesity is a risk factor and acts on persons who are otherwise in the 'no ASHD' category. Given these data, the point estimate $\hat{\delta}$ (3.1) of the relative difference is 0.072, with an estimated standard error of 0.026. This indicates that there is an additional 7.2% risk of developing ASHD among patients who would not have developed ASHD if their weights had been in the lower weight category. Applying (3.3), (3.5), and (3.7) gives 95% confidence intervals for δ of [0.021, 0.123], [0.020, 0.122], and [0.021, 0.123], respectively. Since we have such a large number of subjects in this example, all the interval estimators (3.3), (3.5), and (3.7) are appropriate for use.

Example 3.2 Consider the hypothetical data (Fleiss, 1981, p. 101) from a clinical trial comparing two treatments. Assume that the improvement rate π_1 in treatment group 1 is higher than the improvement rate π_0 in treatment group 0. Suppose there are 56 out of 70 subjects showing improvement ($\hat{\pi}_1 = 0.80$) in the treatment group. By contrast, there are only 48 out of 80 subjects showing improvement ($\hat{\pi}_0 = 0.60$) in the control group. Given these data, the point estimate $\hat{\delta}$ is 0.50, with an estimated standard error of 0.138. Applying (3.3), (3.5), and (3.7) gives 95% confidence intervals for δ of [0.230, 0.770], [0.142, 0.709], and [0.178, 0.744]. We can see that the interval estimate (3.3) is shifted to the right compared to the interval estimate (3.5). In this case, where the number of subjects is not large, we should apply (3.3) with caution.

3.2 A SERIES OF INDEPENDENT BINOMIAL SAMPLING PROCEDURES

Suppose that we employ pre-stratified sampling in a study with S strata. For each stratum $s(s = 1, 2, \ldots, S)$, we independently sample n_{is} subjects from the exposed $(i = 1)$ and the non-exposed $(i = 0)$ populations, respectively. Let π_{is} denote the probability that a randomly selected subject from the ith population in the sth stratum is a case. Assume that exposure to the risk factor under investigation tends to increase the risk of the disease of interest – that is, we assume that $\pi_{1s} > \pi_{0s}$. The relative difference in the sth stratum is then equal to $\delta_s = (\pi_{1s} - \pi_{0s})/(1 - \pi_{0s}) = 1 - \phi_s$, where $\phi_s = (1 - \pi_{1s})/(1 - \pi_{0s})$. Let X_{is} denote the number of cases among the n_{is} subjects in the ith group of the sth stratum. Under the above assumptions, the random variables X_{is} independently follow the binomial distribution with parameters n_{is} and π_{is}. Therefore, the joint probability mass function of the random vector $\mathbf{X}' = (\mathbf{X}_1', \mathbf{X}_0')$, where $\mathbf{X}_i' = (X_{i1}, X_{i2}, \ldots, X_{iS})$ for $i = 1, 0$, is

$$f_{\mathbf{X}}(\mathbf{x}|\mathbf{n}, \boldsymbol{\pi}) = \prod_{s=1}^{S} \prod_{i=0}^{1} \binom{n_{is}}{x_{is}} (\pi_{is})^{x_{is}} (1 - \pi_{is})^{n_{is} - x_{is}}. \quad (3.8)$$

where $x_{is} = 0, 1, 2, \ldots, n_{is}$, $\mathbf{n}' = (n_{11}, n_{12}, \ldots, n_{1S}, n_{01}, \ldots, n_{0S})$, and $\boldsymbol{\pi}' = (\pi_{11}, \pi_{12}, \ldots, \pi_{1S}, \pi_{01}, \ldots, \pi_{0S})$.

3.2.1 Asymptotic interval estimators

Under distribution (3.8), the MLE of the relative difference δ_s in the sth stratum is $\hat{\delta}_s = 1 - \hat{\phi}_s$ for $\hat{\pi}_{1S} > \hat{\pi}_{0S}$, where $\hat{\phi}_s = (1 - \hat{\pi}_{1s})/(1 - \hat{\pi}_{0s})$, and $\hat{\pi}_{is} = X_{is}/n_{is}$ for

$i = 1, 0$, and $s = 1, \ldots, S$. In this section, we assume that $\delta_1 = \delta_2 = \ldots = \delta_S$. Let δ_c denote this common value. We can easily see that this is equivalent to saying that $\phi_1 = \phi_2 = \ldots, = \phi_S = \phi_c$, where $\phi_s = 1 - \delta_s$ and $\phi_c = 1 - \delta_c$.

First, consider the situation in which there is a reasonably large sample size in each stratum. If the parameters π_{is} were known, we might estimate δ_c with an optimally weighted average of the stratum-specific estimates $\hat{\phi}_s$ (**Exercise 2.6**): $1 - (\sum_s W_s\hat{\phi}_s)/\sum_s W_s$, where $W_s = 1/\text{Var}(\hat{\phi}_s)$, and $\text{Var}(\hat{\phi}_s) = \phi_c^2[\pi_{1s}/(n_{1s}(1-\pi_{1s})) + \pi_{0s}/(n_0(1-\pi_{0s}))]$. Note that because ϕ_c is a constant, we can easily see that $1 - (\sum_s W_s\hat{\phi}_s)/\sum_s W_s = 1 - (\sum_s W_s^*\hat{\phi}_s)/\sum_s W_s^*$, where $W_s^* = 1/\text{Var}(\log(\hat{\phi}_s))$, and $\text{Var}(\log(\hat{\phi}_s)) = \pi_{1s}/(n_{1s}(1-\pi_{1s})) + \pi_{0s}/(n_{0s}(1-\pi_{0s}))$. Thus, we consider the following weighted least-squares estimator when the parameters π_{is} are unknown:

$$\hat{\delta}_{\text{WLS}} = 1 - \left(\sum_s \hat{W}_s^*\hat{\phi}_s\right) \bigg/ \sum_s \hat{W}_s^* \tag{3.9}$$

where $\hat{W}_s^* = 1/\widehat{\text{Var}}(\log(\hat{\phi}_s))$, and $\widehat{\text{Var}}(\log(\hat{\phi}_s)) = \hat{\pi}_{1s}/(n_{1s}(1-\hat{\pi}_{1s})) + \hat{\pi}_{0s}/(n_{0s}(1-\hat{\pi}_{0s}))$. Note that whenever $\hat{\delta}_{\text{WLS}} < 0$, we set $\hat{\delta}_{\text{WLS}} = 0$. Note also that since $\hat{\phi}_s$ is a ratio of two proportions, its sampling distribution may be skewed. When developing an interval estimator of δ_c, we may use the logarithmic transformation $\log(1 - x)$ to improve the normal approximation of $\hat{\delta}_{\text{WLS}}$ (3.9). Because the variance $\text{Var}((\sum_s W_s^*\log(\hat{\phi}_s))/(\sum_s W_s^*))$ equals $1/\sum_s W_s^*$, an asymptotic $100(1-\alpha)$ percent confidence interval for δ_c is given by

$$\left[\max\left\{1 - \exp\left(\frac{\sum_s \hat{W}_s^*\log(\hat{\phi}_s)}{\sum_s \hat{W}_s^*} + \frac{Z_{\alpha/2}}{\sqrt{\sum_s \hat{W}_s^*}}\right), 0\right\},\right.$$

$$\left. 1 - \exp\left(\frac{\sum_s \hat{W}_s^*\log(\hat{\phi}_s)}{\sum_s \hat{W}_s^*} - \frac{Z_{\alpha/2}}{\sqrt{\sum_s \hat{W}_s^*}}\right)\right]. \tag{3.10}$$

Note that when π_{is} is close to 0 or 1, the probability of obtaining $\hat{\pi}_{is} = 0$ or 1 may not be negligible. Thus, we may have the difficulty in calculating $\hat{W}_s^*$ when using (3.9) and (3.10). Following Rothman and Boice (1979) and Tarone (1981), we may consider the following Mantel–Haenszel type estimator (Mantel and Haenszel, 1959) of the relative difference,

$$\hat{\delta}_{\text{MH}} = 1 - \hat{\phi}_{\text{MH}}, \tag{3.11}$$

where $\hat{\phi}_{\text{MH}} = (\sum_s(n_{1s} - x_{1s})n_{0s}/n_{.s})/(\sum_s(n_{0s} - x_{0s})n_{1s}/n_{.s})$ and $n_{.s} = n_{1s} + n_{0s}$. Furthermore, using the delta method, we can easily show that the estimated asymptotic variance of $\hat{\phi}_{MH}$ (**Exercise 3.6**) is given by

$$\widehat{\mathrm{Var}}(\hat{\phi}_{MH}) = \left(\sum_s \frac{(1-\hat{\pi}_{0s})n_{0s}n_{1s}}{n_{.s}}\right)^{-2} \left[\sum_s \left(\frac{n_{0s}}{n_{.s}}\right)^2 n_{1s}\hat{\pi}_{1s}(1-\hat{\pi}_{1s}) + \hat{\phi}^2_{MH} \sum_s \left(\frac{n_{1s}}{n_{.s}}\right)^2 n_{0s}\hat{\pi}_{0s}(1-\hat{\pi}_{0s})\right]. \quad (3.12)$$

To produce an interval estimator of δ_c, we may again use the logarithmic transformation of $\hat{\phi}_{MH}$. Because the estimated asymptotic variance $\widehat{\mathrm{Var}}(\log(\hat{\phi}_{MH}))$ can be approximated by $\widehat{\mathrm{Var}}(\hat{\phi}_{MH})/\hat{\phi}^2_{MH}$, we obtain an asymptotic $100(1-\alpha)$ percent confidence interval for δ_c given by

$$[\max\{1-\hat{\phi}_{MH}\exp(Z_{\alpha/2}\sqrt{\widehat{\mathrm{Var}}(\hat{\phi}_{MH})/\hat{\phi}^2_{MH}}), 0\}, \\ 1-\hat{\phi}_{MH}\exp(-Z_{\alpha/2}\sqrt{\widehat{\mathrm{Var}}(\hat{\phi}_{MH})/\hat{\phi}^2_{MH}})]. \quad (3.13)$$

When calculating (3.9)–(3.13), if any observed cell frequency in stratum s is 0, then we will recommend using the adjustment procedure for sparse data by adding 0.50 to each cell in that particular stratum. On the basis of Monte Carlo simulation, Lui (2002) finds that the coverage probability of (3.10) can be much smaller than the desired confidence level, especially for a small stratum size. When the underlying common δ_c is small (say, 0.20), using (3.13) is preferable to the other estimators considered by Lui (2002). On the other hand, when n_{is} is small (say, 5) and the number of strata S is large (say, 20), the coverage probability of (3.13) tends to be smaller than the desired confidence level, and using the interval estimator with limits $\hat{\delta}_{MH} \pm Z_{\alpha/2}\sqrt{\widehat{\mathrm{Var}}(\hat{\phi}_{MH})}$ – subject to their falling in the interval [0, 1] – is probably best.

3.2.2 Test for the homogeneity of relative difference

Before using the summary estimators discussed in the previous subsection, it is important to examine whether the underlying assumption that $\delta_1 = \delta_2 = \ldots = \delta_S = \delta_c$ holds in the data. Thus, we discuss testing the homogeneity of the relative difference in this section. Note that the relative difference δ_s is constant across strata if and only if ϕ_s is constant across strata. Note further that ϕ_s is a ratio of two proportions. Thus, all the procedures considered by Lui and Kelly (2000) for testing the homogeneity of the relative risk can be applied to test the homogeneity of the relative difference, with some slight modifications.

Following Fleiss (1981), we can apply the WLS test procedure with the logarithmic transformation of $\hat{\delta}_s$,

$$T_{WLS} = \sum_s \hat{W}^*_s \log^2(\hat{\phi}_s) - \left(\sum_s \hat{W}^*_s \log(\hat{\phi}_s)\right)^2 \Big/ \sum_s \hat{W}^*_s. \quad (3.14)$$

Under the null hypothesis $H_0 : \delta_1 = \delta_2 = \ldots = \delta_S (= \delta_c)$, the test statistic (3.14) asymptotically has the χ^2 distribution with $S - 1$ degrees of freedom as all n_{is} are large. Thus, we will reject H_0 at level α if $T_{\text{WLS}} > \chi^2_{S-1,\alpha}$, where $\chi^2_{S-1,\alpha}$ is the upper 100αth percentile of the chi-squared distribution with $S - 1$ degrees of freedom.

When the number of strata S is large, we may consider the following test statistic (Lipsitz *et al.*, 1998):

$$Z_{\text{WLS}} = [T_{\text{WLS}} - (S-1)]/\sqrt{2(S-1)}. \tag{3.15}$$

Note that if $H_0 : \delta_1 = \delta_2 = \ldots = \delta_S$ were not valid, we would expect a large value of T_{WLS}. This suggests that we reject H_0 at level α when $Z_{\text{WLS}} > Z_\alpha$.

When approximating a χ^2 random variable by a normal distribution, Fisher (1928) suggests using the logarithmic transformation. When all n_{is} and S are large, this leads us to consider the test statistic

$$Z_{\text{LWLS}} = \left[\frac{1}{2}\log\left(\frac{T_{\text{WLS}}}{S-1}\right) + \frac{1}{2(S-1)}\right] \Big/ \sqrt{\frac{1}{2(S-1)}}. \tag{3.16}$$

We will reject H_0 at level α if $Z_{\text{LWLS}} > Z_\alpha$.

Using Monte Carlo simulations,Lui and Kelly (2000) find that test procedures (3.14) and (3.16) tend to be conservative and test procedure (3.15) is generally preferable to (3.14) and (3.16). They also find that another test procedure may be slightly better than test procedure (3.15) with respect to controlling Type I error; the procedure is somewhat tedious, and we refer readers to Lui and Kelly (2000) for details.

Example 3.3 Consider the all-cause mortality data for the aspirin ($i = 1$) and placebo ($i = 0$) groups of post-myocardial infarction patients (Table 2.1). Because we expect that taking aspirin may protect a fraction δ of those who would have died if they had not taken aspirin, we assume that $\pi_{1s} > \pi_{0s}$, where π_{1s} and π_{0s} denote the survival rates in the aspirin and the placebo groups, respectively. Also, as noted by Canner (1987), the baseline imbalance of medical conditions between the aspirin and placebo groups in the sixth trial may cause the unexpected finding that the survival rate in the aspirin is lower than that in the placebo group. Thus, we exclude this trial from consideration. On the basis of the data of the first five trials, we first test the homogeneity of the relative difference $\delta_s (= 1 - (1 - \pi_{1s})/(1 - \pi_{0s}))$. Applying test statistics (3.14)–(3.16), we obtain *p*-values of 0.95, 0.88, and 0.98, respectively. Thus, the assumption that the relative difference is constant over the first five trials should be reasonable. In fact, the MLEs $\hat{\delta}_s (s = 1, 2, \ldots, 5)$ are 0.26, 0.17, 0.30, 0.18, and 0.18, respectively. Applying (3.9) and (3.11), we obtain $\hat{\delta}_{\text{WLS}} = 0.210$ and $\hat{\delta}_{\text{MH}} = 0.213$. Furthermore, using (3.10) and (3.13), we obtain 95% confidence intervals of [0.087, 0.320] and [0.089, 0.321], respectively. Because the lower limits of these resulting interval estimates are both slightly larger than 0, we conclude that there is significant evidence at the 5% level that taking aspirin provides a protection effect for the all-cause mortality in post-myocardial infarction patients.

3.3 INDEPENDENT CLUSTER SAMPLING

Suppose that in cluster randomization trials we randomly assign n_i clusters of $m_{ij}(j = 1, 2, \ldots, n_i)$ subjects to receive the experimental $(i = 1)$ and the standard (or placebo) treatments $(i = 0)$, respectively. Define the random variable $X_{ijk} = 1$ if the response on the kth subject $(k = 1, 2, \ldots, m_{ij})$ in cluster j of treatment i is positive, and $X_{ijk} = 0$ otherwise. Then, the total number $X_{ij.} = \sum_k X_{ijk}$ of responses in cluster j from population i follows a binomial distribution with parameters m_{ij} and p_{ij}, where p_{ij} denotes the response probability $P(X_{ijk} = 1)$. To account for the intraclass correlation between subject responses within clusters, we assume further that p_{ij} independently follows the beta distribution beta(α_i, β_i) with mean $\pi_i = \alpha_i/T_i$ and variance $\pi_i(1 - \pi_i)/(T_i + 1)$, where $\alpha_i > 0, \beta_i > 0$, and $T_i = \alpha_i + \beta_i$ (Johnson and Kotz, 1970). Under these model assumptions, the probability of a positive response for a randomly selected subject from group i is then $E(Y_{ijk}) = \pi_i$ and the intraclass correlation between X_{ijk} and $X_{ijk'}$ for $k \neq k'$ is $\rho_i = 1/(T_i + 1)$ (**Exercise 1.7**). Without loss of generality, we assume that the experimental treatment tends to increase the probability of a positive response as compared with the standard treatment. In other words, we assume that $\pi_1 > \pi_0$.

Define $\hat{\pi}_i = \sum_j X_{ij.}/m_{i.}$, where $m_{i.} = \sum_j m_{ij}$. Note that $\hat{\pi}_i$ is, in fact, the sample proportion of subjects with positive response in treatment $i(i = 1, 0)$. Note also that $\hat{\pi}_i$ is an unbiased consistent estimator of π_i with variance $\text{Var}(\hat{\pi}_i)$ equal to $\pi_i(1 - \pi_i)f(\mathbf{m}_i, \rho_i)/m_{i.}$, where $\mathbf{m}_i' = (m_{i1}, m_{i2}, \ldots, m_{in_i})$ and $f(\mathbf{m}_i, \rho_i)$ is the variance inflation factor due to the intraclass correlation ρ_i and equals $\sum_j m_{ij}[1 + (m_{ij} - 1)\rho_i]/m_{i.}$ (**Exercise 1.8**). To estimate ρ_i we may use the traditional intraclass correlation estimator $\hat{\rho}_i$ (2.19) (Elston, 1977; Lui *et al.*, 1996).

To estimate the relative difference δ, we may substitute $\hat{\pi}_i$ for π_i and obtain the estimator $\hat{\delta} = (\hat{\pi}_1 - \hat{\pi}_0)/(1 - \hat{\pi}_0)$. Using the delta method, we obtain an estimated asymptotic variance of $\hat{\delta}$ given by (**Exercise 3.7**) $\widehat{\text{Var}}(\hat{\delta}) = \hat{\phi}^2\{\hat{\pi}_1 f(\mathbf{m}_1, \hat{\rho}_1)/[m_{1.}(1 - \hat{\pi}_1)] + \hat{\pi}_0 f(\mathbf{m}_0, \hat{\rho}_0)/[m_{0.}(1 - \hat{\pi}_0)]\}$, where $\hat{\phi} = (1 - \hat{\pi}_1)/(1 - \hat{\pi}_0)$. Thus, we obtain an asymptotic $100(1 - \alpha)$ percent confidence interval for δ given by

$$[\max\{\hat{\delta} - Z_{\alpha/2}\sqrt{\widehat{\text{Var}}(\hat{\delta})}, 0\}, \min\{\hat{\delta} + Z_{\alpha/2}\sqrt{\widehat{\text{Var}}(\hat{\delta})}, 1\}]. \tag{3.17}$$

When both the sample size n_i and the probability of positive response π_i are small, the sampling distribution of $\hat{\delta}$ may be skewed and hence interval estimator (3.17) may not perform well. Following a similar idea to that for deriving interval estimator (3.5), we may consider the logarithmic transformation $\log(1 - x)$.

Thus, we obtain an asymptotic $100(1-\alpha)$ percent confidence interval for δ given by

$$[1-\min\{\hat{\phi}\exp(Z_{\alpha/2}\sqrt{\widehat{\mathrm{Var}}(\log(\hat{\phi}))}), 1\}, 1-\hat{\phi}\exp(-Z_{\alpha/2}\sqrt{\widehat{\mathrm{Var}}(\log(\hat{\phi}))})], \tag{3.18}$$

where $\widehat{\mathrm{Var}}(\log(\hat{\phi})) = \hat{\pi}_1 f(\mathbf{m}_1, \hat{\rho}_1)/[m_{1.}(1-\hat{\pi}_1)] + \hat{\pi}_0 f(\mathbf{m}_0, \hat{\rho}_0)/[m_{0.}(1-\hat{\pi}_0)]$.

Furthermore, when both n_i are large, we have the probability $P(\{[(1-\hat{\pi}_1)-\phi(1-\hat{\pi}_0)]/\sqrt{\mathrm{Var}((1-\hat{\pi}_1)-\phi(1-\hat{\pi}_0))}\}^2 \leq Z^2_{\alpha/2}) \doteq 1-\alpha$. This leads us to consider the following quadratic equation in ϕ:

$$A^\dagger\phi^2 - 2B^\dagger\phi + C^\dagger \leq 0, \tag{3.19}$$

where $A^\dagger = (1-\hat{\pi}_0)^2 - Z^2_{\alpha/2}\hat{\pi}_0(1-\hat{\pi}_0)f(\mathbf{m}_0, \hat{\rho}_0)/m_{0.}$ $B^\dagger = (1-\hat{\pi}_1)(1-\hat{\pi}_0)$, and $C^\dagger = (1-\hat{\pi}_1)^2 - Z^2_{\alpha/2}\hat{\pi}_1(1-\hat{\pi}_1)f(\mathbf{m}_1, \hat{\rho}_1)/m_{1.}$. If $A^\dagger > 0$ and $B^{\dagger 2} - A^\dagger C^\dagger > 0$, then an approximate $100(1-\alpha)$ percent confidence interval for δ would be given by

$$[1-\min\{(B^\dagger + \sqrt{B^{\dagger 2} - A^\dagger C^\dagger})/A^\dagger, 1\}, 1-\max\{(B^\dagger - \sqrt{B^{\dagger 2} - A^\dagger C^\dagger})/A^\dagger, 0\}]. \tag{3.20}$$

When applying interval estimators (3.17)–(3.20), whenever $\hat{\pi}_i = 0$ or 1, we recommend use of $(X_{i..} + 0.5)/(m_{i.} + 1)$ to estimate π_i in $\widehat{\mathrm{Var}}(\hat{\delta})$ and $\widehat{\mathrm{Var}}(\log(\hat{\delta}))$. Note also that when $m_{ij} = 1$ for all i and j, then interval estimators (3.17), (3.18), and (3.20) reduce to (3.3), (3.5), and (3.7), respectively.

Example 3.4 Consider the data (Table 1.1) taken from a study of the effect of education on behavior change with regard to employing solar protection (Mayer *et al.*, 1997). Suppose that the educational intervention increases the proportion of children who apply solar protection (Girgis *et al.*, 1994) and hence $\pi_1 = \pi_0 + \delta(1-\pi_0)$, where π_1 and π_0 denote the proportion of children who employ adequate solar protection in the intervention and the control groups, respectively. Because we randomly assign classes to one of these two comparison groups, we may reasonably assume that the intraclass correlations for the intervention and placebo groups are equal (i.e., $\rho_1 = \rho_0 = \rho_c$). We use $\hat{\rho}_c = (m_{1.}\hat{\rho}_1 + m_{0.}\hat{\rho}_0)/(m_{1.} + m_{0.})$ to estimate this common intraclass correlation ρ_c and obtain $\hat{\rho}_c = 0.30$. The point estimate $\hat{\delta}$ is 0.317. When we apply (3.17), (3.18), and (3.20), we obtain 95% confidence intervals for δ of [0.024, 0.610], [0.000, 0.555], and [0.000, 0.582], respectively. Applying (3.17) tends to produce an interval estimate that is shifted to the right compared to (3.18) and (3.20). This is consistent with the finding noted in Example 3.2. Because the lower limits of interval estimators (3.18) and (3.20) do not exclude 0, we conclude that there is no significant evidence at the 5% level that the educational program affects children's behavior in terms of having an adequate level of solar protection.

3.4 PAIRED-SAMPLE DATA

To increase the efficiency of comparing two treatments in randomized controlled trials, we may match subjects with respect to characteristics strongly associated with both the treatments under comparison and the outcomes of interest to form matched pairs. Within each pair, we randomly assign one subject to receive the experimental treatment, and the other subject to receive the standard treatment. We then compare the response rates between these two treatments. Let π_{ij} (for $i = 1, 0$ and $j = 1, 0$) denote the corresponding cell probability in the following table:

		Standard treatment response		
		Yes	No	
Experimental treatment response	Yes	π_{11}	π_{10}	$\pi_{1.}$
	No	π_{01}	π_{00}	$\pi_{0.}$
		$\pi_{.1}$	$\pi_{.0}$	

where $\pi_{i.} = \pi_{i1} + \pi_{i0}$, $\pi_{.j} = \pi_{1j} + \pi_{0j}$, and $0 < \pi_{ij} < 1$. Note that the probabilities of response for the experimental and standard treatments are simply the marginal probabilities $\pi_{1.}$ and $\pi_{.1}$, respectively.

We assume that $\pi_{1.} > \pi_{.1}$. As noted by Fleiss (1981), this assumption is tenable if the experimental treatment is a combination of the standard treatment with some beneficial compound or is at a higher dosage level than the standard treatment. The relative difference δ is then the proportion of patients who fail to respond to the standard treatment, but who are expected to respond to the experimental treatment. In other words, $\delta = (\pi_{1.} - \pi_{.1})/(1 - \pi_{.1})$. Thus $\delta = 1 - \phi$, where $\phi = (1 - \pi_{1.})/(1 - \pi_{.1})$.

Let N_{ij} denote the number of pairs from a random sample of n matched pairs that fall in cell (i, j) with probability π_{ij}. The random vector $(N_{11}, N_{10}, N_{01}, N_{00})'$ then follows the multinomial distribution (2.25) with parameters n and the probability vector $(\pi_{11}, \pi_{10}, \pi_{01}, \pi_{00})'$. The sample proportions $\hat{\pi}_{ij} = N_{ij}/n$, for $i, j = 0, 1$, are unbiased consistent estimators of π_{ij}. Furthermore, we can easily show that estimators $\hat{\pi}_{1.} = (N_{11} + N_{10})/n$ and $\hat{\pi}_{.1} = (N_{11} + N_{01})/n$ are also unbiased consistent estimators of parameters $\pi_{1.}$ and $\pi_{.1}$, respectively. The MLE of δ for $\hat{\pi}_{1.} > \hat{\pi}_{.1}$ is $\hat{\delta} = (\hat{\pi}_{1.} - \hat{\pi}_{.1})/(1 - \hat{\pi}_{.1})$.

By using the multivariate central limit theorem, as $n \to \infty$, we may claim that $\sqrt{n}((\hat{\pi}_{1.}, \hat{\pi}_{.1})' - (\pi_{1.}, \pi_{.1})')$ has an asymptotic normal distribution with mean vector $(0, 0)'$ and covariance matrix $\mathbf{\Sigma}$, where $\mathbf{\Sigma}$ is a 2×2 matrix with diagonal terms equal to $\pi_{1.}(1 - \pi_{1.})$ and $\pi_{.1}(1 - \pi_{.1})$, respectively, and with both off-diagonal terms equal to $\pi_{11}\pi_{00} - \pi_{10}\pi_{01}$ (**Exercise 3.8**). Using the delta method (Anderson, 1958; Appendix), we may claim that $\sqrt{n}(\hat{\phi} - \phi)$, where $\hat{\phi} = (1 - \hat{\pi}_{1.})/(1 - \hat{\pi}_{.1})$, has an asymptotic normal distribution with mean 0 and variance $n\text{Var}(\hat{\phi}) = (1 - \pi_{1.})(1 - \pi_{.1})\{[(1 - \pi_{.1})\pi_{1.} - (\pi_{11}\pi_{00} - \pi_{10}\pi_{01})] + [(1 - \pi_{1.})\pi_{.1} - (\pi_{11}\pi_{00} - \pi_{10}\pi_{01})]\}/(1 - \pi_{.1})^4$ (**Exercise 3.9**). We can estimate this variance $\text{Var}(\hat{\phi})$ by simply substituting the corresponding unbiased consistent estimate $\hat{\pi}_{ij}$ for π_{ij} to obtain $\widehat{\text{Var}}(\hat{\phi}) = (N_{10} + N_{01})(N_{10} + N_{00})(N_{01} + N_{00})/(N_{10} + N_{00})^4$. Therefore, an asymptotic $100(1 - \alpha)$ percent confidence interval for $\delta(= 1 - \phi)$ is

$$[1 - \min\{\hat{\phi} + Z_{\alpha/2}\sqrt{\widehat{\text{Var}}(\hat{\phi})}, 1\}, 1 - \max\{\hat{\phi} - Z_{\alpha/2}\sqrt{\widehat{\text{Var}}(\hat{\phi})}, 0\}]. \quad (3.21)$$

Except for the trivial adjustment to ensure that the resulting confidence limits fall in the range [0, 1] of δ, the interval estimator (3.21) is actually identical to that given in Fleiss (1981, p. 118). Since the sampling distribution of $1 - \hat{\phi}$ may be skewed when n is small, we consider using the logarithmic transformation to improve the normal approximation of $\hat{\phi}$.

Using of the delta method, as $n \to \infty$, we can show that $\sqrt{n}(\log(\hat{\phi}) - \log(\phi))$ has an asymptotic normal distribution with mean 0 and variance $\pi_{1.}/(1 - \pi_{1.}) + \pi_{.1}/(1 - \pi_{.1}) - 2(\pi_{11}\pi_{00} - \pi_{10}\pi_{01})/((1 - \pi_{1.})(1 - \pi_{.1}))$ (**Exercise 3.10**). Substituting the corresponding unbiased consistent estimate $\hat{\pi}_{ij}$ for π_{ij}, we obtain an estimated variance $n\widehat{\text{Var}}(\log(\hat{\phi})) = (N_{11} + N_{10})/(N_{01} + N_{00}) + (N_{11} + N_{01})/(N_{10} + N_{00}) - 2(N_{11}N_{00} - N_{10}N_{01})/((N_{01} + N_{00})(N_{10} + N_{00}))$. As noted earlier, since the range of $\log(\phi)$ is $-\infty < \log(\phi) < 0$, an asymptotic $100(1 - \alpha)$ percent confidence interval for δ is then

$$[1 - \exp(\min\{\log(\hat{\phi}) + Z_{\alpha/2}\sqrt{\widehat{\text{Var}}(\log(\hat{\phi}))}, 0\}), 1 - \hat{\phi}\exp(-Z_{\alpha/2}\sqrt{\widehat{\text{Var}}(\log(\hat{\phi}))})]. \quad (3.22)$$

On the basis of Monte Carlo simulation, Lui (1998a) notes that when n is small, the estimated coverage probability when using interval estimator (3.21) tends to be less than the desired confidence level. Lui (1998a) also finds that applying (3.22) not only improves the coverage probability of (3.21) but also gives an interval estimator that is shorter on average than (3.21) in a variety of situations. When n is large, however, (3.21) and (3.22) are essentially equivalent.

Example 3.5 To illustrate the use of (3.21) and (3.22), consider first a controlled comparative trial with a small risk difference $\Delta(= \pi_{1.} - \pi_{.1})$. Suppose that we obtain the data: $n_{11} = 1, n_{10} = 7, n_{01} = 6$, and $n_{00} = 6$. The estimated risk difference $\hat{\Delta} = \hat{\pi}_{1.} - \hat{\pi}_{.1} = \hat{\pi}_{10} - \hat{\pi}_{01}$ is 0.05. The 95% confidence limits for δ are [0.0, 0.60] and [0.0, 0.48], corresponding to use of (3.21) and (3.22), respectively. As noted elsewhere (Lui, 1998a), the interval estimator (3.21) in this case is less precise than (3.22) with respect to the average length of the confidence interval.

By contrast, consider a controlled trial with a large risk difference Δ. Suppose that we obtain the data: $n_{11} = 1, n_{10} = 10, n_{01} = 1$, and $n_{00} = 8$. The estimate $\hat{\Delta}$ is then 0.45. Applying (3.21) and (3.22), we obtain 95% confidence limits for δ of [0.25, 0.76] and [0.17, 0.70], respectively. Although the average lengths of these two resulting confidence intervals are essentially equal, Lui (1998a) notes that the coverage probability of the former tends to be less than the desired confidence level in this case, where the sample size n is small and Δ is large.

3.5 INDEPENDENT INVERSE SAMPLING

When the underlying disease is rare or when the data arrive sequentially, we may consider use of inverse sampling (Haldane, 1945) to ensure that we can collect an appropriate number of cases in our sample. Suppose that we independently continue sampling subjects until we obtain a predetermined number x_i of cases from the population with exposure ($i = 1$) and from the population with non-exposure ($i = 0$) to a risk factor, respectively. Let Y_i denote the number of non-cases collected before obtaining x_i cases in group $i (i = 1, 0)$. Then Y_i follows the negative binomial distribution (1.13) with parameters x_i and π_i. We assume that $\pi_1 > \pi_0$.

Under independent negative binomial sampling, we can easily show that the MLE of π_i is $\hat{\pi}_i = x_i/N_i$, where $N_i = x_i + Y_i$, with asymptotic variance $\pi_i^2(1 - \pi_i)/x_i$ (**Exercise 1.11**). This implies that the MLE of δ for $\hat{\pi}_1 > \hat{\pi}_0$ is $\hat{\delta} = (\hat{\pi}_1 - \hat{\pi}_0)/(1 - \hat{\pi}_0)$. Applying the delta method again, we obtain an asymptotic $100(1 - \alpha)$ percent confidence interval for δ given by

$$[\max\{\hat{\delta} - Z_{\alpha/2}\sqrt{\widehat{\text{Var}}(\hat{\delta})}, 0\}, \min\{\hat{\delta} + Z_{\alpha/2}\sqrt{\widehat{\text{Var}}(\hat{\delta})}, 1\}], \tag{3.23}$$

where $\widehat{\text{Var}}(\hat{\delta}) = \hat{\phi}^2\{\hat{\pi}_1^2/[x_1(1 - \hat{\pi}_1)] + \hat{\pi}_0^2/[x_0(1 - \hat{\pi}_0)]\}$ and $\hat{\phi} = (1 - \hat{\pi}_1)/(1 - \hat{\pi}_0)$.

Because the sampling distribution of $\hat{\delta}$ may be skewed when x_i is not large, we can apply an idea similar to that used to derive (3.4) using a logarithmic transformation. Our asymptotic $100(1 - \alpha)$ percent confidence interval for δ is now

$$[1 - \min\{\hat{\phi}\exp(Z_{\alpha/2}\sqrt{\widehat{\text{Var}}(\log(\hat{\phi}))}), 1\}, 1 - \hat{\phi}\exp(-Z_{\alpha/2}\sqrt{\widehat{\text{Var}}(\log(\hat{\phi}))})], \tag{3.24}$$

where $\widehat{\text{Var}}(\log(\hat{\phi})) = \hat{\pi}_1^2/[x_1(1 - \hat{\pi}_1)] + \hat{\pi}_0^2/[x_0(1 - \hat{\pi}_0)]$.

Recall that, for $x_i > 1$, the unbiased estimator of π_i under the negative binomial distribution (1.13) is $\hat{\pi}_i^{(u)} = (x_i - 1)/(N_i - 1)$. Therefore, as both x_i are large, we have $P(\{[(1 - \hat{\pi}_1^{(u)}) - \phi(1 - \hat{\pi}_0^{(u)})]/\sqrt{\text{Var}((1 - \hat{\pi}_1^{(u)}) - \phi(1 - \hat{\pi}_0^{(u)}))}\}^2 \leq Z_{\alpha/2}^2) \doteq 1 - \alpha$. Because we can estimate the variance $\text{Var}(1 - \hat{\pi}_i^{(u)})$ by the unbiased estimator $\hat{\pi}_i^{(u)}(1 - \hat{\pi}_i^{(u)})/(N_i - 2)$ (1.18), these lead to the following quadratic equation in ϕ:

$$A^{\ddagger}\phi^2 - 2B^{\ddagger}\phi + C^{\ddagger} \leq 0, \tag{3.25}$$

where $A^{\ddagger} = (1 - \hat{\pi}_0^{(u)})^2 - Z_{\alpha/2}^2\hat{\pi}_0^{(u)}(1 - \hat{\pi}_0^{(u)})/(N_0 - 2)$, $B^{\ddagger} = (1 - \hat{\pi}_1^{(u)})(1 - \hat{\pi}_0^{(u)})$, and $C^{\ddagger} = (1 - \hat{\pi}_1^{(u)})^2 - Z_{\alpha/2}^2\hat{\pi}_1^{(u)}(1 - \hat{\pi}_1^{(u)})/(N_1 - 2)$. If $A^{\ddagger} > 0$ and $B^{\ddagger 2} - A^{\ddagger}C^{\ddagger} > 0$, then an asymptotic $100(1-\alpha)$ percent confidence interval for δ would be given by

$$[1-\min\{(B^{\ddagger}+\sqrt{B^{\ddagger 2}-A^{\ddagger}C^{\ddagger}})/A^{\ddagger},1\},1-\max\{(B^{\ddagger}-\sqrt{B^{\ddagger 2}-A^{\ddagger}C^{\ddagger}})/A^{\ddagger},0\}]. \tag{3.26}$$

Note that (3.23), (3.24), and (3.26) are derived on the basis of large-sample theory. When x_i is small, these interval estimators may not be adequate for use. In the following section, we discuss the derivation of the exact $100(1-\alpha)$ percent confidence interval on the basis of the conditional distribution, given the marginal $Y_1+Y_0=y.$ fixed. This exact conditional confidence interval can be applied even when the number of cases x_i is as small as 1.

As shown elsewhere (Lui, 1995), the conditional distribution of Y_1, given a fixed total number of non-cases $y.=y_1+y_0$, is then (**Exercise 3.11**)

$$P(Y_1=y_1|y.,x_1,x_0,\phi)=\frac{\binom{y_1+x_1-1}{y_1}\binom{y.-y_1+x_0-1}{y.-y_1}\phi^{y_1}}{\sum_{y=0}^{y.}\binom{y+x_1-1}{y}\binom{y.-y+x_0-1}{y.-y}\phi^{y}}, \tag{3.27}$$

where $\phi=(1-\pi_1)/(1-\pi_0)$, $y_1=0,1,\ldots,y.$.

On the basis of the conditional distribution (3.27), note that the conditional MLE $\hat{\delta}_{\text{cond}}$ of δ is $1-\hat{\phi}_{\text{cond}}$, where $\hat{\phi}_{\text{cond}}$ is the conditional MLE of ϕ satisfying the equation $y_1=E(Y_1|y.,x_1,x_0,\phi)$ (**Exercise 3.13**). Furthermore, we obtain the estimated asymptotic conditional variance $\widehat{\text{Var}}(\hat{\phi}_{\text{cond}})=\hat{\phi}^2_{\text{cond}}/\text{Var}(Y_1|y.,x_1,x_0,\hat{\phi}_{\text{cond}})$ based on the inverse of Fisher's information matrix (Appendix). A discussion of the sufficient and necessary conditions for the unique existence of the conditional MLE of ϕ can be found in **Exercise 3.14**. These considerations lead us to obtain an asymptotic $100(1-\alpha)$ percent conditional confidence interval for δ given by

$$[1-\min\{\hat{\phi}_{\text{cond}}+Z_{\alpha/2}\sqrt{\widehat{\text{Var}}(\hat{\phi}_{\text{cond}})},1\},1-\max\{\hat{\phi}_{\text{cond}}-Z_{\alpha/2}\sqrt{\widehat{\text{Var}}(\hat{\phi}_{\text{cond}})},0\}]. \tag{3.28}$$

Note that $\sum_{y=0}^{y_1}P(Y=y|y.,x_1,x_0,\phi)$ is a decreasing function of ϕ. Thus, we can obtain an exact $100(1-\alpha)$ percent confidence interval $[\phi_l,\phi_u]$ for ϕ by solving the following two equations for ϕ_l and ϕ_u (Casella and Berger, 1990):

$$\sum_{y=y_1}^{y.}P(Y=y|y.,x_1,x_0,\phi_l)=\alpha/2$$

$$\sum_{y=0}^{y_1}P(Y=y|y.,x_1,x_0,\phi_u)=\alpha/2. \tag{3.29}$$

If y_1 is 0 or the solution ϕ_l in (3.29) is less than 0, then we will set the lower limit ϕ_l to 0. Similarly, if $y_1=y.$ or the solution ϕ_u is greater than 1, we will set the upper limit ϕ_u to 1. The $100(1-\alpha)$ percent confidence interval for δ is then

$$[1-\phi_u, 1-\phi_l]. \tag{3.30}$$

When π_0 is small (i.e., the underlying disease is rare in the unexposed group), the relative difference $\delta \doteq \pi_1 - \pi_0$. Thus, in this situation, all interval estimators considered here for the relative difference can also be used to produce confidence intervals for the risk difference $\Delta = \pi_1 - \pi_0$. When there are multiple centers in a study, we may wish to employ pre-stratified inverse sampling. All the discussions on both point and interval estimation of the common relative difference for a series of independent negative binomial sampling based on the conditional approach, as well as test procedures for testing the homogeneity of the relative difference across strata, appear elsewhere (Lui, 1997, 1998b, 2000). Other discussions on estimation of the rate ratio and the relative difference for paired-sample data under inverse sampling can also be found elsewhere (Lui, 2001).

Example 3.6 Suppose that we are studying the association between maternal age ($\leq$20 years; >20 years) and birthweight ($\leq$2500 grams; >2500 grams). From each of the two maternal age categories, suppose further that we independently employ inverse sampling to sample subjects from medical records in a hospital. Assume that we obtain $80 (= Y_1)$ and $190 (= Y_0)$ women whose babies weigh in excess of 2500 grams before we obtain exactly the predetermined numbers $x_1 = 20$ and $x_0 = 10$ of women whose babies weigh 2500 grams or less, respectively. Given these data, the MLE $\hat{\delta}$ is 0.158. Applying (3.23), (3.24), and (3.26), we obtain asymptotic 95% confidence intervals for δ of [0.071, 0.245], [0.067, 0.240], and [0.067, 0.239], respectively. Furthermore, on the basis of the conditional distribution (3.27), we obtain the conditional MLE $\hat{\delta}_{cond} = 0.161$, which is slightly different from the unconditional MLE $\hat{\delta}(= 0.158)$. The asymptotic 95% conditional confidence interval using (3.28) is [0.074, 0.248], while the exact 95% confidence interval using (3.30) is [0.076, 0.251]. Because the lower limits are all above 0, we may conclude that there is a significant association at the 5% level between maternal age and low birthweight.

EXERCISES

3.1. Using the delta method, show that an estimated asymptotic variance of the relative difference estimate $\hat{\delta}$ (3.1) is given by $\widehat{\text{Var}}(\hat{\delta}) = \hat{\phi}^2\{\hat{\pi}_1/[n_1(1-\hat{\pi}_1)] + \hat{\pi}_0/[n_0(1-\hat{\pi}_0)]\}$ under independent binomial sampling.

3.2. On the basis of the finding in **Exercise 3.1** and using the delta method, show that an estimated asymptotic variance of $\log(\hat{\phi})$ is given by $\widehat{\text{Var}}(\log(\hat{\phi})) = \hat{\pi}_1/[n_1(1-\hat{\pi}_1)] + \hat{\pi}_0/[n_0(1-\hat{\pi}_0)]$.

3.3. Under independent binomial sampling, (a) show that the expectation $E((1-\hat{\pi}_1)-\phi(1-\hat{\pi}_0))=0$, where $\hat{\pi}_i = X_i/n_i$. (b) Show that the variance $\mathrm{Var}((1-\hat{\pi}_1)-\phi(1-\hat{\pi}_0))$ is given by $\pi_1(1-\pi_1)/n_1+\phi^2\pi_0(1-\pi_0)/n_0$. (c) Show that the inequality $\{[(1-\hat{\pi}_1)-\phi(1-\hat{\pi}_0)]/\sqrt{\mathrm{Var}((1-\hat{\pi}_1)-\phi(1-\hat{\pi}_0))}\}^2 \leq Z^2_{\alpha/2}$ holds if and only if equation (3.6) is true.

3.4. Consider the data on the 4-year incidence of arteriosclerotic heart disease (ASHD) among male patients in the Framingham epidemiologic study of heart disease (Dawber *et al.*, 1957). We observed 52 new cases of ASHD among 898 patients aged 45–62 years, and 13 new cases among 1078 patients aged 30–44 years. Suppose that we are interested in using the relative difference to measure the effect of age on the risk of developing ASHD between these two age categories.
(a) What is the point estimate $\hat{\delta}$ (3.1)?
(b) What are the asymptotic 95% confidence intervals for δ using interval estimators (3.3), (3.5), and (3.7)?

3.5. When the ratios π_{1s}/π_{0s} are constant across all strata s, where $s=1,2,\ldots,S$, can we claim that the ratios $(1-\pi_{1s})/(1-\pi_{0s})$ are also constant across all strata?

3.6. In Section 3.2, show that the asymptotic variance of $\hat{\phi}_{MH}$ is given by

$$\widehat{\mathrm{Var}}(\hat{\phi}_{MH}) = \left[\sum_s \left(\frac{n_{0s}}{n_{.s}}\right)^2 n_{1s}\hat{\pi}_{1s}(1-\hat{\pi}_{1s}) + \hat{\phi}^2_{MH}\sum_s\left(\frac{n_{1s}}{n_{.s}}\right)^2 n_{0s}\hat{\pi}_{0s}(1-\hat{\pi}_{0s})\right] \Bigg/ \left(\sum_s \frac{(1-\hat{\pi}_{0s})n_{0s}n_{1s}}{n_{.s}}\right)^2.$$

3.7. Using the delta method, show that an estimated asymptotic variance of the estimator $\hat{\delta}$ under independent cluster sampling discussed in Section 3.3 is given by $\widehat{\mathrm{Var}}(\hat{\delta}) = (\hat{\phi})^2\{\hat{\pi}_1 f(\mathbf{m}_1,\hat{\rho}_1)/[m_{1.}(1-\hat{\pi}_1)]+\hat{\pi}_0 f(\mathbf{m}_0,\hat{\rho}_0)/[m_{0.}(1-\hat{\pi}_0)]\}$, where $\hat{\phi}=(1-\hat{\pi}_1)/(1-\hat{\pi}_0)$.

3.8. Show that $\mathrm{Var}(\hat{\pi}_{1.}) = \pi_{1.}(1-\pi_{1.})/n$ and $\mathrm{Cov}(\hat{\pi}_{1.},\hat{\pi}_{.1}) = (\pi_{11}\pi_{00}-\pi_{10}\pi_{01})/n$, where the π_{ij} are defined in Section 3.4.

3.9. Using the delta method, show that $\sqrt{n}(\hat{\phi}-\phi)$, defined in Section 3.4, where $\hat{\phi}=(1-\hat{\pi}_{1.})/(1-\hat{\pi}_{.1})$, has the asymptotic normal distribution with mean 0 and variance $n\mathrm{Var}(\hat{\phi}) = (1-\pi_{1.})(1-\pi_{.1})\{[(1-\pi_{.1})\pi_{1.}-(\pi_{11}\pi_{00}-\pi_{10}\pi_{01})]+[(1-\pi_{1.})\pi_{.1}-(\pi_{11}\pi_{00}-\pi_{10}\pi_{01})]\}/(1-\pi_{.1})^4$.

3.10. Show that $\sqrt{n}(\log(\hat{\phi})-\log(\phi))$ has the asymptotic normal distribution with mean 0 and variance $\pi_{1.}/(1-\pi_{1.})+\pi_{.1}/(1-\pi_{.1})-2(\pi_{11}\pi_{00}-\pi_{10}\pi_{01})/((1-\pi_{1.})(1-\pi_{.1}))$.

3.11. Suppose that Y_i independently follows the negative binomial distribution (1.13) with parameters x_i and π_i. Show that the conditional distribution of Y_1, given that the total number of non-cases $y_. = y_1 + y_0$ is fixed, is given in (3.27).

3.12. When the relative difference δ equals 0, show that the conditional mean $E(Y_1|y_.) = x_1 y_./x_.$, where $x_. = x_1 + x_0$, and the conditional variance $\mathrm{Var}(Y_1|y_.) = x_1 x_0 y_.(x_. + y_.)/[x_.^2(x_. + 1)]$, where the expectation is taken with respect to the conditional probability mass function (3.27) (Barton, 1967; Kudô and Tarumi, 1978).

3.13. On the basis of (3.27), show that the conditional MLE $\hat{\delta}_{\text{cond}}$ of δ is $1 - \hat{\phi}_{\text{cond}}$, where $\hat{\phi}_{\text{cond}}$ will be the solution ϕ satisfying the equation $y_1 = E(Y_1|y_., x_1, x_0, \phi)$ if it exists in the range $0 < \phi < 1$. Furthermore, the conditional asymptotic variance of this estimator, given by the Cramér–Rao lower bound (Casella and Berger, 1990), is simply $\phi^2/\mathrm{Var}(Y_1|y_., x_1, x_0, \phi)$.

3.14. Show that $E(Y_1|y_., x_1, x_0, \phi)$ is a strictly increasing function of ϕ over the range [0, 1]. Furthermore, note that $\lim_{\phi \to 0} E(Y_1|y_., x_1, x_0, \phi) = 0$. Thus, show that if $0 < y_1 < E(Y_1|y_., x_1, x_0, \phi = 1) = x_1 y_./x_.$, then the conditional MLE $\hat{\delta}_{\text{cond}}$ will exist and be unique.

3.15. Consider the data consisting of 192 females with breast cancer in Table 2.3 (Newman, 2001, p. 126). The data are stratified by various stages of breast cancer. As noted elsewhere (Newman, 2001, p. 98), we may assume that the death rate in the patients with a low level of estrogen receptor is higher than that in patients with a high level of estrogen receptor. Suppose that the relative difference is constant across various stages of breast cancer (i.e., $\delta_1 = \delta_2 = \ldots = \delta_S = \delta_c$).
(a) What are the point estimates of the common relative difference δ_c using (3.9) and (3.11)?
(b) What are the 95% confidence intervals for δ_c using (3.10) and (3.13)?
(c) If we test the assumption that $\delta_1 = \delta_2 = \ldots = \delta_S$, what are the p-values using (3.14)–(3.16)?

3.16. Suppose that we employ a matched-pair design to compare the response rate of an experimental treatment ($i = 1$) with that of a placebo group ($i = 0$). We assume that the response rate π_1 in the experimental treatment group is higher than the response rate π_0 in the placebo group. Suppose further that we obtain the following hypothetical data: $n_{11} = 10$, $n_{10} = 20$, $n_{01} = 5$, and $n_{00} = 15$, where the n_{ij} are defined in Section 3.4.
(a) What is the MLE $\hat{\delta}$ of the relative difference δ?
(b) What are the 95% confidence intervals for δ using interval estimators (3.21) and (3.22)?

3.17. Suppose that in Example 3.6 we independently employ inverse sampling to sample subjects from each of the two maternal age groups. Suppose further that we obtain 40($= Y_1$) and 90($= Y_0$) women whose babies weigh in excess of 2500 grams before we obtain exactly the predetermined numbers $x_1 = 10$ and

$x_0 = 10$ of women whose babies have low birthweight (2500 grams or less), respectively.

(a) What are the unconditional and conditional MLEs $\hat{\delta}$ and $\hat{\delta}_{\text{cond}}$?

(b) What are the asymptotic 95% confidence intervals for δ using (3.23), (3.24), and (3.26)?

(c) What are the asymptotic and exact 95% conditional confidence intervals for δ using (3.28) and (3.30)?

REFERENCES

Anderson, T. W. (1958) *An Introduction to Multivariate Statistical Analysis*. Wiley, New York.

Barton, D. E. (1967) Comparison of sample sizes in inverse binomial sampling. *Technometrics*, **9**, 337–339.

Bishop, Y. M. M., Fienberg, S. E. and Holland, P. W. (1975) *Discrete Multivariate Analysis, Theory and Practice*. MIT Press, Cambridge, MA.

Canner, P. L. (1987) An overview of six clinical trials of aspirin in coronary disease. *Statistics in Medicine*, **6**, 255–263.

Casella, G. and Berger, R. L. (1990) *Statistical Inference*. Duxbury Press, Belmont, CA.

Dawber, T. R., Moore, F. E. and Mann, G. V. (1957) Coronary heart disease in the Framingham Study. II. *American Journal of Public Health*, **47**, 4–24.

Elston, R. C. (1977) Response to query: estimating 'inheritability' of a dichotomous trait. *Biometrics*, **33**, 232–233.

Fisher, R. A. (1928) On a distribution yielding the error functions of several well known statistics. In J. C. Fields (ed.), *Proceedings of the International Mathematical Congress*, Vol. 2. University of Toronto Press, Toronto, pp. 805–813.

Fleiss, J. L. (1981) *Statistical Methods for Rates and Proportions*. Wiley, New York.

Girgis, A., Sanson-Fisher, R. W. and Watson, A. (1994) A workplace intervention for increasing outdoor workers' use of solar protection. *American Journal of Public Health*, **84**, 77–81.

Haldane, J. B. S. (1945) On a method of estimating frequencies. *Biometrika*, **33**, 222–225.

Hutton, J. L. (2000) Number needed to treat: properties and problems. *Journal of Royal Statistical Society, Series A*, **163**, 403–419.

Johnson, N. L. and Kotz, S. (1970) *Distributions in Statistics: Continuous Univariate Distributions 2*. Wiley, New York.

Kudô, A. and Tarumi, T. (1978) 2 × 2 tables emerging out of different chance mechanisms. *Communications in Statistics – Theory and Methods*, A**7**, 977–986.

Laupacis, A., Sackett, D. L. and Roberts, R. S. (1988) An assessment of clinically useful measures of the consequences of treatment. *New England Journal of Medicine*, **318**, 1728–1733.

Lipsitz, S. R., Dear, K. B. G., Laird, N. M. and Molenberghs, G. (1998) Tests for homogeneity of the risk difference when data are sparse. *Biometrics*, **54**, 148–160.

Lui, K.-J. (1995) Notes on conditional confidence limits under inverse sampling. *Statistics in Medicine*, **14**, 2051–2056.

Lui, K.-J. (1997) Conditional maximum likelihood estimate and exact test of the common relative difference in combination of 2 × 2 tables under inverse sampling. *Biometrical Journal*, **39**, 215–225.

Lui, K.-J. (1998a) A note on interval estimation of the relative difference in data with matched pairs. *Statistics in Medicine*, **17**, 1509–1515.

Lui, K.-J. (1998b) Tests of homogeneity for the relative difference between strata under inverse sampling. *Biometrical Journal*, **40**, 227–235.

Lui, K.-J. (2000) Asymptotic conditional test procedures for relative difference under inverse sampling. *Computational Statistics and Data Analysis*, **34**, 335–343.
Lui, K.-J. (2001) Estimation of rate ratio and relative difference in matched pairs under inverse sampling. *Environmetrics*, **12**, 539–546.
Lui, K.-J. (2002) Interval estimation of relative difference in a series of independent 2 × 2 tables. Unpublished manuscript. Department of Mathematical and Computer Sciences, San Diego State University.
Lui, K.-J. and Kelly, C. (2000) Tests for homogeneity of the risk ratio in a series of 2 × 2 tables. *Statistics in Medicine*, **19**, 2919–2932.
Lui, K.-J., Cumberland, W. G., and Kuo, L. (1996) An interval estimate for the intraclass correlation in beta-binomial sampling. *Biometrics*, **52**, 412–425.
Mantel, N. and Haenszel, W. (1959) Statistical aspects of the analysis of data from retrospective studies of disease. *Journal of the National Cancer Institute*, **22**, 719–748.
Mayer, J., Slymen, D. J., Eckhardt, L., *et al.* (1997) Reducing ultraviolet radiation exposure in children. *Preventive Medicine*, **26**, 516–522.
Newman, S. C. (2001) *Biostatistical Methods in Epidemiology*. Wiley, New York.
Rothman, K. J. and Boice, J. D. (1979) *Epidemiologic Analysis with a Programmable Calculator*. NIH Publication No. 79–1649, Washington DC.
Sheps, M. C. (1958) Shall we count the living or the dead? *New England Journal of Medicine*, **259**, 1210–1214.
Sheps, M. C. (1959) An examination of some methods of comparing several rates or proportions. *Biometrics*, **15**, 87–97.
Tarone, R. E. (1981) On summary estimators of relative risk. *Journal of Chronic Diseases*, **34**, 463–468.
Walter, S. D. (1975) The distribution of Levin's measure of attributable risk. *Biometrika*, **62**, 371–374.

4

Relative Risk

To quantify the strength of the association between a given disease and a suspected risk factor in etiological studies, the relative risk (RR) between the exposed and the non-exposed groups is certainly one of the most important indices (Fleiss, 1981). In cohort studies, in which we follow into the future two groups of disease-free subjects, distinguished by the presence or absence of a suspected antecedent risk factor, the RR, calculated as the ratio of the probabilities of developing the disease of interest between the exposed and non-exposed groups, is called the incidence RR. When employing a cohort design to study the effect of a risk factor on the incidence rate of a rare chronic disease, some of the subjects may drop out during the lengthy follow-up period. To account for variations in length of follow-up time between groups, we may often consider the incidence rate expressed as the number of cases over the number of person-years at risk rather than the proportion of cases among the number of subjects in the study. Thus, we may also calculate the incidence RR as the ratio of the former between the exposed and non-exposed groups. By contrast, in prevalence studies, in which we simultaneously classify subjects according to the status of the disease and the exposure, the RR, calculated as the ratio of the proportions of cases between the exposed and the unexposed, is called the prevalence RR. When the incidence rate and the duration of the underlying disease are stable over a period of time, and the duration of the disease is constant between two comparison populations, the prevalence RR and the incidence RR are equivalent (Mausner and Bahn, 1974). Note that if the effect due to a risk factor is beneficial (e.g., in the vaccine trials of poliomyelitis described in Chapter 3), then the RR of being a case between the exposed (i.e., vaccinated) group and the non-exposed (i.e., unvaccinated) group will be less than 1. In this case, we frequently use $1 - \text{RR}$, which is called the relative difference (Chapter 3) or relative risk reduction (Laupacis *et al.*, 1988), to measure the efficacy of a vaccine. Note that, by definition, $\text{RR} > 0$. When there is no association between the risk factor and the disease, $\text{RR} = 1$.

In this chapter, we first discuss estimation of the RR under independent binomial sampling for the case of no confounders. We then extend discussion to the situation in which stratified analysis is applied to control confounders or the data

Statistical Estimation of Epidemiological Risk K-J. Lui
 ISBN: 0-470-85071-X (HB)

are collected by use of a multicenter study design. We further discuss estimation of the RR under independent cluster sampling. This is useful for the situation in which subjects are naturally grouped into clusters in epidemiological investigations (Lui *et al.*, 2000) or treatments are compared in cluster randomization trials (Donner *et al.*, 1981). We also discuss estimation of the RR in paired-sample data, which may arise when the matched-pairs design is used to increase the efficiency or the validity of a study (Fleiss, 1981). We further discuss estimation of the RR under inverse sampling. Finally, we discuss estimation of the RR under Poisson sampling when a cohort design is employed to study a rare chronic disease and the follow-up time may vary between different comparison groups (Colditz *et al.*, 1990).

4.1 INDEPENDENT BINOMIAL SAMPLING

Suppose that we wish to assess the effect of a suspected risk factor on the development of a disease. Let π_i denote the probability of having the disease of interest for two populations, distinguished by exposure ($i = 1$) or non-exposure ($i = 0$) to a risk factor. Suppose further that we take an independent random sample of size n_i from each of these two populations and obtain X_i cases. The RR between the exposed and the non-exposed groups is defined as $\theta = \pi_1/\pi_0$. First, under independent binomial sampling (1.1), we can show that the maximum likelihood estimator of π_i is $\hat{\pi}_i = X_i/n_i$ and hence the MLE of θ is $\hat{\theta} = \hat{\pi}_1/\hat{\pi}_0$. Furthermore, by the delta method (Bishop *et al.*, 1975; see Appendix), the asymptotic variance of $\hat{\theta}$ is $\mathrm{Var}(\hat{\theta}) = \theta^2[(1-\pi_1)/(n_1\pi_1) + (1-\pi_0)/(n_0\pi_0)]$ (**Exercise 4.1**). This leads to an asymptotic $100(1-\alpha)$ percent confidence interval for the RR given by

$$[\max\{\hat{\theta} - Z_{\alpha/2}\sqrt{\widehat{\mathrm{Var}}(\hat{\theta})}, 0\}, \hat{\theta} + Z_{\alpha/2}\sqrt{\widehat{\mathrm{Var}}(\hat{\theta})}], \tag{4.1}$$

where $\widehat{\mathrm{Var}}(\hat{\theta}) = \hat{\theta}^2[(1-\hat{\pi}_1)/(n_1\hat{\pi}_1) + (1-\hat{\pi}_0)/(n_0\hat{\pi}_0)]$, and Z_α is the upper 100αth percentile of the standard normal distribution. Note that the sampling distribution of $\hat{\theta}(= \hat{\pi}_1/\hat{\pi}_0)$, a ratio of two proportions, may be skewed, especially when both the sample size n_i and the underlying probability π_i are small. To improve the normal approximation of (4.1), Katz *et al.* (1978) propose using the logarithmic transformation. This produces the following asymptotic $100(1-\alpha)$ percent confidence interval for the RR (**Exercise 4.2**):

$$[\hat{\theta}\exp(-Z_{\alpha/2}\sqrt{\widehat{\mathrm{Var}}(\log(\hat{\theta}))}), \hat{\theta}\exp(Z_{\alpha/2}\sqrt{\widehat{\mathrm{Var}}(\log(\hat{\theta}))})], \tag{4.2}$$

where $\widehat{\mathrm{Var}}(\log(\hat{\theta})) = (1-\hat{\pi}_1)/(n_1\hat{\pi}_1) + (1-\hat{\pi}_0)/(n_0\hat{\pi}_0)$.

Following a principle analogous to that used in Fieller's theorem (Casella and Berger, 1990), we consider $Z = \hat{\pi}_1 - \theta\hat{\pi}_0$ to avoid inference based on a ratio of two sample proportions. We can easily see that $E(Z) = 0$. By the central limit theorem, we have the probability $P([(\hat{\pi}_1 - \theta\hat{\pi}_0)/\sqrt{\mathrm{Var}(\hat{\pi}_1 - \theta\hat{\pi}_0)}]^2 \leq Z^2_{\alpha/2}) \doteq 1 - \alpha$ as both n_i are large. This leads us to consider the following quadratic equation in θ (**Exercise 4.3**):

$$A\theta^2 - 2B\theta + C \leq 0, \tag{4.3}$$

where $A = \hat{\pi}_0^2 - Z_{\alpha/2}^2 \hat{\pi}_0(1 - \hat{\pi}_0)/n_0 B = \hat{\pi}_1\hat{\pi}_0$, and $C = \hat{\pi}_1^2 - Z_{\alpha/2}^2 \hat{\pi}_1(1 - \hat{\pi}_1)/n_1$. If $A > 0$ and $B^2 - AC > 0$, then an asymptotic $100(1 - \alpha)$ percent confidence interval for the RR would be given by

$$[\max\{(B - \sqrt{B^2 - AC})/A, 0\}, (B + \sqrt{B^2 - AC})/A]. \tag{4.4}$$

To reduce the skewness of the sampling distribution of $(\hat{\pi}_1 - \theta\hat{\pi}_0)/\sqrt{\text{Var}(\hat{\pi}_1 - \theta\hat{\pi}_0)}$, Bailey (1987) suggests using $\hat{\pi}_1^{1/3} - (\theta\hat{\pi}_0)^{1/3}$ instead of $\hat{\pi}_1 - \theta\hat{\pi}_0$ as for deriving (4.4). Using the delta method again, we may consider the following quadratic equation in $\theta^{1/3}$ (**Exercise 4.4**):

$$\mathcal{A}\theta^{2/3} - 2\mathcal{B}\theta^{1/3} + \mathcal{C} \leq 0, \tag{4.5}$$

where $\mathcal{A} = \hat{\pi}_0^{2/3} - Z_{\alpha/2}^2(1 - \hat{\pi}_0)/(9n_0\hat{\pi}_0^{1/3})$, $\mathcal{B} = (\hat{\pi}_1\hat{\pi}_0)^{1/3}$, and $\mathcal{C} = \hat{\pi}_1^{2/3} - Z_{\alpha/2}^2$ $(1 - \hat{\pi}_1)/(9n_1\hat{\pi}_1^{1/3})$. If $\mathcal{A} > 0$ and $\mathcal{B}^2 - \mathcal{AC} > 0$, then an asymptotic $100(1 - \alpha)$ percent confidence interval for the RR would be

$$[\max\{((\mathcal{B} - \sqrt{\mathcal{B}^2 - \mathcal{AC}})/\mathcal{A})^3, 0\}, ((\mathcal{B} + \sqrt{\mathcal{B}^2 - \mathcal{AC}})/\mathcal{A})^3]. \tag{4.6}$$

Note that in application of (4.1), (4.2), (4.4), and (4.6), substituting $\hat{\pi}_i$ for π_i in $\text{Var}(\hat{\pi}_i)$ is obviously inappropriate when either of the $\hat{\pi}_i$ is 0 or 1. To alleviate this concern, we substitute $(X_i + 0.5)/(n_i + 1)$ for $\hat{\pi}_i$ whenever this occurs.

Example 4.1 Consider the all-cause mortality data of the randomized clinical trial comparing aspirin with placebo in post-myocardial infarction patients (Elwood *et al.*, 1974). There are $(X_1 =)$ 49 cases out of $n_1 = 615$ patients in the aspirin group, while there are $(X_0 =)$ 67 cases out of $(n_0 =)$ 624 patients in the placebo group. The MLE of the RR between the groups of aspirin and placebo is 0.742 (= (49/615)/(67/624)). Thus, the corresponding estimate of relative difference (discussed in Chapter 3), that is commonly used to measure the efficacy of a treatment, is 0.258 (= 1 − 0.742). Applying (4.1), (4.2), (4.4), and (4.6), we obtain 95% confidence intervals for the RR of [0.481, 1.003], [0.522, 1.054], [0.512, 1.053], and [0.520, 1.052]. We can see that interval estimator (4.1) is shifted to the left compared to the others, while interval estimator (4.4), using the idea of Fieller's theorem, appears to have the greatest length. However, because all these interval estimates cover 1, there is no significant evidence at the 5% level against the RR of all-cause mortality between taking aspirin and placebo being equal to 1.

Example 4.2 Dyspepsia is very common in the general population and hence an empirical therapy such as taking cisapride without prior diagnostic procedures is often recommended for patients. Consider the data obtained from a placebo-controlled clinical trial studying the effect of cisapride on dyspepsia (Chung, 1993; Hartung and Knapp, 2001). We have 4 failures among 14 patients receiving cisapride, versus 12 failures among 15 patients receiving the placebo. The MLE of the RR of failure between cisapride and placebo is 0.357. The estimated relative difference is 0.643. Applying (4.1), (4.2), (4.4), and (4.6), we obtain 95% confidence intervals for the RR of [0.048, 0.666], [0.150, 0.849], [0.061, 0.702], and [0.132, 0.779], respectively. As in the previous example, we can see that the interval estimate (4.1) is again shifted to the left compared to the others. Since the number of failures in the cisapride treatment is small, we may want to apply (4.1) with caution in this case.

4.2 A SERIES OF INDEPENDENT BINOMIAL SAMPLING PROCEDURES

Suppose that there are S strata formed by either centers in a multicenter trial or studies in a meta-analysis. For each stratum $s (s = 1, 2, \ldots, S)$, we independently sample n_{is} subjects from the exposed ($i = 1$) and the non-exposed ($i = 0$) populations, respectively. Suppose we obtain X_{is} cases among n_{is} subjects. Let π_{is} denote the probability that a randomly selected subject from the ith population in the sth stratum is a case. The RR in the sth stratum is defined as $\theta_s = \pi_{1s}/\pi_{0s}$. Under the above assumptions, the random variables X_{is} independently follow the binomial distribution (1.1) with parameters n_{is} and π_{is}. Therefore, the joint probability mass function of the random vector $\mathbf{X}' = (\mathbf{X}'_1, \mathbf{X}'_0)$, where $\mathbf{X}'_i = (X_{i1}, X_{i2}, \ldots, X_{iS})$, is

$$f_{\mathbf{X}}(\mathbf{x}|\mathbf{n}, \boldsymbol{\pi}_0, \boldsymbol{\theta}) = \prod_{s=1}^{S} \binom{n_{1s}}{x_{1s}} (\theta_s \pi_{0s})^{x_{1s}} (1 - \theta_s \pi_{0s})^{n_{1s}-x_{1s}} \binom{n_{0s}}{x_{0s}} \times (\pi_{0s})^{x_{0s}} (1 - \pi_{0s})^{n_{0s}-x_{0s}}, \tag{4.7}$$

where $x_{is} = 0, 1, 2, \ldots, n_{is}$, $\mathbf{n}' = (n_{11}, n_{12}, \ldots, n_{1S}, n_{01}, \ldots, n_{0S})$, and $\boldsymbol{\pi}'_0 = (\pi_{01}, \ldots, \pi_{0S})$, and $\boldsymbol{\theta}' = (\theta_1, \ldots, \theta_S)$.

4.2.1 Asymptotic interval estimators

The MLE of the RR in the sth stratum is simply $\hat{\theta}_s = \hat{\pi}_{1s}/\hat{\pi}_{0s}$, where $\hat{\pi}_{is} = X_{is}/n_{is}$ for $i = 1, 0$ and $s = 1, \ldots, S$. In the discussion of this section, we assume that $\theta_1 = \theta_2 = \ldots = \theta_S$. We denote this common value of the RR by θ_c.

First, consider the situation in which we take a reasonably large sample from each stratum. If the parameters π_{is} were known, we might apply precision-weighting of the stratum-specific RR estimates, $\sum_s W_s \hat{\theta}_s / \sum_s W_s$, to estimate θ_c, where $W_s = 1/\text{Var}(\hat{\theta}_s)$ and $\text{Var}(\hat{\theta}_s) = \theta_c^2[(1 - \pi_{1s})/(n_{1s}\pi_{1s}) + (1 - \pi_{0s})/(n_0\pi_{0s})]$. Note that since θ_c is a constant, we have $\sum_s W_s \hat{\theta}_s / \sum_s W_s = \sum_s W_s^* \hat{\theta}_s / \sum_s W_s^*$, where $W_s^* = 1/\text{Var}(\log(\hat{\theta}_s))$ and $\text{Var}(\log(\hat{\theta}_s)) = (1 - \pi_{1s})/(n_{1s}\pi_{1s}) + (1 - \pi_{0s})/(n_{0s}\pi_{0s})$. If parameters π_{js} are unknown, we can substitute $\hat{\pi}_{is}$ for π_{is} and obtain the estimated weight $\hat{W}_s^* = 1/$

$\widehat{\text{Var}}(\log(\hat{\theta}_s))$, where $\widehat{\text{Var}}(\log(\hat{\theta}_s)) = (1-\hat{\pi}_{1s})/(n_{1s}\hat{\pi}_{1s}) + (1-\hat{\pi}_{0s})/(n_{0s}\hat{\pi}_{0s})$. This leads to the weighted least-squares estimator (Gart, 1962),

$$\hat{\theta}_{\text{WLS}} = \sum_s \hat{W}_s^* \hat{\theta}_s / \sum_s \hat{W}_s^*. \tag{4.8}$$

As noted in Section 4.1, the sampling distribution $\hat{\theta}_s$ may be skewed. To improve the normal approximation of $\hat{\theta}_{\text{WLS}}$ (4.8), we again consider the logarithmic transformation: $\sum_s W_s^* \log(\hat{\theta}_s)/\sum_s W_s^*$. It is easy to show that the variance $\text{Var}(\sum_s W_s^* \log(\hat{\theta}_s)/\sum_s W_s^*) = 1/\sum_s W_s^*$. Substituting $\hat{\pi}_{is}$ for π_{is}, we obtain an asymptotic $100(1-\alpha)$ percent confidence interval for θ_c given by

$$\left[\exp\left(\frac{\sum_s \hat{W}_s^* \log(\hat{\theta}_s)}{\sum_s \hat{W}_s^*} - \frac{Z_{\alpha/2}}{\sqrt{\sum_s \hat{W}_s^*}}\right), \exp\left(\frac{\sum_s \hat{W}_s^* \log(\hat{\theta}_s)}{\sum_s \hat{W}_s^*} + \frac{Z_{\alpha/2}}{\sqrt{\sum_s \hat{W}_s^*}}\right)\right]. \tag{4.9}$$

Applying estimators (4.8) and (4.9) requires us to take a reasonably large sample from each stratum. Furthermore, when the underlying disease rate π_{is} is low, the probability of obtaining $\hat{\pi}_{is} = 0$ may not be negligible and so estimators (4.8) and (4.9) are not applicable. To circumvent this limitation, Tarone (1981) proposes the Mantel–Haenszel type estimator (Mantel and Haenszel, 1959) to estimate the underlying common RR (Rothman and Boice, 1979),

$$\hat{\theta}_{\text{MH}} = \left(\sum_s x_{1s}n_{0s}/n_{.s}\right) \Big/ \left(\sum_s x_{0s}n_{1s}/n_{.s}\right), \qquad n_{.s} = n_{1s} + n_{0s}. \tag{4.10}$$

Using the delta method, we can easily show that the asymptotic variance of $\hat{\theta}_{\text{MH}}$ can be approximated by (**Exercise 4.6**)

$$\widehat{\text{Var}}(\hat{\theta}_{\text{MH}}) = \left(\sum_s \frac{x_{0s}n_{1s}}{n_{.s}}\right)^{-2} \left[\sum_s \left(\frac{n_{0s}}{n_{.s}}\right)^2 n_{1s}\hat{\pi}_{is}(1-\hat{\pi}_{is}) + \hat{\theta}_{\text{MH}}^2 \sum_s \left(\frac{n_{1s}}{n_{.s}}\right)^2 n_{0s}\hat{\pi}_{0s}(1-\hat{\pi}_{0s})\right]. \tag{4.11}$$

If the underlying probabilities π_{is} were so small that $1-\hat{\pi}_{is} \doteq 1$, $\widehat{\text{Var}}(\hat{\theta}_{\text{MH}})$ (4.11) would reduce to the estimated variance given by Tarone (1981). Considering use of the logarithmic transformation, we can easily show that the estimated asymptotic variance $\widehat{\text{Var}}(\log(\hat{\theta}_{\text{MH}}))$ is given by $\widehat{\text{Var}}(\hat{\theta}_{\text{MH}})/\hat{\theta}_{\text{MH}}^2$. Thus, an asymptotic $100(1-\alpha)$ percent confidence interval for θ_c is given by

$$[\hat{\theta}_{\text{MH}} \exp(-Z_{\alpha/2}\sqrt{\widehat{\text{Var}}(\hat{\theta}_{\text{MH}})/\hat{\theta}_{\text{MH}}^2}), \hat{\theta}_{\text{MH}} \exp(Z_{\alpha/2}\sqrt{\widehat{\text{Var}}(\hat{\theta}_{\text{MH}})/\hat{\theta}_{\text{MH}}^2})]. \tag{4.12}$$

4.2.2 Test for the homogeneity of risk ratio

Before employing the estimators described in Section 4.2.1, it is important to examine whether the underlying assumption that $\theta_1 = \theta_2 = \ldots = \theta_S (= \theta_c)$ holds in the data. In this subsection, we briefly discuss testing the homogeneity of the RR.

First, we consider using the WLS test procedure based on $\log(\hat{\theta}_s)$ (Fleiss, 1981):

$$T_{\mathrm{WLS}} = \sum_s \hat{W}_s^* \log^2(\hat{\theta}_s) - \left(\sum_s \hat{W}_s^* \log(\hat{\theta}_s)\right)^2 \Big/ \sum_s \hat{W}_s^*. \tag{4.13}$$

Under the null hypothesis $H_0 : \theta_1 = \theta_2 = \ldots = \theta_S (= \theta_c)$, test statistic (4.13) asymptotically has the χ^2 distribution with $S - 1$ degrees of freedom as all the n_{is} are large. Thus, we will reject H_0 at level α if $T_{\mathrm{WLS}} > \chi^2_{S-1,\alpha}$, where $\chi^2_{S-1,\alpha}$ is the upper 100αth percentile of the chi-squared distribution with $S - 1$ degrees of freedom.

Following Lipsitz *et al.* (1998), we consider the following statistic when the number of strata S is large:

$$Z_{\mathrm{WLS}} = [T_{\mathrm{WLS}} - (S - 1)]/\sqrt{2(S - 1)}. \tag{4.14}$$

Note that if H_0 were not valid, we would expect a large value of T_{WLS}. This suggests that we reject H_0 at level α when $Z_{\mathrm{WLS}} > Z_\alpha$.

When approximating a χ^2 random variable by a normal distribution, Fisher (1928) suggests use of the logarithmic transformation. When all n_{is} and S are large, we consider the test statistic

$$Z_{\mathrm{LWLS}} = \left[\frac{1}{2}\log\left(\frac{T_{\mathrm{WLS}}}{S - 1}\right) + \frac{1}{2(S - 1)}\right] \Big/ \sqrt{\frac{1}{2(S - 1)}}. \tag{4.15}$$

We will reject H_0 at level α if $Z_{\mathrm{LWLS}} > Z_\alpha$.

Lui and Kelly (2000) compare the three simple test procedures above with three others that involve quite tedious and intensive calculations in a variety of situations. They note that when the stratum size is small or moderate and the number of strata is large, test procedure (4.13) is generally too conservative and hence loses power. Test procedure (4.15) does not improve the performance of (4.13). Except for a few cases, test procedure (4.14) is generally preferable to (4.13). When the number of strata S is large (at least 10), readers may wish to use the test procedure with the weight depending only on the n_{ij}, as described by Lui and Kelly (2000). We do not present this test procedure here due to its complexity.

Example 4.3 Consider comparing the incidence rates of non-melanoma skin cancer for women in Dallas-Fort Worth and Minneapolis-St. Paul (Scotto *et al.*, 1974; Gart, 1979; Tarone, 1981). Table 4.1 summarizes these data for different age categories ($S = 8$). Applying test statistics (4.13)–(4.15) to test the

Table 4.1 The number of non-melanoma skin cancer patients/the population size for women in Dallas-Fort Worth and Minneapolis-St. Paul.

Age Category	Dallas-Fort Worth	Minneapolis-St. Paul
15–24	4/181 343	1/172 675
25–34	38/146 207	16/123 065
35–44	119/121 374	30/96 216
45–54	221/111 353	71/92 051
55–64	259/83 004	102/72 159
65–74	310/55 932	130/54 722
75–84	226/29 007	133/32 195
>85	65/7538	40/8328

Sources: Scotto *et al.* (1974).

homogeneity of the RR across age categories, we obtain p-values of 0.33, 0.39, and 0.30, respectively. These results suggest that there is no evidence against the homogeneity of RR. The MLEs $\hat{\theta}_s$ of the RR of non-melanoma skin cancer for different age categories are 3.81, 2.00, 3.14, 2.57, 2.21, 2.33, 1.89 and 1.80; these estimates are generally not much different from one another. When using $\hat{\theta}_{WLS}$ (4.8) and $\hat{\theta}_{MH}$ (4.10), we obtain summary estimates of RR of 2.25 and 2.24, respectively. Thus, the risk of developing non-melanoma skin cancer for women in Dallas-Fort Worth is approximately 2.25 times the incidence rate in Minneapolis-St. Paul. Using (4.9) and (4.12), we obtain 95% confidence intervals of [2.007, 2.465] and [2.024, 2.483], respectively. Because the lower limits are greater than 1, we conclude that there is significant evidence at the 5% level that the risk of developing the non-melanoma skin cancer in Dallas-Fort Worth is higher than that in Minneapolis-St. Paul.

4.3 INDEPENDENT CLUSTER SAMPLING

Suppose that there are two populations distinguished by exposure ($i = 1$) and non-exposure ($i = 0$) to a risk factor. Suppose further that from population i we take a random sample of n_i clusters, each with m_{ij} subjects ($j = 1, 2, \ldots, n_i$). We define the random variable $X_{ijk} = 1$ if the kth subject ($k = 1, 2, \ldots, m_{ij}$) in cluster j of group i is a case, and $X_{ijk} = 0$ otherwise. Then the total number $X_{ij.} = \sum_k X_{ijk}$ of cases in cluster j from population i follows a binomial distribution with parameters m_{ij} and p_{ij}, where p_{ij} denotes the probability of having the disease of interest. Note that the responses X_{ijk} within clusters may be correlated. To account for this intraclass correlation, we assume that p_{ij} independently follows the beta distribution beta(α_i, β_i) with mean π_i $(= \alpha_i/(\alpha_i + \beta_i))$ and variance $\pi_i(1 - \pi_i)/(T_i + 1)$, where $\alpha_i > 0$, $\beta_i > 0$, and $T_i = \alpha_i + \beta_i$ (Johnson and Kotz, 1970; Lui, 1991). Thus, the probability that a randomly selected subject from group i is a case is $E(X_{ijk}) = \pi_i$. Furthermore, the intraclass correlation between X_{ijk} and $X_{ijk'}$, within clusters can be shown to equal $\rho_i = 1/(T_i + 1)$ (see **Exercise 1.7**).

Define $\hat{\pi}_i = \sum_j \sum_k X_{ijk}/m_{i.}$, where $m_{i.} = \sum_j m_{ij}$. Note that $\hat{\pi}_i$ is an unbiased estimator of π_i with variance $\mathrm{Var}(\hat{\pi}_i)$ equal to $\pi_i(1-\pi_i)f(\mathbf{m}_i, \rho_i)/m_{i.}$ (**Exercise 1.8**), where $\mathbf{m}_i' = (m_{i1}, m_{i2}, \ldots, m_{in_i})$, and $f(\mathbf{m}_i, \rho_i)$ is the variance inflation factor due to the intraclass correlation ρ_i and equals $\sum_j m_{ij}[1 + (m_{ij} - 1)\rho_i]/m_{i.}$. The properties of the variance inflation factor are discussed in Chapter 1. To estimate ρ_i we may apply the traditional intraclass correlation estimator $\hat{\rho}_i$ (2.19) given in Chapter 2 (Fleiss, 1986; Lui *et al.*, 1996; Elston, 1977).

To derive an interval estimator of $\theta (= \pi_1/\pi_0)$, we first note that the asymptotic variance of the estimator $\hat{\theta} = \hat{\pi}_1/\hat{\pi}_0$ is $\mathrm{Var}(\hat{\theta}) = \theta^2[(1-\pi_1)f(\mathbf{m}_1, \rho_1)/(m_{1.}\pi_1) + (1-\pi_0)f(\mathbf{m}_0, \rho_0)/(m_{0.}\pi_0)]$. Thus, we obtain an asymptotic $100(1-\alpha)$ percent confidence interval for the RR given by

$$[\max\{\hat{\theta} - Z_{\alpha/2}\sqrt{\widehat{\mathrm{Var}}(\hat{\theta})}, 0\}, \quad \hat{\theta} + Z_{\alpha/2}\sqrt{\widehat{\mathrm{Var}}(\hat{\theta})}], \tag{4.16}$$

where $\widehat{\mathrm{Var}}(\hat{\theta}) = \hat{\theta}^2[(1-\hat{\pi}_1)f(\mathbf{m}_1, \hat{\rho}_1)/(m_{1.}\hat{\pi}_1) + (1-\hat{\pi}_0)f(\mathbf{m}_0, \hat{\rho}_0)/(m_{0.}\hat{\pi}_0)]$. When using the logarithmic transformation on $\hat{\theta}$, we obtain an asymptotic $100(1-\alpha)$ percent confidence interval for the RR given by

$$[\hat{\theta}\exp(-Z_{\alpha/2}\sqrt{\widehat{\mathrm{Var}}(\log(\hat{\theta}))}), \quad \hat{\theta}\exp(Z_{\alpha/2}\sqrt{\widehat{\mathrm{Var}}(\log(\hat{\theta}))})], \tag{4.17}$$

where $\widehat{\mathrm{Var}}(\log(\hat{\theta})) = (1-\hat{\pi}_1)f(\mathbf{m}_1, \hat{\rho}_1)/(m_{1.}\hat{\pi}_1) + (1-\hat{\pi}_0)f(\mathbf{m}_0, \hat{\rho}_0)/(m_{0.}\hat{\pi}_0)$.

Following a similar principle to that used in Fieller's theorem (Casella and Berger, 1990), we consider the probability $P([(\hat{\pi}_1 - \theta\hat{\pi}_0)/\sqrt{\mathrm{Var}(\hat{\pi}_1 - \theta\hat{\pi}_0)}]^2 \le Z^2_{\alpha/2}) \doteq 1 - \alpha$ when all m_{ij} are large. This leads us to consider the following quadratic equation in θ:

$$A^\dagger\theta^2 - 2B^\dagger\theta + C^\dagger \le 0, \tag{4.18}$$

where $A^\dagger = \hat{\pi}_0^2 - Z^2_{\alpha/2}\hat{\pi}_0(1-\hat{\pi}_0)f(\mathbf{m}_0, \hat{\rho}_0)/m_0$. $B^\dagger = \hat{\pi}_1\hat{\pi}_0$, and $C^\dagger = \hat{\pi}_1^2 - Z^2_{\alpha/2}\hat{\pi}_1(1-\hat{\pi}_1)f(\mathbf{m}_1, \hat{\rho}_1)/m_1$. If $A^\dagger > 0$ and $B^{\dagger 2} - A^\dagger C^\dagger > 0$, then an asymptotic $100(1-\alpha)$ percent confidence interval for the RR will be given by

$$[\max\{(B^\dagger - \sqrt{B^{\dagger 2} - A^\dagger C^\dagger})/A^\dagger, 0\}, \quad (B^\dagger + \sqrt{B^{\dagger 2} - A^\dagger C^\dagger})/A^\dagger]. \tag{4.19}$$

To reduce the possible skewness of the sampling distribution of $(\hat{\pi}_1 - \theta\hat{\pi}_0)/\sqrt{\mathrm{Var}(\hat{\pi}_1 - \theta\hat{\pi}_0)}$, Bailey (1987) proposes considering $\hat{\pi}_1^{1/3} - (\theta\hat{\pi}_0)^{1/3}$. This leads us to consider the following quadratic equation in $\theta^{1/3}$:

$$\mathcal{A}^\dagger(\theta)^{2/3} - 2\mathcal{B}^\dagger(\theta)^{1/3} + \mathcal{C}^\dagger \le 0, \tag{4.20}$$

where $\mathcal{A}^\dagger = \hat{\pi}_0^{2/3} - Z^2_{\alpha/2}(1-\hat{\pi}_0)f(\mathbf{m}_0, \hat{\rho}_0)/(9m_{0.}\hat{\pi}_0^{1/3})$, $\mathcal{B}^\dagger = (\hat{\pi}_1\hat{\pi}_0)^{1/3}$, and $\mathcal{C}^\dagger = \hat{\pi}_1^{2/3} - Z^2_{\alpha/2}(1-\hat{\pi}_1)f(\mathbf{m}_1, \hat{\rho}_1)/(9m_{1.}\hat{\pi}_1^{1/3})$. If $\mathcal{A}^\dagger > 0$ and $\mathcal{B}^{\dagger 2} - \mathcal{A}^\dagger\mathcal{C}^\dagger > 0$, then an approximate $100(1-\alpha)$ percent confidence interval for the RR will be

$$[\max\{((\mathcal{B}^\dagger - \sqrt{\mathcal{B}^{\dagger 2} - \mathcal{A}^\dagger\mathcal{C}^\dagger})/\mathcal{A}^\dagger)^3, 0\}, ((\mathcal{B}^\dagger + \sqrt{\mathcal{B}^{\dagger 2} - \mathcal{A}^\dagger\mathcal{C}^\dagger})/\mathcal{A}^\dagger)^3]. \tag{4.21}$$

Note that in application of (4.16), (4.17), (4.19), and (4.21), when either of the $\hat{\pi}_i$ is 0 or 1, we recommend using the *ad hoc* adjustment procedure for sparse data involving adding 0.50 to each cell and using $(X_{i..} + 0.5)/(m_{i.} + 1)$ to estimate π_i. Note also that if we replaced $f(\mathbf{m}_i, \hat{\rho}_i)$ by 1 due to either $\rho_i = 0$ or $m_{ij} = 1$ for all i and j, then interval estimators (4.16), (4.17), (4.19), and (4.21) would reduce to (4.1), (4.2), (4.4), and (4.6), respectively.

Lui *et al.* (2000) evaluate and compare the performance of interval estimators (4.16), (4.17), (4.19), and (4.21). They note that the coverage probability of (4.16) tends to be less than the desired confidence level when the underlying common intraclass correlation ρ_c $(= \rho_1 = \rho_0)$ is large, while those of the other three interval estimators generally agree reasonably well in a variety of situations. However, because interval estimator (4.17) is often shorter on average than those of (4.19) and (4.21) in a variety of situations, the former is recommended for general use.

Example 4.4 To illustrate the application of the proposed interval estimators (4.16), (4.17), (4.19), and (4.21), consider the data (Table 1.1) taken from a study of an educational intervention with emphasis on behavior change (Mayer *et al.*, 1997). There were 132 children, in 58 classes of size 1–6; for each class it is known how many children had an adequate level of solar protection. From the data, the proportions of children in the intervention ($i = 1$) and control ($i = 0$) groups who do not have an adequate level of solar protection are $\hat{\pi}_1 = 27/64$ and $\hat{\pi}_0 = 42/68$, respectively. The point estimate of the RR is thus $\hat{\theta} = \hat{\pi}_1/\hat{\pi}_0 = 0.683$. Assuming $\rho_1 = \rho_0 = \rho_c$, we obtain a point estimate for the common intraclass correlation of $\hat{\rho}_c(= (m_{1.}\hat{\rho}_1 + m_{0.}\hat{\rho}_0.)/m_{..}$, where $m_{..} = m_{1.} + m_{0.}) = 0.30$. Applying interval estimators (4.16), (4.17), (4.19), and (4.21), we obtain 95% confidence intervals for the RR of [0.390, 0.976], [0.445, 1.049], [0.418, 1.026], and [0.438, 1.037], respectively. Since the number of classes in this example is not large, Lui *et al.* (2000) note that using (4.16) can often give too short an interval. Because the other three 95% confidence intervals include 1, there is no significant evidence at the 5% level that the educational intervention affects whether children employ an adequate level of solar protection.

4.4 PAIRED-SAMPLE DATA

Suppose that we want to measure the strength of the association between a risk factor and a disease of interest. Denote the possibly distinct levels of a combination of matching variables by $z_1, z_2, \ldots, z_K$. Let E denote the exposure status: $E = 1$ if a randomly selected subject is exposed to the risk factor, and $E = 0$ otherwise. Similarly, let D denote the disease status: $D = 1$ if a randomly selected subject is a case, and $D = 0$ otherwise. Let p_k denote the conditional probability $\mathrm{P}(Z = z_k|E = 1) > 0$ of a subject randomly selected from the exposure group having matching covariate level z_k, where $\sum_k p_k = 1$. We assume that the probability

$P(D = 1|E, z_k)$ of a randomly selected subject being a case, given the exposure status E and matching covariate level z_k, is given by $\exp(-(\beta_0 + \beta_1 E + \beta_2 z_k))$, where β_1 and β_2 denote the effects due to the exposure and matching covariate Z, and the parameter space is $\{(\beta_0, \beta_1, \beta_2)|\beta_0 + \beta_1 E + \beta_2 z_k > 0$ for $E = 0, 1$, and $Z = z_1, z_2, \ldots, z_K\}$. Consider a commonly used matched design, in which we take a random sample of n subjects from the exposure population. For each of these n sampled subjects with matching covariate level $z^* \in \{z_1, z_2, \ldots, z_K\}$ we then find a subject with the same matching covariate level z^* from the non-exposed population. Note that within each pair, the RR is simply equal to $\exp(-\beta_1)$. For clarity, we use the following 2×2 table to summarize the data structure:

		Non-exposed population Case	Non-case	
Exposed population	Case	π_{11}	π_{10}	$\pi_{1.}$
	Non-case	π_{01}	π_{00}	$\pi_{0.}$,
		$\pi_{.1}$	$\pi_{.0}$	

where $0 < \pi_{ij} < 1 (i = 0, 1$, and $j = 0, 1)$ denotes the corresponding cell probability. Under the above model assumptions, $\pi_{ij} = \sum_k P(D = i|E = 1, z_k)P(D = j|E = 0, z_k)p_k$. We define $\pi_{i.} = \pi_{i1} + \pi_{i0}, \pi_{.j} = \pi_{1j} + \pi_{0j}$. We can show that the RR $(= \exp(-\beta_1))$ is equal to $\theta = \pi_{1.}/\pi_{.1}$ (**Exercise 4.8**).

Let N_{ij} denote the number of pairs out of these n pairs that fall in cell (i, j) with probability π_{ij}, where $\sum_i \sum_j N_{ij} = n$. Then the random vector $(N_{11}, N_{10}, N_{01}, N_{00})'$ follows the multinomial distribution (2.25) with parameters n and $(\pi_{11}, \pi_{10}, \pi_{01}, \pi_{00})'$. The MLE of π_{ij} is $\hat{\pi}_{ij} = N_{ij}/n$. Similarly, the MLEs of $\pi_{1.}$ and $\pi_{.1}$ are $\hat{\pi}_{1.} = (N_{11} + N_{10})/n$ and $\hat{\pi}_{.1} = (N_{11} + N_{01})/n$, respectively.

Using the multivariate central limit theorem, as n goes to ∞, we can show that $\sqrt{n}((\hat{\pi}_{1.}, \hat{\pi}_{.1})' - (\pi_{1.}, \pi_{.1})')$ has the asymptotic normal distribution with mean vector $(0, 0)'$ and covariance matrix $\boldsymbol{\Sigma}$, where $\boldsymbol{\Sigma}$ is a 2×2 matrix with diagonal terms equal to $\pi_{1.}(1 - \pi_{1.})$ and $\pi_{.1}(1 - \pi_{.1})$, respectively, and with both off-diagonal terms equal to $\pi_{11}\pi_{00} - \pi_{10}\pi_{01}$. Let $f(X_1, X_0)$ denote the function $\log(X_1/X_0)$. Using the delta method (see Appendix) and the function $f_1(X_1, X_0)$, we obtain that $\sqrt{n}[\log(\hat{\theta}) - \log(\theta)]$, where $\hat{\theta} = \hat{\pi}_{1.}/\hat{\pi}_{.1}$, has the asymptotic normal distribution with mean 0 and variance $n\mathrm{Var}(\log(\hat{\theta})) = (1 - \pi_{1.})/\pi_{1.} + (1 - \pi_{.1})/\pi_{.1} - 2(\pi_{11}\pi_{00} - \pi_{10}\pi_{01})/(\pi_{1.}\pi_{.1})$ (**Exercise 4.9**). Substituting the corresponding unbiased consistent estimate $\hat{\pi}_{ij}$ for π_{ij}, we obtain the estimated asymptotic variance $\widehat{\mathrm{Var}}(\log(\hat{\theta})) = (N_{01} + N_{00})/[n(N_{11} + N_{10})] + (N_{10} + N_{00})/[n(N_{11} + N_{01})] - 2(N_{11}N_{00} - N_{10}N_{01})/[n(N_{11} + N_{10})(N_{11} + N_{01})]$. Therefore, an asymptotic $100(1 - \alpha)$ percent confidence interval for the RR is given by

$$[\hat{\theta} \exp(-Z_{\alpha/2}\sqrt{\widehat{\mathrm{Var}}(\log(\hat{\theta}))}), \quad \hat{\theta} \exp(Z_{\alpha/2}\sqrt{\widehat{\mathrm{Var}}(\log(\hat{\theta}))})]. \tag{4.22}$$

Note that testing the null hypothesis $H_0 : \pi_{1.} = \pi_{.1}$ is equivalent to testing the null hypothesis $H_0 : \theta = 1$. When testing equality in dichotomous data with matched pairs, Lui (2001a) notes that applying the statistic $\log(\hat{\theta})$ for the purpose of hypothesis testing can slightly improve the power of McNemar's test without essentially damaging the extent of agreement between the actual Type I error and the nominal α level.

Example 4.5 Consider the data given in Rosner, (1990, pp. 342–343) comparing two treatments for a rare form of cancer. We matched patients on age, sex, and clinical condition to form pairs. Within each pair, we randomly assign one patient to receive chemotherapy and the other to receive surgery. The patients are followed for 5 years, with survival as the outcome variable. We obtain $n_{11} = 510$ (the number of pairs where both patients die), $n_{10} = 5$ (the number of pairs where the patient receiving chemotherapy dies and the patient receiving surgery survives), $n_{01} = 16$ (the number of pairs where the patient receiving chemotherapy survives and the patient receiving surgery dies), and $n_{00} = 90$ (the number of pairs where both patients survive). Given these data, the MLE $\hat{\theta}$ of the RR is simply 0.979. Applying interval estimator (4.22), we obtain a 95% confidence interval for the RR of [0.962, 0.996]. Since the upper limit of this confidence interval is less than 1, there is significant evidence at the 5% level that the death rate for chemotherapy is lower than that for surgery.

4.5 INDEPENDENT INVERSE SAMPLING

Recall that the MLE $\hat{\theta}$ of the RR $(= \pi_1/\pi_0)$ under independent binomial sampling is $\hat{\pi}_1/\hat{\pi}_0$. Because there is a positive probability that the MLE $\hat{\pi}_0$ is 0, the MLE $\hat{\theta}$ has an infinitely large bias with no finite variance under independent binomial sampling. Furthermore, when the disease of interest is rare in the unexposed, the probability that $\hat{\pi}_0$ equals 0 can be non-negligible. Although we can always apply the *ad hoc* procedure for sparse data of adding a small positive constant to alleviate this concern, such an adjustment cannot eliminate the bias. On the other hand, we can easily apply inverse sampling to avoid this theoretical limitation inherent in binomial sampling (Lui, 1996).

4.5.1 Uniformly minimum variance unbiased estimator of relative risk

Assume that we employ independent inverse sampling (Haldane, 1945a, 1945b) from the exposed $(i = 1)$ and the non-exposed populations $(i = 0)$, in which we continue sampling subjects until we obtain a predetermined number x_i of cases. Let Y_1 denote the number of non-cases accumulated in the exposed sample before obtaining exactly the first x_1 cases from the exposed population. Similarly, let Y_0 denote the number of non-cases accumulated in the unexposed sample before obtaining exactly the first x_0 cases from the unexposed population. Then the random variables Y_i(for $i = 1, 0$) independently follow the negative binomial distribution (Haldane, 1945a, 1945b) with mean $x_i(1 - \pi_i)/\pi_i$ and variance $x_i(1 - \pi_i)/\pi_i^2$ (Casella and Berger, 1990). The joint probability mass function (Y_1, Y_0) is then given by

$$\binom{Y_1 + x_1 - 1}{Y_1} (\theta\pi_0)^{x_1}(1 - \theta\pi_0)^{Y_1} \binom{Y_0 + x_0 - 1}{Y_0} \pi_0{}^{x_0}(1 - \pi_0)^{Y_0}, \quad (4.23)$$

where $Y_i = 0, 1, 2, \ldots$ and $i = 1, 0$. Note that the MLE of $\theta(= \pi_1/\pi_0)$ under (4.23) is simply $\hat{\pi}_1/\hat{\pi}_0$, where $\hat{\pi}_i = x_i/(x_i + Y_i)$. Following Feller (1968, pp. 241, 493) and Best (1974), we can show that the MLE $\hat{\pi}_1/\hat{\pi}_0$ is a biased estimator of θ with the bias $\mathrm{E}(\hat{\pi}_1/\hat{\pi}_0 - \theta)$ given by

$$\left[\sum_{k=1}^{x_1-1}(-1)^{k-1}\frac{x_1}{x_1 - k}\left(\frac{\pi_1}{1 - \pi_1}\right)^k + x_1\left(\frac{-\pi_1}{1 - \pi_1}\right)^{x_1}\log(\pi_1) - \pi_1\right]\pi_0^{-1}. \quad (4.24)$$

In fact, for $x_1 \geq 2$, the uniformly minimum variance unbiased estimator (UMVUE) of θ is simply given by

$$\hat{\pi}_1^{(\mathrm{u})}/\hat{\pi}_0, \quad (4.25)$$

where $\hat{\pi}_1^{(\mathrm{u})} = (x_1 - 1)/(x_1 + Y_1 - 1)$ with variance $\mathrm{Var}(\hat{\pi}_1^{(\mathrm{u})}/\hat{\pi}_0)$ given in **Exercise 4.10**. The UMVUE of this variance (**Exercise 4.11**) for $x_1 > 2$ is

$$\widehat{\mathrm{Var}}(\hat{\pi}_1^{(\mathrm{u})}/\hat{\pi}_0) = \widehat{\mathrm{Var}}(\hat{\pi}_1^{(\mathrm{u})})\widehat{\mathrm{Var}}(\hat{\pi}_0^{-1}) + \widehat{\mathrm{Var}}(\hat{\pi}_1^{(\mathrm{u})})\widehat{\pi_0^{-2}} + \widehat{\mathrm{Var}}(\hat{\pi}_0^{-1})\widehat{\pi_1^2}, \quad (4.26)$$

where

$$\begin{aligned}
\widehat{\mathrm{Var}}(\hat{\pi}_1^{(\mathrm{u})}) &= \hat{\pi}_1^{(\mathrm{u})}(1 - \hat{\pi}_1^{(\mathrm{u})})/(x_1 + Y_1 - 2), \\
\widehat{\mathrm{Var}}(\hat{\pi}_0^{-1}) &= Y_0(Y_0 + x_0)/[x_0^2(x_0 + 1)], \\
\widehat{\pi_0^{-2}} &= (x_0 + Y_0 + 1)(x_0 + Y_0)/[x_0(x_0 + 1)], \\
\widehat{\pi_1^2} &= (x_1 - 1)(x_1 - 2)/[(x_1 + Y_1 - 1)(x_1 + Y_1 - 2)].
\end{aligned}$$

Lui (1996) demonstrates that applying the UMVUE (4.25) may significantly reduce the mean-squared error of applying the MLE $\hat{\pi}_1/\hat{\pi}_0$ when the number of index cases x_i in both groups is small.

Example 4.6 Consider the example given by Bennett (1981, p. 71) concerning the study of the association between a certain disease and two plants. We have the data $x_1 = x_0 = 30$, $y_1 = 120$, and $y_0 = 225$. On the basis of these data, the MLE and the UMVUE of θ are 1.700 and 1.654, respectively; the MLE is slightly larger than the UMVUE. The UMVUE $\widehat{\mathrm{Var}}(\hat{\pi}_1^{(\mathrm{u})}/\hat{\pi}_0)$ (4.26) of the variance is 0.152.

4.5.2 Interval estimators of relative risk

Note that because the sampling distribution of the UMVUE $\hat{\pi}_1^{(u)}/\hat{\pi}_0$ (4.25) could be skewed if either x_i were small, we do not recommend use of the UMVUE and its estimated standard error to form the confidence limits. When x_i is large, note also that the random variable $Y_i/x_i + 1$ is asymptotically normally distributed with mean $1/\pi_i$ and variance $(1 - \pi_i)/(x_i\pi_i^2)$. Using the delta method (Bishop *et al.*, 1975; Agresti, 1990), the random variable $\log(Y_i/x_i + 1)$ is also asymptotically normally distributed with mean $-\log(\pi_i)$ and variance $(1 - \pi_i)/x_i$. Therefore, an asymptotic $100(1 - \alpha)$ percent confidence interval for θ is given by

$$\left[\frac{Y_0/x_0 + 1}{Y_1/x_1 + 1}\exp(-Z_{\alpha/2}\sqrt{(1 - \hat{\pi}_1)/x_1 + (1 - \hat{\pi}_0)/x_0}),\right.$$
$$\left.\frac{Y_0/x_0 + 1}{Y_1/x_1 + 1}\exp(Z_{\alpha/2}\sqrt{(1 - \hat{\pi}_1)/x_1 + (1 - \hat{\pi}_0)/x_0})\right]. \qquad (4.27)$$

Note that when both π_i are small, because $2(x_i + Y_i)\pi_i$ approximately follows a χ^2 distribution with $2x_i$ degrees of freedom (**Exercise 1.18**), the ratio $\{[2(x_1 + Y_1)\pi_1]/2x_1\}/\{[2(x_0 + Y_0)\pi_0]/2x_0\}$ approximately follows an F distribution with $2x_1$ and $2x_0$ degrees of freedom. Thus, an approximate $100(1 - \alpha)$ percent confidence interval for θ (Bennett, 1981) is given by (**Exercise 4.12**):

$$\left[\frac{x_1(x_0 + Y_0)}{x_0(x_1 + Y_1)}F_{2x_1,2x_0,1-\alpha/2}, \frac{x_1(x_0 + Y_0)}{x_0(x_1 + Y_1)}F_{2x_1,2x_0,\alpha/2}\right], \qquad (4.28)$$

where F_{f_1,f_2},α is the upper 100αth percentile of the F distribution with f_1 and f_2 degrees of freedom. Given reasonably large numbers x_i (≥ 20), Lui (1995a) applies Monte Carlo simulation and notes that interval estimator (4.28) can be conservative and hence lose efficiency when the π_i are not small. On the other hand, (4.27) is derived on the basis of large-sample theory. Therefore, when neither x_i is large, if the underlying probabilities π_i were both small, then (4.28) based on the F distribution may be considered.

Note that the relative difference δ discussed in Chapter 3 is equal to $1 - \phi$, where $\phi = (1 - \pi_1)/(1 - \pi_0)$, which is a ratio of two proportions. Therefore, we can easily derive the exact confidence interval of the RR on the basis of the conditional distribution of the negative binomial distribution as for the relative difference presented in Section 3.5 (Lui, 1995b). Similarly, the discussions on estimation of the common relative difference under a series of independent negative binomial sampling procedures, as well as procedures for testing the homogeneity of the relative difference across strata, can also be modified to accommodate the situation where the RR is our primary parameter of interest under pre-stratified sampling (Lui, 1997, 1998, 2000). A discussion on both point and interval estimation of RR and RD for paired-sample data under inverse sampling is given by Lui (2001b).

Example 4.7 Consider the numerical data given in Example 4.6. We have $x_1 = x_0 = 30$, $y_1 = 120$, and $y_0 = 225$. Suppose that we are interested in obtaining an interval estimate for the RR ($= \theta = \pi_1/\pi_0$). The estimators (4.27) and (4.28) give 95% confidence intervals for the RR of [1.069, 2.704] and [1.020, 2.834], respectively. We can see that the latter is longer than the former. Note that since the MLE estimates $\hat{\pi}_1 = 0.20$ and $\hat{\pi}_0 = 0.12$, both of which are not small, as noted before (Lui, 1995a), interval estimator (4.28) may be conservative.

4.6 INDEPENDENT POISSON SAMPLING

When studying the effect of a risk factor on the incidence rate of a chronic disease in a cohort design, we often apply the Poisson distribution to model the incidence data (Breslow, 1984). A few notes on the motivation behind assuming the Poisson model can be found in Section 2.6. In this section, we focus discussion on estimation of the incidence RR, $\theta^* = \lambda_1/\lambda_0$, where λ_1 and λ_0 are the incidence rates in the exposed and non-exposed groups, respectively. Suppose that we obtain X_i cases among n_i^* person-years at risk in group $i(i = 1, 0)$. We assume that X_1 and X_0 independently follow Poisson distributions with parameters $n_1^*\lambda_1$ and $n_0^*\lambda_0$, respectively. On the basis of the likelihood (2.40), we can show that the MLE of θ^* is simply $\hat{\theta}^* = \hat{\lambda}_1/\hat{\lambda}_0$, where $\hat{\lambda}_i = X_i/n_i^*$. When the disease of interest is rare, the sampling distribution of $\hat{\theta}^*$ may be skewed. Thus, we may wish to apply the logarithmic transformation to improve the normal approximation of $\hat{\theta}^*$. Using the delta method (**Exercise 4.15**), we can show that the estimated asymptotic variance $\widehat{\text{Var}}(\log(\hat{\theta}^*)) = 1/X_1 + 1/X_2$. Therefore, an asymptotic $100(1-\alpha)$ percent confidence interval for θ^* is simply given by

$$[\hat{\theta}^* \exp(-Z_{\alpha/2}\sqrt{\widehat{\text{Var}}(\log(\hat{\theta}^*))}),\ \hat{\theta}^* \exp(Z_{\alpha/2}\sqrt{\widehat{\text{Var}}(\log(\hat{\theta}^*))})]. \tag{4.29}$$

To avoid our inference depending on the possibly skewed distribution of $\hat{\theta}^*$ (i.e., the ratio of two random variables), we may also use the principle of Fieller's theorem by considering $\mathcal{Z} = \hat{\lambda}_1 - \theta^*\hat{\lambda}_0$. We can easily see that the expectation $E(\mathcal{Z}) = 0$. Furthermore, the variance $\text{Var}(\mathcal{Z})$ is $\lambda_1/n_1^* + (\theta^*)^2\lambda_0/n_0^*$. We are thus led to consider the following quadratic equation:

$$\mathfrak{A}(\theta^*)^2 - 2\mathfrak{B}\theta^* + \mathfrak{C} \leq 0, \tag{4.30}$$

where $\mathfrak{A} = \hat{\lambda}_0^2 - Z_{\alpha/2}^2\hat{\lambda}_{0s}/n_0^*$, $\mathfrak{B} = \hat{\lambda}_1\hat{\lambda}_0$, and $\mathfrak{C} = \hat{\lambda}_1^2 - Z_{\alpha/2}^2\hat{\lambda}_1/n_1^*$. If $\mathfrak{A} > 0$ and $\hat{\lambda}_1 > 0$, then we can show that $\mathfrak{B}^2 - \mathfrak{A}\mathfrak{C} > 0$ (**Exercise 4.16**). In this case, the two distinct roots of (4.30) for which equality holds exist. Thus, an asymptotic $100(1-\alpha)$ percent confidence interval for θ^* is given by

$$\left[\max\left\{\frac{\mathfrak{B} - \sqrt{\mathfrak{B}^2 - \mathfrak{A}\mathfrak{C}}}{\mathfrak{A}}, 0\right\}, \frac{\mathfrak{B} + \sqrt{\mathfrak{B}^2 - \mathfrak{A}\mathfrak{C}}}{\mathfrak{A}}\right]. \tag{4.31}$$

Note that when redefining the parameters in the model, we can easily see that the likelihood (2.40) depends on θ^* and a nuisance parameter. To eliminate this nuisance parameter, we may consider the conditional distribution $P(X_1 = x_1|x_., \theta^*)$ of X_1, given $x_. = X_1 + X_0$ fixed. As shown in **Exercise 4.17**, this conditional distribution follows the binomial distribution

$$P(X_1 = x_1|x_., \theta^*) = \binom{x_.}{x_1} \psi^{x_1}(1-\psi)^{x_.-x_1}, \tag{4.32}$$

where $\psi = n_1^*\theta^*/(n_1^*\theta^* + n_0^*)$, and $x_1 = 0, 1, \ldots, x_.$. On the basis of (4.32), one can easily show that the conditional MLE of θ^* is the same as the unconditional MLE $\hat{\theta}^* = \hat{\lambda}_1/\hat{\lambda}_0$ with the asymptotic conditional variance equal to (**Exercise 4.17**)

$$\widehat{\text{Var}}(\hat{\theta}^*|x_.) = (\hat{\theta}^*)^2/(x_.\hat{\psi}(1-\hat{\psi})), \tag{4.33}$$

where $\hat{\psi} = n_1^*\hat{\theta}^*/(n_1^*\hat{\theta}^* + n_0^*)$. Since the sampling distribution $\hat{\theta}^*$ may, as noted previously, be skewed, we consider using the logarithmic transformation. We obtain an asymptotic $100(1-\alpha)$ percent confidence interval for θ^* given by

$$[\hat{\theta}^* \exp(-Z_{\alpha/2}\sqrt{\widehat{\text{Var}}(\log(\hat{\theta}^*)|x_.)}), \hat{\theta}^* \exp(Z_{\alpha/2}\sqrt{\widehat{\text{Var}}(\log(\hat{\theta}^*)|x_.)})], \tag{4.34}$$

where $\widehat{\text{Var}}(\log(\hat{\theta}^*)|x_.) = 1/(x_.\hat{\psi}(1-\hat{\psi}))$. When the number of person-years n_i^* (or equivalently, the expected number of cases in group i) is not large, interval estimators (4.29), (4.31), and (4.34), all derived from large-sample theory, may not perform well. In this case, we may wish to find the exact confidence interval for θ^*. On the basis of the exact confidence interval (1.6) for the parameter ψ of the binomial distribution and the monotonic transformation $T(x) = n_0^*x/(n_1^*(1-x))$, we may obtain an exact $100(1-\alpha)$ percent confidence interval for θ^* given by (**Exercise 4.17**)

$$[n_0^*x_1/(n_1^*(x_0+1)F_{2(x_0+1),2x_1,\alpha/2}), \quad n_0^*(x_1+1)F_{2(x_1+1),2x_0,\alpha/2}/(n_1^*x_0)]. \tag{4.35}$$

Example 4.8 Consider the data regarding the study of the effect of hormone use on the risk of breast cancer in menopausal women aged 39 and 44 years. We have $X_1 = 12$, $n_1^* = 10199$, $x_0 = 5$, and $n_0^* = 4722$ (Table 2.4). From these data, the MLE $\hat{\theta}^* = 1.111$. Applying (4.29), (4.31), (4.34), and (4.35), we obtain 95% confidence intervals for θ^* of [0.391, 3.154], [0.394, 9.198], [0.391, 3.154], and [0.364, 4.026]. We can see that in this case using (4.31), based on Fieller's theorem, can cause loss of efficiency as compared with the other three interval estimates. Furthermore, it is not surprising to observe that (4.35), using the exact method that assures that the coverage probability is equal to or greater than the desired confidence level, is wider than (4.29) or (4.34). Because all the lower limits of these interval estimates are less than 1, there is no significant evidence at the 5% level that hormone use can increase the risk of breast cancer in women aged between 39 and 44 years.

4.7 STRATIFIED POISSON SAMPLING

When assessing the incidence RR between the exposed ($i = 1$) and non-exposed ($i = 0$) groups, we may need to apply stratified analysis to adjust for the effects of confounding variables. For example, consider the data in Table 2.4, in which age is a confounder when we study association between hormone use and the risk of breast cancer. Suppose that from each stratum $s(s = 1, 2, \ldots, S)$ we obtain X_{is} cases among n_{is}^* person-years at risk in group i. We assume that the random variables $X_{is}(= 0, 1, 2, \ldots)$ independently follow Poisson distributions with parameters $n_{is}^*\lambda_{is}$, where λ_{is} is the underlying disease rate for stratum s in group i. Note that the assumption of independence between the X_{is} may still be reasonable even when the same person may contribute observation time to several contiguous age categories (Breslow and Day, 1987). The incidence RR in stratum s, denoted by θ_s^*, is simply equal to $\lambda_{1s}/\lambda_{0s}$. Under stratified Poisson sampling (2.44), we can easily show that the MLE of θ_s^* is $\hat{\theta}_s^* = \hat{\lambda}_{1s}/\hat{\lambda}_{0s}$, where $\hat{\lambda}_{is} = X_{is}/n_{is}^*$. In the following discussion, we assume that $\theta_1^* = \theta_2^* = \ldots = \theta_S^*$ and denote this common RR by θ_c^*. We focus our discussion on estimation of θ_c^*.

First, consider the most commonly used WLS interval estimator (Greenland and Robins, 1985; Rosner, 2000; Newman, 2001). An approximate $100(1 - \alpha)$ percent confidence interval for θ_c^* is given by

$$\left[\exp\left(\frac{\sum \hat{W}_s \log(\hat{\theta}_s^*)}{\sum \hat{W}_s} - \frac{Z_{\alpha/2}}{\sqrt{\sum \hat{W}_s}}\right),\ \exp\left(\frac{\sum \hat{W}_s \log(\hat{\theta}_s^*)}{\sum \hat{W}_s} + \frac{Z_{\alpha/2}}{\sqrt{\sum \hat{W}_s}}\right)\right], \tag{4.36}$$

where $\hat{W}_s = (1/X_{1s} + 1/X_{0s})^{-1}$. Note that $\hat{W}_s$ is actually the inverse of the estimated asymptotic variance $\widehat{\mathrm{Var}}(\log(\hat{\theta}_s^*))$ (**Exercise 4.15**).

Based on stratified Poisson sampling, the likelihood (2.44) depends on θ_c^* and on $S - 1$ other nuisance parameters. We may easily eliminate these nuisance parameters by considering the conditional distribution $P(X_{1s} = x_{1s}|x_{.s}; \theta_c^*)$ of X_{1s}, given $x_{.s} = X_{1s} + X_{2s}$ fixed. As shown in **Exercise 4.17**, this conditional distribution follows the binomial distribution

$$P(X_{1s} = x_{1s}|x_{.s}; \theta_c^*) = \binom{x_{.s}}{x_{1s}} \psi_s^{x_{1s}}(1 - \psi_s)^{x_{.s} - x_{1s}}, \tag{4.37}$$

where $\psi_s = n_{1s}^*\theta_c^*/(n_{1s}^*\theta_c^* + n_{0s}^*)$ and $x_{1s} = 0, 1, \ldots, x_{.s}$. Thus, the conditional likelihood of θ_c^*, given the fixed vector $\mathbf{x}_. = (x_{.1}, x_{.2}, \ldots, x_{.S})'$, is

$$\prod_s P(X_{1s} = x_{1s}|x_{.s}; \theta_c^*), \tag{4.38}$$

which depends on θ_c^* only. On the basis of (4.38), we can obtain the conditional MLE $\hat{\theta}_c^*$ by solving for θ_c^* for the following equation (**Exercise 4.18**; Rothman and

Boice, 1979):

$$x_{1.} = \sum_s \mathrm{E}(X_{1s}|x_{.s}, \theta_c^*) \tag{4.39}$$

where $x_{1.} = \sum_s X_{1s}$, and $\mathrm{E}(X_{1s}|x_{.s}; \theta_c^*)$ denotes the conditional expectation of X_{1s}, given $X_{.s} = x_{.s}$, and is equal to $x_{.s}\psi_s$. Lui (2003) notes that a sufficient and necessary condition for the existence of a unique finite conditional MLE $\hat{\theta}_c^*(> 0)$ is that both $x_{1.}$ and $x_{0.}$ $\left(= \sum_s X_{0s}\right)$ are positive. Using the logarithmic transformation, the asymptotic conditional variance of $\hat{\theta}_c^*$ is given by (**Exercise 4.18**)

$$\mathrm{Var}(\log(\hat{\theta}_c^*)|\mathbf{x}_.) = 1 \Big/ \sum_s x_{.s}\psi_s(1 - \psi_s). \tag{4.40}$$

Thus, an asymptotic $100(1 - \alpha)$ percent confidence interval for θ_c^* is

$$[\hat{\theta}_c^* \exp(-Z_{\alpha/2}\sqrt{\widehat{\mathrm{Var}}(\log(\hat{\theta}_c^*)|\mathbf{x}_.)}), \hat{\theta}_c^* \exp(Z_{\alpha/2}\sqrt{\widehat{\mathrm{Var}}(\log(\hat{\theta}_c^*)|\mathbf{x}_.)})] \tag{4.41}$$

where $\widehat{\mathrm{Var}}(\log(\hat{\theta}_c^*)|\mathbf{x}_.) = 1/\left(\sum_s x_{.s}\hat{\psi}_s(1 - \hat{\psi}_s)\right)$ and $\hat{\psi}_s = n_{1s}^*\hat{\theta}_c^*/(n_{1s}^*\hat{\theta}_c^* + n_{0s}^*)$.

Note that the application of (4.41) requires the use of iterative numerical procedure to obtain the conditional MLE $\hat{\theta}_c^*$. A commonly used point estimator of θ_c^* which does not involve an iterative procedure is the Mantel–Haenszel estimator (Rothman and Boice, 1979),

$$\hat{\theta}_{\mathrm{MH}}^* = \frac{\sum_s X_{1s}n_{0s}^*/n_{.s}^*}{\sum_s X_{0s}n_{1s}^*/n_{.s}^*}, \tag{4.42}$$

where $n_{.s}^* = n_{1s}^* + n_{0s}^*$. Furthermore, using the delta method, we can show that the estimated asymptotic conditional variance of $\log(\hat{\theta}_{\mathrm{MH}}^*)$ using the logarithmic transformation is (Breslow, 1984)

$$\widehat{\mathrm{Var}}(\log(\hat{\theta}_{\mathrm{MH}}^*)|\mathbf{x}_.) = \frac{\sum_s n_{1s}^*n_{0s}^*x_{.s}/(n_{.s}^*)^2}{\left[(\hat{\theta}_{\mathrm{MH}})^{0.5}\left(\sum_s n_{0s}^*n_{1s}^*x_{.s}/(n_{.s}^*(\hat{\theta}_{\mathrm{MH}}^*n_{1s}^* + n_{0s}^*))\right)\right]^2}. \tag{4.43}$$

Thus, an asymptotic $100(1 - \alpha)$ percent confidence interval for θ_c^* based on (4.42) and (4.43) is given by

$$[\hat{\theta}_{\mathrm{MH}}^* \exp(-Z_{\alpha/2}\sqrt{\widehat{\mathrm{Var}}(\log(\hat{\theta}_{\mathrm{MH}}^*)|\mathbf{x}_.)}), \hat{\theta}_{\mathrm{MH}}^* \exp(Z_{\alpha/2}\sqrt{\widehat{\mathrm{Var}}(\log(\hat{\theta}_{\mathrm{MH}}^*)|\mathbf{x}_.)})]. \tag{4.44}$$

Applying the score test (Cox and Oakes, 1984; Lawless, 1982; see also the Appendix) to the conditional likelihood, we have the probability $\mathrm{P}((x_{1.} -$

$\sum_s x_{.s}\psi_s)^2 - Z^2_{\alpha/2}\sum_s x_{.s}\psi_s(1-\psi_s) \leq 0) \doteq 1-\alpha$, as both n^*_{is} are large. Thus, we obtain an asymptotic $100(1-\alpha)$ percent confidence interval for θ^*_c given by

$$[\theta^*_l, \theta^*_u], \tag{4.45}$$

where θ^*_l and θ^*_u are the smaller and the larger roots of

$$\left(x_{1.} - \sum_s x_{.s}\psi_s\right)^2 - Z^2_{\alpha/2}\left(\sum_s x_{.s}\psi_s(1-\psi_s)\right) = 0. \tag{4.46}$$

We can again use trial and error to find the two roots of equation (4.46).

The interval estimators (4.36), (4.41), (4,44), and (4.45) are derived under the assumption that the underlying incidence RR is constant across strata. Thus, before using any of these estimators, it is advisable to examine whether this assumption is satisfied by the data. We may apply the asymptotic likelihood ratio test (Rothman and Boice, 1979) and reject the hypothesis $H_0 : \theta^*_1 = \theta^*_2 = \ldots = \theta^*_S$ at level α if

$$2\left\{\sum_s [x_{1s}\log(\hat{\theta}^*_s) - x_{.s}\log(n^*_{1s}\hat{\theta}^*_s + n^*_{0s})]\right.$$
$$\left. - \sum_s [x_{1s}\log(\hat{\theta}^*_c) - x_{.s}\log(n^*_{1s}\hat{\theta}^*_c + n^*_{0s})]\right\} > \chi^2_{S-1,\alpha}. \tag{4.47}$$

Recently, Lui (2003) has evaluated and compared eight interval estimators of θ^*_c in a variety of situations. These include (4.36), (4.41), (4.44), (4.45), and four other interval estimators. Lui notes that the coverage probability of the WLS estimator (4.36) tends to be less than the desired 95% confidence level, especially when we have a large number of strata with a small expected total number of cases per stratum and the underlying θ^*_c is far away from 1 (i.e., $\theta^*_c \leq 1/8$ or $\theta^*_c \geq 8$). Lui further notes that (4.41), (4.44), and (4.45) can actually perform reasonably well in a variety of situations, although (4.44), using the Mantel–Haenszel statistic, is likely less efficient than (4.41) and (4.46). Readers may wish to read this paper for other findings and discussions.

Example 4.9 To illustrate the use of (4.36), (4.41), (4.44), and (4.45), consider the data in Table 2.4 studying the association between estrogen replacement therapy and the risk of breast cancer in menopausal women (Colditz *et al.*, 1990). Applying (4.47) to test the homogeneity of θ^*_s across strata, we obtain a p-value of 0.315. Thus, there is no significant evidence against the assumption that the RR of breast cancer between current users and those who have never used estrogen replacement therapy is constant across different age categories. Furthermore, given these data, we obtain a conditional MLE and a Mantel–Haenszel estimate of $\hat{\theta}^*_c = 1.401$ and $\hat{\theta}^*_{MH} = 1.398$, respectively. Using interval estimators (4.36), (4.41), (4.44), and (4.45), we obtain 95% confidence intervals for θ^*_c of [1.175,

1.692], [1.167, 1.683], [1.164, 1.679], and [1.167, 1.682], respectively. Because the lower limits are all above 1, there is significant evidence at the 5% level that hormone use in menopausal women tends to increase the risk of breast cancer.

EXERCISES

4.1. Under independent binomial sampling, show that the asymptotic variance of the MLE $\hat{\theta} = \hat{\pi}_1/\hat{\pi}_0$ is $\mathrm{Var}(\hat{\theta}) = (\pi_1/\pi_0)^2[(1-\pi_1)/(n_1\pi_1) + (1-\pi_0)/(n_0\pi_0)]$, where $\hat{\pi}_i = X_i/n_i$ for $i = 1, 0$.

4.2. On the basis of the result in **Exercise 4.1**, show that the asymptotic variance of $\log(\hat{\theta})$ is $\mathrm{Var}(\log(\hat{\theta})) = (1-\pi_1)/(n_1\pi_1) + (1-\pi_0)/(n_0\pi_0)$.

4.3. Show that the inequality $[(\hat{\pi}_1 - \theta\hat{\pi}_0)/\sqrt{\mathrm{Var}(\hat{\pi}_1 - \theta\hat{\pi}_0)}]^2 \leq Z^2_{\alpha/2}$ in Section 4.1 can be rewritten as $A\theta^2 - 2B\theta + C \leq 0$, where $A = \hat{\pi}_0^2 - Z^2_{\alpha/2}\hat{\pi}_0(1 - \hat{\pi}_0)/n_0$, $B = \hat{\pi}_1\hat{\pi}_0$, and $C = \hat{\pi}_1^2 - Z^2_{\alpha/2}\hat{\pi}_1(1-\hat{\pi}_1)/n_1$.

4.4. (a) In Section 4.1, what is the asymptotic asymptotic variance of $\hat{\pi}_1^{1/3} - (\theta\hat{\pi}_0)^{1/3}$? (Hint: use the delta method.) (b) Show that $[(\hat{\pi}_1^{1/3} - (\theta\hat{\pi}_0)^{1/3})/\mathrm{Var}(\hat{\pi}_1^{1/3} - (\theta\hat{\pi}_0)^{1/3})]^2 \leq Z^2_{\alpha/2}$ is equivalent to the following quadratic equation in $\theta^{1/3}$: $\mathcal{A}(\theta)^{2/3} - 2\mathcal{B}(\theta)^{1/3} + \mathcal{C} \leq 0$, where $\mathcal{A} = \hat{\pi}_0^{2/3} - Z^2_{\alpha/2}(1 - \hat{\pi}_0)/(9n_0\hat{\pi}_0^{1/3})$, $\mathcal{B} = (\hat{\pi}_1\hat{\pi}_0)^{1/3}$, and $\mathcal{C} = \hat{\pi}_1^{2/3} - Z^2_{\alpha/2}(1-\hat{\pi}_1)/(9n_1\hat{\pi}_1^{1/3})$.

4.5. Consider the data taken from the German-Austrian Multicenter Study (Breddin *et al.*, 1979) comparing deaths from post-myocardial infarction between the placebo and aspirin groups. We have 32 deaths out of 309 patients in the placebo group ($i = 0$), and 27 deaths out of 317 patients in the aspirin group ($i = 1$). What is the MLE of the RR ($= \pi_1/\pi_0$)? What are the corresponding 95% confidence intervals for the RR using (4.1), (4.2), (4.4), and (4.6)?

4.6. Using the delta method, show that in Section 4.2, an asymptotic variance $\mathrm{Var}(\hat{\theta}_{MH})$ of $\hat{\theta}_{MH}$ is given by $[\sum_s (n_{0s}/n_{.s})^2 n_{1s}\pi_{1s}(1-\pi_{1s}) + \theta_c^2 \sum_s (n_{1s}/n_{.s})^2 n_{0s}\pi_{0s}(1-\pi_{0s})]/(\sum_s((n_{0s}n_{1s})/n_{.s})\pi_{0s})^2$. If π_{is} is small ($\doteq 0$), then this asymptotic variance reduces to $[\sum_s (n_{0s}/n_{.s})^2 n_{1s}\pi_{1s} + \theta_c^2 \sum_s (n_{1s}/n_{.s})^2 n_{0s}\pi_{0s}]/(\sum_s((n_{0s}n_{1s})/n_{.s})\pi_{0s})^2$.

4.7. Consider the all-cause mortality data (Table 2.1) from the first five ($s = 1, 2, \ldots, 5$) randomized trials comparing aspirin with placebo in post-myocardial infarction patients (Canner, 1987). (a) What are the p-values for testing the homogeneity of the mortality RR between aspirin and placebo using (4.13)–(4.15)? (b) What are the MLEs $\hat{\theta}_s$ of the mortality RR for these trials? (c) What are the summary estimates $\hat{\theta}_{WLS}$ (4.8) and $\hat{\theta}_{MH}$ (4.10)? (d) What are the 95% confidence intervals for θ_c using (4.9) and (4.12)? Recall that on the basis of using the data from a single randomized trial (Elwood *et al.*, 1974) considered in Example 4.1, we do not find significant evidence of the protection effect due to the aspirin. By contrast, when combining data from the five trials, we do find a minor reduction in all-cause mortality due to aspirin.

4.8. Show that the marginal RR $\theta = \pi_{1.}/\pi_{.1} = \exp(-\beta_1)$ under the exponential model assumptions as described in Section 4.4.

4.9. Show that in Section 4.4, the asymptotic variance $\mathrm{Var}(\log(\hat{\theta}))$ is given by $(1-\pi_{1.})/(n\pi_{1.}) + (1-\pi_{.1})/(n\pi_{.1}) - 2(\pi_{11}\pi_{00} - \pi_{10}\pi_{01})/(n\pi_{1.}\pi_{.1})$, where $\hat{\theta} = \hat{\pi}_{1.}/\hat{\pi}_{.1}$.

4.10. Show that in Section 4.5, the variance $\mathrm{Var}(\hat{\pi}_1^{(u)}/\hat{\pi}_0) = \mathrm{Var}(\hat{\pi}_1^{(u)})\mathrm{Var}(\hat{\pi}_0^{-1}) + \pi_0^{-2}\mathrm{Var}(\hat{\pi}_1^{(u)}) + \pi_1^2\mathrm{Var}(\hat{\pi}_0^{-1})$, where $\hat{\pi}_1^{(u)} = (x_1 - 1)/(x_1 + Y_1 - 1)$ and $\hat{\pi}_0 = x_0/(x_0 + Y_0)$. Thus, the variance $\mathrm{Var}(\hat{\pi}_1^{(u)}/\hat{\pi}_0)$ can be expressed in closed form. (Hint: Best (1974) shows that $\mathrm{Var}(\hat{\pi}_1^{(u)}) = (x_1 - 1)(1 - \pi_1)[\sum_{k=2}^{x_1-1} (-\pi_1/(1-\pi_1))^k/(x_1 - k) - (-\pi_1/(1-\pi_1))^{x_1}\log(\pi_1)] - \pi_1^2$. Furthermore, we can show that $\mathrm{Var}(\hat{\pi}_0^{-1}) = (1-\pi_0)/(x_0\pi_0^2)$.)

4.11. For $x_1 > 2$ in Section 4.5, show that the variance estimator (4.26) is an unbiased estimator of $\mathrm{Var}(\hat{\pi}_1^{(u)}/\hat{\pi}_0)$.

4.12. When π_i is small, as noted in **Exercise 1.18**, $2(x_i + Y_i)\pi_i$ approximately follows a χ^2 distribution with $2x_i$ degrees of freedom. Thus, the ratio $\{[2(x_1 + Y_1)\pi_1]/2x_1\}/\{[2(x_0 + Y_0)\pi_0]/2x_0\}$ follows an F distribution with $2x_1$ and $2x_0$ degrees of freedom. Show how to apply this result to derive an approximate $100(1-\alpha)$ percent confidence interval for the RR as given in (4.28).

4.13. Consider the data on the opinions of a random sample of 1600 voters on the President's performance in two surveys (Agresti, 1990, p. 350). There are $(n_{11} =)794$ people indicating approval in both surveys, $(n_{10} =)150$ people indicating approval in the first survey but indicating disapproval in the second survey, $(n_{01} =)86$ people indicating disapproval in the first survey but indicating approval in the second survey, and $(n_{00} =)570$ people indicating disapproval in both surveys. When we compare the approval rate $\pi_{1.}$ of the first survey with the approval rate $\pi_{.1}$ of the second survey, what is the MLE of the ratio $\pi_{1.}/\pi_{.1}$? What is the 95% confidence interval for $\pi_{1.}/\pi_{.1}$? (c) What conclusion can you draw from these data?

4.14. Suppose that we want to compare the cancer prevalence rates between a mining town ($i = 1$) and a control town ($i = 0$). Suppose further that we employ inverse sampling to collect subjects from population i until we obtain $x_i = 10$ cases, and we obtain $y_1 = 490$ and $y_2 = 4990$. What are the MLE and the UMVUE of the prevalence RR? What are the 95% confidence intervals for RR using (4.27) and (4.28)?

4.15. Suppose that X_i independently follows the Poisson distribution with parameter $n_i^*\lambda_i (i = 1, 0)$. Show that the estimated asymptotic variance $\widehat{\mathrm{Var}}(\log(\hat{\theta}^*))$ is given by $1/X_1 + 1/X_2$, where $\hat{\theta}^* = \hat{\lambda}_1/\hat{\lambda}_0$ and $\hat{\lambda}_i = X_i/n_i^*$.

4.16. Define $\mathfrak{A} = \hat{\lambda}_0^2 - Z_{\alpha/2}^2\hat{\lambda}_0/n_0^*$, $\mathfrak{B} = \hat{\lambda}_1\hat{\lambda}_0$, and $\mathfrak{C} = \hat{\lambda}_1^2 - Z_{\alpha/2}^2\hat{\lambda}_1/n_1^*$. Show that if $\mathfrak{A} > 0$ and $\hat{\lambda}_1 > 0$, then the inequality $\mathfrak{B}^2 - \mathfrak{A}\mathfrak{C} > 0$ will hold.

4.17. Suppose that $X_i (i = 1, 0)$ independently follows the Poisson distribution with mean $n_i^*\lambda_i$. (a) Show that $X_1 + X_0$ follows the Poisson distribution with mean $n_1^*\lambda_1 + n_0^*\lambda_0$. (b) Show that the conditional probability mass function $f(X_1 = x_1|X_1 + X_0 = x_.)$ follows the binomial distribution with parameter $x_.$ and $n_1^*\lambda_1/(n_1^*\lambda_1 + n_0^*\lambda_0)(= n_1^*\theta^*/(n_1^*\theta^* + n_0))$, where $\theta^* = \lambda_1/\lambda_0$. (c) Show that the conditional MLE of θ^* on the basis of (4.32) is the same as the unconditional MLE $\hat{\theta}^* = \hat{\lambda}_1/\hat{\lambda}_0$ with the estimated asymptotic conditional variance given by $\widehat{\text{Var}}(\theta^*|x_.) = (\hat{\theta}^*)^2/(x_.\hat{\psi}(1 - \hat{\psi}))$, where $\hat{\psi} = n_1^*\hat{\theta}^*/(n_1^*\hat{\theta}^* + n_0^*)$.

4.18. On the basis of the likelihood (4.38), show that the conditional MLE $\hat{\theta}_c^*$ can be obtained by solving for θ_c^* the equation $x_{1.} = \sum_s E(X_{1s}|x_{.s}, \theta_c^*)$, where $x_{1.} = \sum_s X_{1s}$ and $E(X_{1s}|x_{.s}, \theta_c^*) = x_{.s}\psi_s$, where $\psi_s = n_{1s}^*\theta_c^*/(n_{1s}^*\theta_c^* + n_{0s}^*)$. Also show that the asymptotic conditional variance $\text{Var}(\log(\hat{\theta}_c^*)|\mathbf{x}_.) = 1/\sum_s x_{.s}\psi_s(1 - \psi_s)$.

4.19. Consider the data concerning estrogen replacement therapy and breast cancer just for women aged between 39 and 54 years in Table 2.4 (Colditz *et al.*, 1990; Rosner, 2000).
(a) What is the p-value using the asymptotic likelihood ratio test statistic to test the homogeneity of the incidence RR across the three strata?
(b) What is the conditional MLE $\hat{\theta}_c^*$ and Mantel–Haenszel estimate $\hat{\theta}_{MH}^*$ of the underlying common RR?
(c) What are the 95% confidence intervals for the underlying common incidence RR using (4.36), (4.41), (4.44), and (4.45)?

REFERENCES

Agresti, A. (1990) *Categorical Data Analysis*. Wiley, New York.
Bailey, B. J. R. (1987) Confidence limits to the risk ratio. *Biometrics*, **43**, 201–205.
Bennett, B. M. (1981) On the use of the negative binomial in epidemiology. *Biometrical Journal*, **23**, 69–72.
Best, D. J. (1974) The variance of the inverse binomial estimator. *Biometrika*, **61**, 385–386.
Bishop, Y. M. M., Fienberg, S. E., and Holland, P. W. (1975) *Discrete Multivariate Analysis, Theory and Practice*. MIT Press, Cambridge, MA.
Breddin, K., Loew, D., Lechner, K., Uberla, K. and Walter, E. (1979) Secondary prevention of myocardial infarction. Comparison of acetylsalicylic acid, phenprocoumon and placebo. A multicenter two-year prospective study. *Thrombosis and Haemostasis*, **40**, 225–236.
Breslow, N. E. (1984) Elementary methods of cohort analysis. *International Journal of Epidemiology*, **13**, 112–115.
Breslow, N. E. and Day, N. E. (1987) *Statistical Methods in Cancer Research*, Vol II. *The Design and Analysis of Cohort Studies*. International Agency for Research on Cancer, Lyon, France.
Canner, P. L. (1987) An overview of six clinical trials of aspirin in coronary disease. *Statistics in Medicine*, **6**, 255–263.

Casella, G. and Berger, R. L. (1990) *Statistical Inference*. Duxbury Press, Belmont, CA.
Chung, J. M. (1993) Cisapride in chronic dyspepsia: results of a double-blind, placebo-controlled trial. *Scandinavian Journal of Gastroenterology*, **28**, 11–14.
Colditz, G. A., Stampfer, M. J., Willett, W. C., Hennekens, C. H., Rosner, B., and Speizer, F. E. (1990) Prospective study of estrogen replacement therapy and risk of breast cancer in postmenopausal women. *Journal of the American Medical Association*, **264**, 2648–2653.
Cox, D. R. and Oakes, D. (1984) *Analysis of Survival Data*. Chapman & Hall, London.
Donner, A., Birkett, N. and Buck, C. (1981) Randomization by cluster sample size requirements and analysis. *American Journal of Epidemiology*, **14**, 906–914.
Elston, R. C. (1977) Response to query: estimating 'inheritability' of a dichotomous trait. *Biometrics*, **33**, 232–233.
Elwood, P. C., Cochrane, A. L., Burr, M. L., Sweetnam, P. M., Williams, G., Welsby, E., Hughes, S. J., and Renton, R. (1974) A randomized controlled trial of acetyl salicylic acid in the secondary prevention of mortality from myocardial infarction. *British Medical Journal*, **1**, 436–440.
Fleiss, J. L. (1981) *Statistical Methods for Rates and Proportions*, 2nd edition. Wiley, New York.
Fleiss, J. L. (1986) *The Design and Analysis of Clinical Experiments*. Wiley, New York.
Feller, W. (1968). *An Introduction to Probability Theory and Its Applications*, Vol. 1, 3rd edition. Wiley, New York.
Fisher, R. A. (1928) On a distribution yielding the error functions of several well known statistics. In J. C. Fields (ed.), *Proceedings of the International Mathematical Congress*, Vol. 2. University of Toronto Press, Toronto, pp. 805–813.
Gart, J. J. (1962) On the combination of relative risks. *Biometrics*, **18**, 601–610.
Gart, J. J. (1979) Statistical analyses of relative risk. *Environmental Health Perspectives*, **32**, 157–167.
Greenland, S. and Robins, J. M. (1985) Estimation of a common effect parameter from sparse followup data. *Biometrics*, **41**, 55–68.
Haldane, J. B. S. (1945a) A labour-saving method of sampling. *Nature*, **155**, 49–50.
Haldane, J. B. S. (1945b) On a method of estimating frequencies. *Biometrika*, **33**, 222–225.
Hartung, J. and Knapp, G. (2001) A refined method for the meta-analysis of controlled clinical trials with binary outcome. *Statistics in Medicine*, **20**, 3875–3899.
Johnson, N. L. and Kotz, S. (1970) *Distributions in Statistics: Continuous Univariate Distributions 2*. Wiley, New York.
Katz, D., Baptista, J., Azen, S. P. and Pike, M. C. (1978) Obtaining confidence intervals for the risk ratio in cohort studies. *Biometrics*, **34**, 469–474.
Laupacis, A., Sackett, D. L. and Roberts, R. S. (1988) An assessment of clinically useful measures of the consequences of treatment. *New England Journal of Medicine*, **218**, 1728–1733.
Lawless, J. F. (1982). *Statistical Models and Methods for Lifetime Data*. Wiley, New York.
Lipsitz, S. R., Dear, K. B. G., Laird, N. M. and Molenberghs, G. (1998) Tests for homogeneity of the risk difference when data are sparse. *Biometrics*, **54**, 148–160.
Lui, K.-J. (1991) Sample size for repeated measurements in dichotomous data. *Statistics in Medicine*, **10**, 463–472.
Lui, K.-J. (1995a) Confidence intervals for the risk ratio in cohort studies under inverse sampling. *Biometrical Journal*, **37**, 965–971.
Lui, K.-J. (1995b) Notes on conditional confidence limits under inverse sampling. *Statistics in Medicine*, **14**, 2051–2056.
Lui, K.-J. (1996) Point estimation of relative risk under inverse sampling. *Biometrical Journal*, **38**, 669–680.
Lui, K.-J. (1997) Conditional maximum likelihood estimate and exact test of the common relative difference in combination of 2 × 2 tables under inverse sampling. *Biometrical Journal*, **39**, 215–225.

Lui, K.-J. (1998) Tests of homogeneity for the relative difference between strata under inverse sampling. *Biometrical Journal*, **40**, 227–235.

Lui, K.-J. (2000) Asymptotic conditional test procedures for relative difference under inverse sampling. *Computational Statistics and Data Analysis*, **34**, 335–343.

Lui, K.-J. (2001a) Notes on testing equality in dichotomous data with matched pairs. *Biometrical Journal*, **43**, 313–321.

Lui, K.-J. (2001b) Estimation of rate ratio and relative difference in matched-pairs under inverse sampling. *Environmetrics*, **12**, 539–546.

Lui, K.-J. (2003) Eight interval estimators of a common rate ratio under stratified Poisson sampling, *Statistics in Medicine*. To appear.

Lui, K.-J. and Kelly, C. (2000) Tests for homogeneity of the risk ratio in a series of 2 × 2 tables. *Statistics in Medicine*, **19**, 2919–2932.

Lui, K.-J., Cumberland, W. G., and Kuo, L. (1996) An interval estimate for the intraclass correlation in beta-binomial sampling. *Biometrics*, **52**: 412–425.

Lui, K.-J., Mayer, J. A., and Eckhardt, L. (2000) Confidence intervals for the risk ratio under cluster sampling based on the beta-binomial model. *Statistics in Medicine* **19**, 2933–2942.

Mantel, N. and Haenszel, W. (1959) Statistical aspects of the analysis of data from retrospective studies of disease. *Journal of the National Cancer Institute*, **22**, 719–748.

Mausner, J. S. and Bahn, A. K. (1974) *Epidemiology, An Introduction Text*. W. B. Saunders, Philadelphia.

Mayer, J., Slymen, D. J., Eckhardt, L., *et al.* (1997). Reducing ultraviolet radiation exposure in children. *Preventive Medicine*, **26**, 516–522.

Newman, S. C. (2001) *Biostatistical Methods in Epidemiology*. Wiley, New York.

Rosner, B. (1990) *Fundamental of Biostatistics*, 3rd edition. PWS-Kent, Boston.

Rosner, B. (2000) *Fundamentals of Biostatistics*, 5th edition. Duxbury Press, Pacific Grove, CA.

Rothman, K. J. and Boice, J. D. (1979) *Epidemiologic Analysis with a Programmable Calculator*. NIH Publication No. 79–1649, Washington, DC.

Scotto, J., Kopf, A. W. and Urbach, F. (1974) Non-melanoma skin cancer among Caucasians in four areas of the United States. *Cancer*, **34**, 1333–1338.

Tarone, R. E. (1981). On summary estimators of relative risk. *Journal of Chronic Diseases*, **34**, 463–468.

5

Odds Ratio

The odds ratio (OR) is one of the most frequently used indices to measure the extent of association between a risk factor and an outcome in epidemiology. When the underlying outcome is rare, it is well known that OR can provide a good approximation to the risk ratio (RR) discussed in Chapter 4. While the RR cannot be estimated in case–control studies without using Bayes' theorem and other external information, the OR can actually be estimated from a cross-sectional study, a cohort study, or a case–control study. Due to this invariance property, we often use the OR to locate possible etiologic causes in a case–control study design for chronic diseases, such as cancers or cardiovascular diseases, which are generally difficult to study without following up a relatively large number of subjects for a long period of time. Furthermore, because the OR can be expressed in terms of model parameters under the log-linear or logistic regression model (Bishop *et al.*, 1975; Agresti, 1990; Hosmer and Lemeshow, 1989), we can easily study this important parameter while controlling the effects of other confounders.

To help readers appreciate the invariance (Cornfield, 1951) of the OR with respect to various study designs, for clarity we may use the following table to summarize the data structure:

		Status of disease		
		Yes	No	
Status of exposure	Yes	π_{11}	π_{10}	$\pi_{1.}$
	No	π_{01}	π_{00}	$\pi_{0.}$
		$\pi_{.1}$	$\pi_{.0}$	

where π_{ij} denotes the corresponding cell probability for i and $j = 1, 0$. We define $\pi_{i.} = \pi_{i1} + \pi_{i0}$ and $\pi_{.j} = \pi_{1j} + \pi_{0j}$. For convenience, we let D and $\overline{D}$ denote the events of being a case and a control, respectively. Similarly, we let E and $\overline{E}$ denote the events of being exposed and not exposed to the risk factor under investigation. In a cohort study, we take independent random samples of subjects

Statistical Estimation of Epidemiological Risk K-J. Lui
 ISBN: 0-470-85071-X (HB)

from the exposed and non-exposed populations, respectively. The proportions of cases calculated from the two samples are unbiased estimators of the conditional probabilities $P(D|E)(=\pi_{11}/\pi_{1.})$ and $P(D|\bar{E})(=\pi_{01}/\pi_{0.})$ of the disease for the exposed and non-exposed populations. Thus, the OR $(=[P(D|E)/(1-P(D|E))]/[P(D|\bar{E})/(1-P(D|\bar{E}))]=\pi_{11}\pi_{00}/(\pi_{10}\pi_{01}))$ of being a case between the exposed and the non-exposed populations can be estimated. On the other hand, in a case–control study, we take independent random samples of subjects from the diseased and non-diseased populations, respectively. The proportions of exposure calculated from the two samples are unbiased estimators of the conditional probabilities $P(E|D)(=\pi_{11}/\pi_{.1})$ and $P(E|\bar{D})(=\pi_{10}/\pi_{.0})$, respectively. Thus, the OR$(=[P(E|D)/(1-P(E|D))]/[P(E|\bar{D})/(1-P(E|\bar{D}))]=\pi_{11}\pi_{00}/(\pi_{10}\pi_{01}))$ of having the exposure between the cases and the controls can also be estimated. Finally, in a cross-sectional study, we take a random sample of subjects from the general population and simultaneously determine the status of the risk factor and the disease for each sampled subject. The sample proportions falling into the corresponding cells can then be used to directly estimate the cell probabilities π_{ij}, and thus the OR$(=\pi_{11}\pi_{00}/(\pi_{10}\pi_{01}))$ between the disease and the risk factor can be estimated as well. The OR therefore not only remains invariant with respect to the cohort study, the case–control study, and the cross-sectional study, but also is estimable by using any of these three designs. Note that the value of the OR does not change when we multiply a row or a column by a positive constant (**Exercise 5.1**). Therefore, in case–control studies, in which we may sample most cases, but only sample a small fraction of the disease-free population, this differential sampling fraction does not cause a bias in estimation of OR. If the underlying disease were rare, as noted previously, the OR would approximately equal the RR$(=P(D|E)/P(D|\bar{E}))$ (**Exercise 5.2**). By definition, the OR is positive. A value greater than 1 (less than 1, equal to 1) simply indicates a positive association (a negative association, no association) between the risk factor under investigation and the disease of interest.

In this chapter, we first discuss estimation of the OR under independent binomial sampling for the situation of no confounders. We then extend this discussion to accommodate the situation in which we can apply stratified analysis to control confounders or we encounter data collected by means of a multicenter study design. We discuss estimation of the OR under independent cluster sampling, in which the sampled units are classes or households rather than individual patients. We also discuss estimation of the OR for one-to-one matched sampling, commonly used in the case–control study design to increase the efficiency or the validity of inference. We further include a brief discussion on using the logistic regression to control the confounders for the cases of both non-matching and matching designs. Finally, we discuss estimation of the OR for case–control studies under independent inverse sampling and for case–control studies with matched pairs under negative multinomial sampling.

5.1 INDEPENDENT BINOMIAL SAMPLING

Suppose that we independently sample n_1 subjects from the case ($j = 1$) population and n_0 subjects from the control ($j = 0$) population, respectively. Suppose further that we obtain X_j among n_j sampled subjects exposed to the risk factor. Then X_j follows the binomial distribution (1.1) with parameters n_j and $\pi_{1|j} = \pi_{1j}/\pi_{.j}$.

5.1.1 Asymptotic interval estimators

Under independent binomial sampling, we can easily see that the MLE of the OR is simply $\hat{\mathcal{O}} = (X_1/n_1)((n_0 - X_0)/n_0)/[(X_0/n_0)((n_1 - X_1)/n_1)] = X_1(n_0 - X_0)/[(n_1 - X_1)X_0]$. Using the delta method (Bishop *et al.*, 1975; see also the Appendix), we obtain an estimated asymptotic variance $\widehat{\text{Var}}(\log(\hat{\mathcal{O}})) = 1/X_1 + 1/X_0 + 1/(n_1 - X_1) + 1/(n_0 - X_0)$ (**Exercise 5.3**). Note that there is a positive probability that the variance $\widehat{\text{Var}}(\log(\hat{\mathcal{O}}))$ is not defined due to either $X_i = 0$ or $X_i = n_i$. To alleviate this concern, we may apply the commonly used adjustment procedure for sparse data of adding 0.50 to each cell. In fact, substituting $(X_j + 0.50)$ and $n_j - X_j + 0.5$ for X_j and $n_j - X_j$, respectively, will generally improve the performance of the interval estimator based on the statistic $\log(\hat{\mathcal{O}})$ (Gart and Zweifel, 1967; Fleiss, 1981; Gart and Thomas, 1972; Fleiss, 1979; Lui and Lin, 2003). Thus, an asymptotic $100(1 - \alpha)$ percent confidence interval for the OR (Woolf, 1955; Gart and Thomas, 1972) is given by

$$[\hat{\mathcal{O}}_{\text{adj}} \exp(-Z_{\alpha/2}\sqrt{\widehat{\text{Var}}(\log(\hat{\mathcal{O}}_{\text{adj}}))}), \hat{\mathcal{O}}_{\text{adj}} \exp(Z_{\alpha/2}\sqrt{\widehat{\text{Var}}(\log(\hat{\mathcal{O}}_{\text{adj}}))})], \quad (5.1)$$

where

$$\hat{\mathcal{O}}_{\text{adj}} = (X_1 + 0.5)(n_0 - X_0 + 0.5)/[(n_1 - X_1 + 0.5)(X_0 + 0.5)],$$

$$\widehat{\text{Var}}(\log(\hat{\mathcal{O}}_{\text{adj}})) = 1/(X_1 + 0.5) + 1/(X_0 + 0.5) + 1/(n_1 - X_1 + 0.5) + 1/(n_0 - X_0 + 0.5).$$

Note that applying (5.1) to determine whether there is an association between a risk factor and a disease does not always lead to the same conclusion as that using the χ^2 test for equality between two comparison groups. Thus, Cornfield (1956) proposes an interval estimator by inverting the acceptance region of a family of χ^2 tests for testing the null hypothesis $H_0 : \mathcal{O} = \mathcal{O}_0$, where $\mathcal{O}_0$ is any given specified value.

Given $\mathcal{O} = \mathcal{O}_0$, we have $X_1(\mathcal{O}_0)(n_0 - X_0(\mathcal{O}_0))/[(n_1 - X_1(\mathcal{O}_0))X_0(\mathcal{O}_0)] = \mathcal{O}_0$, where $X_j(\mathcal{O}_0)$ denotes the expected frequency of subjects with exposure in group j ($j = 1, 0$) when $\mathcal{O} = \mathcal{O}_0$. Therefore, given a value $\mathcal{O}_0 \neq 1$ and a marginal total $x_.(= X_1 + X_0)$ of subjects with exposure, we can obtain the expected frequency

$X_1(\mathcal{O}_0)$ by solving the following quadratic equation in X_1 (**Exercise 5.4**):

$$(\mathcal{O}_0 - 1)X_1^2 - [\mathcal{O}_0(n_1 + x_.) + (n_0 - x_.)]X_1 + \mathcal{O}_0 n_1 x_. = 0. \qquad (5.2)$$

Once we obtain $X_1(\mathcal{O}_0)$, we can then uniquely determine the expected frequency for the other cells. Note that $X_1(\mathcal{O}_0)$ must satisfy the constraint given by $\max\{0, x_. - n_0\} < X_1(\mathcal{O}_0) < \min\{n_1, x_.\}$ so that none of the expected frequencies in any cell is negative. We thus conclude (Fleiss, 1981; **Exercise 5.5**) that

$$X_1(\mathcal{O}_0) = \frac{\mathcal{O}_0(n_1 + x_.) + (n_0 - x_.) - \sqrt{[\mathcal{O}_0(n_1 + x_.) + (n_0 - x_.)]^2 - 4(\mathcal{O}_0 - 1)\mathcal{O}_0 n_1 x_.}}{2(\mathcal{O}_0 - 1)}. \qquad (5.3)$$

When $\mathcal{O}_0 = 1$, the expected frequency $X_1(\mathcal{O}_0)$ derived from (5.2) simplifies to $n_1 x_./n_.$, where $n_. = n_1 + n_0$. Stevens (1951) and Cornfield (1956) show that when the underlying OR is $\mathcal{O}_0$, the conditional distribution of the observed frequency X_1, given $x_.$ fixed, is approximately normally distributed with mean $X_1(\mathcal{O}_0)$ and asymptotic variance **(Exercise 5.6)**

$$\begin{aligned}\mathrm{Var}(X_1|x_., \mathcal{O}_0) = {} & [1/X_1(\mathcal{O}_0) + 1/(x_. - X_1(\mathcal{O}_0)) + 1/(n_1 - X_1(\mathcal{O}_0)) \\ & + 1/(n_0 - x_. + X_1(\mathcal{O}_0))]^{-1}. \end{aligned} \qquad (5.4)$$

When $\mathcal{O}_0 = 1$, the variance (5.4) reduces to $n_1 n_0 x_.(n_. - x_.)/n_.^3$. Thus, we obtain an asymptotic $100(1 - \alpha)$ percent confidence interval for the OR given by

$$[\mathcal{O}_l, \mathcal{O}_u], \qquad (5.5)$$

where the confidence limits $\mathcal{O}_l$ and $\mathcal{O}_u$ are determined by solving the following two equations for $\mathcal{O}$. The lower limit $\mathcal{O}_l$ is the smaller root of $\mathcal{O}$ satisfying

$$(X_1 - X_1(\mathcal{O}) - c_1)^2\ \mathrm{Var}(X_1|x_., \mathcal{O})^{-1} - Z_{\alpha/2}^2 = 0, \qquad (5.6)$$

where the constant c_1 is set to 0.50 when one wishes to apply a continuity correction, or to 0 otherwise. Similarly, the upper limit $\mathcal{O}_u$ is the larger root of $\mathcal{O}$ satisfying

$$(X_1 - X_1(\mathcal{O}) + c_2)^2\ \mathrm{Var}(X_1|x_., \mathcal{O})^{-1} - Z_{\alpha/2}^2 = 0, \qquad (5.7)$$

where the constant c_2 is set to 0.50 for the continuity correction, or to 0 otherwise. Although the above discussion is relevant to case–control studies, all the results presented here apply equally to cohort studies. A discussion on the iterative numerical procedure for solving (5.6) and (5.7) can be found elsewhere (Fleiss, 1979).

Gart and Thomas (1972) as well as Brown (1981) note that interval estimator (5.1) can be much too narrow in the conditional sample space when both n_i are small. Fleiss (1979) notes a close agreement of confidence limits between Cornfield's interval with the continuity correction and the exact method (that will be discussed in the next section). Thus, all these papers suggest using interval estimator (5.5) with the continuity correction. On the other hand, Gart and Thomas (1982) find that Cornfield's method without the continuity correction is the preferred approximate method in the unconditional space. However, none of these papers accounts for efficiency by comparing the average length of the resulting confidence intervals using different methods. Agresti (1999) notes that (5.1) can actually perform well even for small samples unless the underlying OR is large. Lui and Lin (2003) further note that using Cornfield's confidence interval (5.5) with the continuity correction can actually lose substantial efficiency compared to that without the continuity correction. When the sample size per group is not large (≤ 30) and the probability of exposure in the control group is small (say, 0.10) or large (say, 0.90), Lui and Lin (2003) recommend using Cornfield confidence interval without the continuity correction. When the sample size is large (say, 100 per group) or when the probability of exposure in the control is moderate (say, 0.50), they note that (5.1) is probably preferable to (5.5).

There is another popular interval estimator of the OR, Miettinen's (1976) test-based confidence interval, that is simple to use. As demonstrated elsewhere (Brown, 1981), however, this interval estimator does not perform well if the underlying OR is far from 1. Thus, we confine ourselves to a brief outline of the construction of Miettinen's test-based interval estimator in **Exercise 5.8**.

Example 5.1 Consider the case–control study of smoking and oral cancer (Graham *et al.*, 1977; Gart and Thomas, 1982). As reported elsewhere (Gart and Thomas, 1982, p. 461), there were ($X_1 =$)399 cases who had ever smoked among ($n_1 =$) 419 cases with oral cancer. By contrast, there were ($X_0 =$) 414 subjects who had ever smoked among ($n_0 =$)516 controls without oral cancer. On the basis of the data, the MLE $\hat{\mathcal{O}}$ of the OR is 4.92. Applying (5.1) and (5.5) with the continuity correction, we obtain 95% confidence intervals of $\mathcal{O}$ of [2.985, 8.093] and [2.916, 8.362], respectively. Note that the former is completely contained in the latter here. On the other hand, if we applied (5.5) without the continuity correction, we would obtain [2.998, 8.056], which is similar to the result of using (5.1). Because all these lower limits are above 1, we may conclude at the 5% level that smoking may increase the risk of oral cancer.

5.1.2 Exact confidence interval

Note that both interval estimators (5.1) and (5.5) are derived on the basis of large-sample theory. When either of the n_i is small, these two interval estimators are theoretically not valid. Thus, we may consider deriving a $100(1-\alpha)$ percent confidence interval on the basis of the exact conditional distribution of X_1, given a fixed marginal total $x_.$ (Gart, 1970; Gart and Thomas, 1972; **Exercise 5.9**):

$$P(X_1 = x_1 | X_. = x_., \mathcal{O}) = \binom{n_1}{x_1}\binom{n_0}{x_. - x_1}\mathcal{O}^{x_1} \Big/ \sum_x \binom{n_1}{x}\binom{n_0}{x_. - x}\mathcal{O}^{x}, \tag{5.8}$$

where the range of x_1 is $\max\{x_. - n_0, 0\} \leq x_1 \leq \min\{n_1, x_.\}$, and the summation for x is over this range. Note that the conditional probability mass function (5.8) is only a function of the parameter $\mathcal{O}$. On the basis of (5.8), we can obtain the conditional MLE $\hat{\mathcal{O}}_{\text{cond}}$ by finding the root $\mathcal{O}$ of the following equation (**Exercise 5.10**):

$$x_1 = \mathrm{E}(X_1 | x_., \mathcal{O}), \tag{5.9}$$

where $\mathrm{E}(X_1|x_., \mathcal{O})$ denotes the conditional expectation of X_1, given $x_.$ fixed. Furthermore, we can show that the cumulative probability distribution $\sum_{x \leq x_1} \mathrm{P}(X_1 = x|X_. = x_., \mathcal{O})$ is a decreasing function of $\mathcal{O}$ (**Exercise 5.11**). Thus, we obtain an exact $100(1-\alpha)$ percent confidence interval for $\mathcal{O}$ given by (Casella and Berger, 1990)

$$[\mathcal{O}_{\mathrm{l}}^*, \mathcal{O}_{\mathrm{u}}^*], \tag{5.10}$$

where the confidence limits $\mathcal{O}_{\mathrm{l}}^*$ and $\mathcal{O}_{\mathrm{u}}^*$ are determined by solving the following two equations:

$$\sum_{x \geq x_1} \mathrm{P}(X_1 = x|X_. = x_., \mathcal{O}_{\mathrm{l}}^*) = \alpha/2, \qquad \sum_{x \leq x_1} \mathrm{P}(X_1 = x|X_. = x_., \mathcal{O}_{\mathrm{u}}^*) = \alpha/2.$$

Applying (5.10) always ensures that the coverage probability is equal to or greater than the desired $100(1-\alpha)$ percent confidence level and hence (5.10) is valid even when both n_i are small.

Example 5.2 To illustrate the use of estimators (5.9) and (5.10), we consider data on the response of diffuse lymphoma patients to combination chemotherapy by gender (Bishop *et al.*, 1975, p. 148; Skarin *et al.*, 1973). Only one of the thirteen male patients responded, while one out of four females responded. The conditional MLE of the OR is $\hat{\mathcal{O}}_{\text{cond}} = 0.278$. Because the number of patients in this study is so small, we may wish to apply interval estimator (5.10) rather than the asymptotic interval estimators (5.1) and (5.5). Using (5.10), we obtain a 95% confidence interval for the OR of [0.003, 26.115]. The width of the confidence interval suggests that the data considered here cannot provide us with a precise estimate of the OR between the response to combination chemotherapy and gender.

5.2 A SERIES OF INDEPENDENT BINOMIAL SAMPLING PROCEDURES

Assume that there are S strata formed by centers in a multicenter study. From each stratum s ($s = 1, 2, \ldots, S$), we independently sample n_{js} subjects from the case ($j = 1$) and the control ($j = 0$) populations, respectively. Assume further that we obtain X_{js} out of n_{js} sampled subjects exposed to the risk factor. Then X_{js} follows the binomial distribution (1.1) with parameters n_{js} and $\pi_{1|js}$. Let $\pi_{0|js} = 1 - \pi_{1|js}$ for $j = 1, 0$. The OR in stratum s is $\mathcal{O}_s = \pi_{1|1s}\pi_{0|0s}/(\pi_{0|1s}\pi_{1|0s})$. The joint probability mass function for the random vector $\mathbf{X}' = (\mathbf{X}_1', \mathbf{X}_0')$, where

$\mathbf{X}_j' = (X_{j1}, X_{j2}, \ldots, X_{jS})$, is

$$\mathrm{f}_{\mathbf{X}}(\mathbf{x}|\mathbf{n}, \boldsymbol{\pi}) = \prod_{s=1}^{S}\prod_{j=0}^{1}\binom{n_{js}}{x_{js}}(\pi_{1|js})^{x_{js}}(1-\pi_{1|js})^{n_{js}-x_{js}},$$

$$= \prod_{s=1}^{S}\binom{n_{1s}}{x_{1s}}\binom{n_{0s}}{x_{0s}}(\mathcal{O}_s)^{x_{1s}}(\pi_{0|1s})^{n_{1s}}(\pi_{1|0s})^{x_{.s}}(\pi_{0|0s})^{n_{0s}-x_{.s}}, \quad (5.11)$$

where $\mathbf{n}' = (n_{11}, n_{12}, \ldots, n_{1S}, n_{01}, n_{02}, \ldots, n_{0S})$ and $\boldsymbol{\pi}' = (\pi_{1|11}, \pi_{1|12}, \ldots, \pi_{1|1S}, \pi_{1|01}, \pi_{1|02}, \ldots, \pi_{1|0S})$. Based on likelihood (5.1), we can easily show that the MLE of $\mathcal{O}_s$ is simply $\hat{\mathcal{O}}_s = X_{1s}(n_{0s} - X_{0s})/[(n_{1s} - X_{1s})X_{0s}]$. In the following two subsections, we shall assume that $\mathcal{O}_s$ is constant across all strata, and we denote this common OR by $\mathcal{O}_c$. Given $\mathbf{x}_.' = (x_{.1}, x_{.2}, \ldots, x_{.S})$ fixed, we can see that from (5.11), the likelihood function of $\mathcal{O}_c$ depends on only on the statistic $X_{1.} (= \sum_s X_{1s})$.

5.2.1 Asymptotic interval estimators

To estimate $\mathcal{O}_c$ we may first consider using the 'precision weighting' of stratum-specific estimates $\hat{\mathcal{O}}_s : \sum_s W_s\hat{\mathcal{O}}_s/\sum_s W_s$, where $W_s = 1/\mathrm{Var}(\hat{\mathcal{O}}_s)$ and $\mathrm{Var}(\hat{\mathcal{O}}_s) = \mathcal{O}_c^2\{1/[n_{1s}\pi_{1|1s}(1-\pi_{1|1s})] + 1/[n_{0s}\pi_{1|0s}(1-\pi_{1|0s})]\}$ when parameters $\pi_{1|js}$ are known. Note that because $\mathcal{O}_c$ is a constant, $\sum_s W_s\hat{\mathcal{O}}_s/\sum_s W_s = \sum_s W_s^*\hat{\mathcal{O}}_s/\sum_s W_s^*$, where $W_s^* = 1/\mathrm{Var}(\log(\hat{\mathcal{O}}_s))$ and $\mathrm{Var}(\log(\hat{\mathcal{O}}_s)) = 1/[n_{1s}\pi_{1|1s}(1-\pi_{1|1s})] + 1/[n_{0s}\pi_{1|0s}(1-\pi_{1|0s})]$. When parameters $\pi_{1|js}$ are unknown, we can substitute $\hat{\pi}_{1|js} = X_{js}/n_{js}$ for $\pi_{1|js}$ in $\mathrm{Var}(\log(\hat{\mathcal{O}}_s))$ and obtain the consistent point estimator $\sum_s \hat{W}_s^*\hat{\mathcal{O}}_s/\sum_s \hat{W}_s^*$, where $\hat{W}_s^* = 1/\widehat{\mathrm{Var}}(\log(\hat{\mathcal{O}}_s))$, and $\widehat{\mathrm{Var}}(\log(\hat{\mathcal{O}}_s)) = 1/X_{1s} + 1/X_{0s} + 1/(n_{1s} - X_{1s}) + 1/(n_{0s} - X_{0s})$. For interval estimation of $\mathcal{O}_c$, we may apply the logarithmic transformation to improve the normal approximation of $\hat{\mathcal{O}}_s$. This will lead us to obtain the weighted least-squares (WLS) estimator, $\sum_s W_s^*\log(\hat{\mathcal{O}}_s)/\sum_s W_s^*$, if parameters $\pi_{1|js}$ are known. We can easily show that the asymptotic variance $\mathrm{Var}\left(\sum_s W_s^*\log(\hat{\mathcal{O}}_s)/\sum_s W_s^*\right) = 1/\sum_s W_s^*$. Thus, when substituting $\hat{\pi}_{1|js}$ for unknown parameters $\pi_{1|js}$, we obtain an asymptotic $100(1-\alpha)$ percent confidence interval for $\mathcal{O}_c$ (Woolf, 1955) given by

$$\left[\exp\left\{\frac{\sum_s \hat{W}_s^*\log(\hat{\mathcal{O}}_s)}{\sum_s \hat{W}_s^*} - \frac{Z_{\alpha/2}}{\sqrt{\sum_s \hat{W}_s^*}}\right\}, \exp\left\{\frac{\sum_s \hat{W}_s^*\log(\hat{\mathcal{O}}_s)}{\sum_s \hat{W}_s^*} + \frac{Z_{\alpha/2}}{\sqrt{\sum_s \hat{W}_s^*}}\right\}\right]. \quad (5.12)$$

Note that when applying (5.12), we may also consider adding 0.50 to each cell, as we did for (5.1). Note also that the application of both the point estimator $\sum_s \hat{W}_s^* \hat{\mathcal{O}}_s / \sum_s \hat{W}_s^*$ and interval estimator (5.12) requires an adequately large sample taken from each stratum. To alleviate this concern, Mantel and Haenszel (1959) propose the following summary OR estimator:

$$\hat{\mathcal{O}}_{\mathrm{MH}} = \frac{\sum_s X_{1s}(n_{0s} - X_{0s})/n_{.s}}{\sum_s X_{0s}(n_{1s} - X_{1s})/n_{.s}}, \tag{5.13}$$

where $n_{.s} = n_{1s} + n_{0s}$. As noted by Hauck (1989) and Agresti (1990), the estimator $\hat{\mathcal{O}}_{\mathrm{MH}}$ has good asymptotic properties for both asymptotic cases: (a) the number of strata is fixed, but the sample size within each stratum becomes large; and (b) the stratum sizes are fixed, but the number of strata becomes large. Robins *et al.* (1986) provide an estimated asymptotic variance of $\log(\hat{\mathcal{O}}_{\mathrm{MH}})$:

$$\begin{aligned}
&\widehat{\mathrm{Var}}(\log(\hat{\mathcal{O}}_{\mathrm{MH}})) \\
&= \frac{\sum_s (x_{1s} + n_{0s} - X_{0s})(X_{1s}(n_{0s} - X_{0s}))/n_{.s}^2}{2\left(\sum_s X_{1s}(n_{0s} - X_{0s})/n_{.s}\right)^2} \\
&+ \frac{\sum_s [(X_{1s} + n_{0s} - X_{0s})(X_{0s}(n_{1s} - X_{1s})) + (X_{0s} + n_{1s} - X_{1s})(X_{1s}(n_{0s} - X_{0s}))]/n_{.s}^2}{2\left(\sum_s X_{1s}(n_{0s} - X_{0s})/n_{.s}\right)\left(\sum_s X_{0s}(n_{1s} - X_{1s})/n_{.s}\right)} \\
&+ \frac{\sum_s (X_{0s} + n_{1s} - X_{1s})(X_{0s}(n_{1s} - X_{1s}))/n_{.s}^2}{2\left(\sum_s X_{0s}(n_{1s} - X_{1s})/n_{.s}\right)^2}
\end{aligned} \tag{5.14}$$

Thus, we obtain an asymptotic $100(1-\alpha)$ percent confidence interval for $\mathcal{O}_c$ given by

$$[\hat{\mathcal{O}}_{\mathrm{MH}} \exp(-Z_{\alpha/2}\sqrt{\widehat{\mathrm{Var}}(\log(\hat{\mathcal{O}}_{\mathrm{MH}}))}), \quad \hat{\mathcal{O}}_{\mathrm{MH}} \exp(Z_{\alpha/2}\sqrt{\widehat{\mathrm{Var}}(\log(\hat{\mathcal{O}}_{\mathrm{MH}}))})]. \tag{5.15}$$

Gart (1970) extends Cornfield's (1956) confidence interval to accommodate a series of independent 2×2 tables. Given $\mathcal{O} = \mathcal{O}_c$, we have $X_{1s}(\mathcal{O}_c)(n_{0s} - X_{0s}(\mathcal{O}_c))/[(n_{1s} - X_{1s}(\mathcal{O}_c))X_{0s}(\mathcal{O}_c)] = \mathcal{O}_c$, where $X_{js}(\mathcal{O}_c)$ denotes the expected frequency of subjects with exposure from group j ($j = 1, 0$) in stratum s ($s = 1, \ldots, S$). Therefore, given $\mathcal{O}_c \neq 1$ and the marginal totals $x_{.s}$ $(= X_{1s} + X_{0s})$ of subjects with exposure fixed, we obtain the expected frequency

$$X_{1s}(\mathcal{O}_c) = \frac{\mathcal{O}_c(n_{1s}+x_{.s})+(n_{0s}-x_{.s})-\sqrt{[\mathcal{O}_c(n_{1s}+x_{.s})+(n_{0s}-x_{.s})]^2-4(\mathcal{O}_c-1)\mathcal{O}_c n_{1s}x_{.s}}}{2(\mathcal{O}_c-1)}. \quad (5.16)$$

Note that when $\mathcal{O}_c = 1$, $X_{1s}(\mathcal{O}_c)$ simplifies to $n_{1s}x_{.s}/n_{.s}$, where $n_{.s} = n_{1s} + n_{0s}$. Note also that the conditional distribution of the observed frequency X_{1s}, given $x_{.s}$ fixed, is approximately normally distributed with mean $X_{1s}(\mathcal{O}_c)$ and asymptotic variance

$$\mathrm{Var}(X_{1s}|x_{.s},\mathcal{O}_c) = [1/X_{1s}(\mathcal{O}_c)+1/(x_{.s}-X_{1s}(\mathcal{O}_c))+1/(n_{1s}-X_{1s}(\mathcal{O}_c)) + 1/(n_{0s}-x_{.s}+X_{1s}(\mathcal{O}_c))]^{-1}. \quad (5.17)$$

Thus, the statistic $X_{1.} = \sum_s X_{1s}$, asymptotically follows the normal distribution with mean $X_{1.}(\mathcal{O}_c) = \sum_s X_{1s}(\mathcal{O}_c)$ and variance $\mathrm{Var}(X_{1.}|\mathbf{x}_{.},\mathcal{O}_c) = \sum_s \mathrm{Var}(X_{1s}|x_{.s},\mathcal{O}_c)$, where $\mathbf{x'}_{.} = (x_{.1}, x_{.2}, \ldots, x_{.S})$. We thus obtain an asymptotic $100(1-\alpha)$ percent confidence interval for the underlying common OR of

$$[\mathcal{O}_l, \mathcal{O}_u], \quad (5.18)$$

where the confidence limits $\mathcal{O}_l$ and $\mathcal{O}_u$ are determined by solving the following two equations. The lower limit $\mathcal{O}_l$ is the smaller root of $\mathcal{O}$ satisfying

$$(X_{1.} - X_{1.}(\mathcal{O}) - c_1)^2\,\mathrm{Var}(X_{1.}|\mathbf{x}_{.},\mathcal{O})^{-1} - Z^2_{\alpha/2} = 0, \quad (5.19)$$

where the constant c_1 is set to 0.50 when using continuity correction, or to 0 otherwise. The upper limit $\mathcal{O}_u$ is the larger root of $\mathcal{O}$ satisfying

$$(X_{1.} - X_{1.}(\mathcal{O}) + c_2)^2\,\mathrm{Var}(X_{1.}|\mathbf{x}_{.},\mathcal{O})^{-1} - Z^2_{\alpha/2} = 0, \quad (5.20)$$

where the constant c_2 is set to 0.50 for the continuity correction, or to 0 otherwise.

5.2.2 Exact confidence interval

Under the product of independent binomial sampling (5.11), we can easily see that the joint conditional probability mass function $\mathbf{X'}_1 = (X_{11}, X_{12}, \ldots, X_{1S})$, given $\mathbf{x}_{.}$ fixed, is simply

$$f_{\mathbf{X}_1}(x_{11}, x_{12}, \ldots, x_{1S}|\mathbf{x}_{.},\mathcal{O}_c) = \prod_s \mathrm{P}(X_{1s} = x_{1s}|x_{.s},\mathcal{O}_c), \quad (5.21)$$

where $P(X_{1s} = x_{1s}|x_{.s}, \mathcal{O}_c) = \binom{n_{1s}}{x_{1s}}\binom{n_{0s}}{x_{.s} - x_{1s}} \mathcal{O}_c^{x_{1s}} / \sum_x \binom{n_{1s}}{x}\binom{n_{0s}}{x_{.s} - x} \mathcal{O}_c^x$, and the summation is over the range $\max\{x_{.s} - n_{0s}, 0\} \leq x \leq \min\{n_{1s}, x_{.s}\}$. Note that $X_{1.}$ is a conditional sufficient statistic for $\mathcal{O}_c$ based on (5.21). Thus, we may focus our attention on the probability mass function of $X_{1.}$,

$$f_{X_{1.}}(x_{1.}|\mathbf{x}_., \mathcal{O}_c) = \sum_{\mathbf{x}_1 \varepsilon R(x_{1.})} f_{\mathbf{X}_1}(x_{11}, x_{12}, \ldots, x_{1S}|\mathbf{x}_., \mathcal{O}_c), \tag{5.22}$$

where $R(x_{1.}) = \{(x_{11}, x_{12}, \ldots, x_{1S})|x_{11} + x_{12} + \cdots + x_{1S} = x_{1.}$, and $\max\{x_{.s} - n_{0s}, 0\} \leq x_{1s} \leq \min\{n_{1s}, x_{.s}\}\}$. On the basis of distribution (5.22), we may obtain the conditional MLE $\hat{\mathcal{O}}_{\text{cond}}$ by finding the root $\mathcal{O}$ of the equation

$$x_{1.} = \mathrm{E}(X_{1.}|\mathbf{x}_., \mathcal{O}), \tag{5.23}$$

where $\mathrm{E}(X_{1.}|\mathbf{x}_., \mathcal{O}) = \sum_s \mathrm{E}(X_{1s}|x_{.s}, \mathcal{O})$ and $\mathrm{E}(X_{1s}|x_{.s}, \mathcal{O})$ is the conditional expectation of X_{1s}, given $x_{.s}$ fixed. Furthermore, following similar arguments to those for deriving (5.10), we obtain an exact $100(1 - \alpha)$ percent confidence interval for the underlying common OR of

$$[\mathcal{O}_l^*, \mathcal{O}_u^*], \tag{5.24}$$

where the confidence limits $\mathcal{O}_l^*$ and $\mathcal{O}_u^*$ are determined by solving the following two equations:

$$\sum_{x \geq x_{1.}} f_{X_{1.}}(x|\mathbf{x}_., \mathcal{O}_l^*) = \alpha/2, \qquad \sum_{x \leq x_{1.}} f_{X_{1.}}(x|\mathbf{x}_., \mathcal{O}_u^*) = \alpha/2.$$

5.2.3 Test for homogeneity of the odds ratio

The estimators discussed in Sections 5.2.1 and 5.2.2 assume that the underlying OR is constant across all strata. To examine this assumption, $H_0 : \mathcal{O}_1 = \mathcal{O}_2 = \ldots = \mathcal{O}_S$, we may apply the WLS test statistic (Fleiss, 1981) based on the variation of $\log(\hat{\mathcal{O}}_s)$:

$$\sum_s \hat{W}_s^* (\log(\hat{\mathcal{O}}_s))^2 - \frac{\left(\sum_s \hat{W}_s^* \log(\hat{\mathcal{O}}_s)\right)^2}{\sum_s \hat{W}_s^*}. \tag{5.25}$$

Under H_0, test statistic (5.25) asymptotically follows the χ^2 distribution with $S - 1$ degrees of freedom. When H_0 is not true, we expect large values of the test statistic. Thus, we will reject H_0 at level α if the test statistic exceeds $\chi^2_{S-1,\alpha}$, where $\chi^2_{S-1,\alpha}$ is the upper $100\,\alpha$th percentile of the χ^2 distribution with $S - 1$ degrees of freedom.

Following Breslow and Day (1980), we may also consider use of the following statistic for testing the homogeneity of the OR:

$$\sum_s (X_{1s} - X_{1s}(\hat{\mathcal{O}}))^2/\mathrm{Var}(X_{1s}|x_{.s}, \hat{\mathcal{O}}), \tag{5.26}$$

where $\hat{\mathcal{O}}$ can be either the conditional MLE $\hat{\mathcal{O}}_{\text{cond}}$ calculated from equation (5.23) or $\hat{\mathcal{O}}_{\text{MH}}$ (5.13). We will reject H_0 if test statistic (5.26) exceeds $\chi^2_{S-1,\alpha}$. For the situation where the stratum size is small, tests of the homogeneity of the OR for sparse data are discussed elsewhere (Ejigou and McHugh, 1984; Liang and Self, 1985). We refer readers to these papers for details.

Example 5.3 Consider the data in Table 5.1, from Hosmer and Lemeshow (1989). Mothers are cross-classified by their smoker status during pregnancy (1 = Yes, 0 = No), the birthweight of their baby (= 1 for < 2500 g; = 0 for ≥ 2500 g), and their race. Suppose that we are interested in estimating the OR of low birthweight between smoking and non-smoking mothers, while controlling the confounder of race. Applying (5.25) and (5.26) to test the homogeneity of the OR, we obtain p-values of 0.22 and 0.21. Thus, it may be reasonable to assume that the OR is constant across race. Using $\sum_s \hat{W}_s^* \hat{\mathcal{O}}_s / \sum_s \hat{W}_s^*$ and $\hat{\mathcal{O}}_{\text{MH}}$ gives 3.603 and 3.086, respectively. If we added 0.50 to each cell before calculating $\sum_s \hat{W}_s^* \hat{\mathcal{O}}_s / \sum_s \hat{W}_s^*$, we would obtain an estimate of 3.342, which is closer to $\hat{\mathcal{O}}_{\text{MH}}$. Furthermore, applying (5.12), (5.15), and (5.18), we obtain 95% confidence intervals for $\mathcal{O}_c$ of [1.372, 6.323], [1.491, 6.390], and [1.491, 6.244], respectively. Because these lower limits are above 1, we conclude at the 5% level that smoking during pregnancy tends to increase the risk of bearing a baby with low birthweight.

5.3 INDEPENDENT CLUSTER SAMPLING

Suppose that we assign n_1 clusters of $m_{1j}(j = 1, 2, \ldots, n_1)$ subjects to receive the experimental ($i = 1$) treatment and n_0 clusters of $m_{0j}(j = 1, 2, \ldots, n_0)$ subjects to receive the standard or placebo ($i = 0$) treatment, respectively. We define the

Table 5.1 Contingency table for low baby birthweight by smoking stratus of mother during pregnancy stratified by race of mother.

	Race	White		Black		Other	
		Yes	No	Yes	No	Yes	No
	Smoke	1	0	1	0	1	0
Low birthweight	Yes 1	19	4	6	5	5	20
	No 0	33	40	4	11	7	35

Source: Hosmer and Lemeshow (1989).

random variable $X_{ijk} = 1$ if the response of the kth subject ($k = 1, 2, \ldots, m_{ij}$) in cluster j for treatment i is positive, and $X_{ijk} = 0$ otherwise. Then the random variable $X_{ij.} = \sum_k X_{ijk}$, denoting the total number of positive responses in cluster j for treatment i, follows a binomial distribution with parameters m_{ij} and p_{ij}, where p_{ij} denotes the probability of positive response. Note that the responses X_{ijk} within clusters are likely correlated. To account for this intraclass correlation, we assume that the p_{ij} independently follow the beta distribution beta(α_i, β_i) with mean $\pi_i (= \alpha_i/(\alpha_i + \beta_i))$ and variance $\pi_i(1 - \pi_i)/(T_i + 1)$, where $\alpha_i > 0$, $\beta_i > 0$, and $T_i = \alpha_i + \beta_i$ (Johnson and Kotz, 1970; Lui, 1991). Thus, the probability that a subject randomly selected subject for treatment i is positive equals $E(X_{ijk}) = \pi_i$, and the intraclass correlation between X_{ijk} and $X_{ijk'}$ for $k \neq k'$ within clusters is $\rho_i = 1/(T_i + 1)$ (**Exercise 1.7**). The OR of being positive between the experimental and the control treatments is simply equal to $\pi_1(1 - \pi_0)/[\pi_0(1 - \pi_1)]$.

Define $\hat{\pi}_i = \sum_j \sum_k X_{ijk}/m_{i.}$, where $m_{i.} = \sum_j m_{ij}$. Note that $\hat{\pi}_i$ is an unbiased estimator of π_i with variance $\text{Var}(\hat{\pi}_i)$ equal to $\pi_i(1 - \pi_i)f(\mathbf{m}_i, \rho_i)/m_{i.}$ (**Exercise 1.7**), where $\mathbf{m}_i' = (m_{i1}, m_{i2}, \ldots, m_{in_i})$ and $f(\mathbf{m}_i, \rho_i)$ is the variance inflation factor due to the intraclass correlation ρ_i and equals $\sum_j m_{ij}[1 + (m_{ij} - 1)\rho_i]/m_{i.}$. To estimate ρ_i we may apply the traditional intraclass correlation estimator (2.19) in Chapter 2 (Fleiss, 1986; Lui *et al.*, 1996; Elston, 1977).

First, note that a consistent estimator of the OR is simply given by $\hat{\mathcal{O}} = \hat{\pi}_1(1 - \hat{\pi}_0)/[\hat{\pi}_0(1 - \hat{\pi}_1)]$. Using the delta method together with the adjustment procedure of adding 0.50 for sparse data, we obtain the estimated asymptotic variance $\widehat{\text{Var}}(\log(\hat{\mathcal{O}}_{\text{adj}})) = f(\mathbf{m}_1, \hat{\rho}_1)[1/(X_{1..} + 0.5) + 1/(m_{1.} - X_{1..} + 0.5)] + f(\mathbf{m}_0, \hat{\rho}_0)[1/(X_{0..} + 0.5) + 1/(m_{0.} - X_{0..} + 0.5)]$, where $\hat{\mathcal{O}}_{\text{adj}} = (X_{1..} + 0.5)(m_{0.} - X_{0..} + 0.5)/[(m_{1.} - X_{1..} + 0.5)(X_{0..} + 0.5)]$ and $X_{i..} = \sum_j \sum_k X_{ijk}$ for $i = 1, 0$. Thus, we obtain an asymptotic $100(1 - \alpha)$ percent confidence interval for OR given by

$$[\hat{\mathcal{O}}_{\text{adj}} \exp(-Z_{\alpha/2}\sqrt{\widehat{\text{Var}}(\log(\hat{\mathcal{O}}_{\text{adj}}))}), \hat{\mathcal{O}}_{\text{adj}} \exp(Z_{\alpha/2}\sqrt{\widehat{\text{Var}}(\log(\hat{\mathcal{O}}_{\text{adj}}))})]. \quad (5.27)$$

Note that interval estimator (5.1) is a special case of (5.27) when $m_{ij} = 1$ for all $i (i = 1, 0)$ and $j (j = 1, 2, \ldots, n_i)$. Thus, as for (5.1), this estimator, though simple to use, is not likely to perform well when the $m_{i.}$ are small.

Following Cornfield (1956), we can generalize interval estimator (5.5) to accommodate the case of cluster sampling. Given $\mathcal{O} = \mathcal{O}_0$ and $X_{...} = x_{...}$ (where $x_{...} = x_{1..} + x_{0..}$) we can calculate the expected frequency $X_{1..}(\mathcal{O}_0)$ from the equation $X_{1..}(\mathcal{O}_0)(m_{0.} - x_{...} + X_{1..}(\mathcal{O}_0))/[(x_{...} - X_{1..}(\mathcal{O}_0))(m_{1.} - X_{1..}(\mathcal{O}_0)] = \mathcal{O}_0$. This is equivalent to saying that, given $\mathcal{O}_0 \neq 1$, we can obtain the expected frequency $X_{1..}(\mathcal{O}_0)$ by solving the following quadratic equation for $X_{1..}$:

$$(\mathcal{O}_0 - 1)X_{1..}^2 - [\mathcal{O}_0(m_{1.} + x_{...}) + (m_{0.} - x_{...})]X_{1..} + \mathcal{O}_0 m_{1.} x_{...} = 0. \quad (5.28)$$

Note that, given $\mathcal{O}_0$ and $x_{...}$, the solution $X_{1..}(\mathcal{O}_0)$ must also satisfy the constraint $\max\{0, x_{...} - m_{0.}\} \leq X_{1..} \leq \min\{m_{1.}, x_{...}\}$. Therefore, the solution $X_{1..}(\mathcal{O}_0)$

is given by

$$X_{1..}(\mathcal{O}_0) = \frac{\mathcal{O}_0(m_{1.}+x_{...})+(m_{0.}-x_{...})-\sqrt{[\mathcal{O}_0(m_{1.}+x_{...})+(m_{0.}-x_{...})]^2-4(\mathcal{O}_0-1)\mathcal{O}_0 m_{1.}x_{...}}}{2(\mathcal{O}_0-1)}. \tag{5.29}$$

When $\mathcal{O}_0 = 1$, the expected frequency $X_{1..}(\mathcal{O}_0)$ is $m_{1.}x_{...}/(m_{1.}+m_{0.})$. Furthermore, when $\mathcal{O}_0$ is the underlying value, we can show that the observed frequency $X_{1..}$, given $x_{...}$, approximately follows a normal distribution with mean $X_{1..}(\mathcal{O}_0)$ and asymptotic variance

$$\begin{aligned}\mathrm{Var}(X_{1..}|x_{...},\mathcal{O}_0,\rho_1,\rho_0) = \{&[1/X_{1..}(\mathcal{O}_0)+1/(m_{1.}-X_{1..}(\mathcal{O}_0))](1/f(\mathbf{m}_1,\rho_1))\\ &+[1/(x_{...}-X_{1..}(\mathcal{O}_0))+1/(m_{0.}-x_{...}\\ &+X_{1..}(\mathcal{O}_0))](1/f(\mathbf{m}_0,\rho_0))\}^{-1}.\end{aligned} \tag{5.30}$$

Thus, an asymptotic $100(1-\alpha)$ percent confidence interval for OR is given by

$$[\mathcal{O}_\mathrm{l}, \mathcal{O}_\mathrm{u}], \tag{5.31}$$

where $\mathcal{O}_\mathrm{l}$ and $\mathcal{O}_\mathrm{u}$ are determined by the following two equations. The lower limit $\mathcal{O}_\mathrm{l}$ is the smaller root $\mathcal{O}$ of the equation

$$(X_{1..}-X_{1..}(\mathcal{O})-c_1)^2\,\mathrm{Var}(X_{1..}|x_{...},\mathcal{O},\hat{\rho}_1,\hat{\rho}_0)^{-1}-Z^2_{\alpha/2}=0, \tag{5.32}$$

where the constant c_1 is set to 0.50 when using continuity correction, or to 0 otherwise. The upper limit $\mathcal{O}_\mathrm{u}$ is the larger root $\mathcal{O}$ of the equation

$$(X_{1..}-X_{1..}(\mathcal{O})+c_2)^2\,\mathrm{Var}(X_{1..}|x_{...},\mathcal{O},\hat{\rho}_1,\hat{\rho}_0)^{-1}-Z^2_{\alpha/2}=0. \tag{5.33}$$

where the constant c_2 is set to 0.50 for continuity correction, or to 0 otherwise. Note that when $m_{ij} = 1$ for all i and j, (5.31) reduces to (5.5).

Example 5.4 Consider the data (Table 1.1) from a study of an educational intervention with emphasis on behavior change (Mayer *et al.*, 1997). There were 132 children divided into 58 classes, and for each class it is known how many children had an inadequate level of solar protection. Given these data, we obtain the point estimate $\hat{\mathcal{O}} = 0.452$ of having an inadequate level of solar protection when comparing the educational intervention group ($i = 1$) with the control group ($i = 0$). In Example 2.5 we estimated the common intraclass correlation for the two comparison groups to be $\hat{\rho}_c = 0.30$. Applying interval estimators (5.27), (5.31) with continuity correction, and (5.31) without continuity correction, we obtain 95% confidence intervals of [0.193, 1.084], [0.179, 1.138], and [0.190, 1.072], respectively. Because all the interval estimators cover 1, there is no significant evidence at the 5% level that the intervention program increases the proportion of children who employ an adequate level of solar protection.

Furthermore, we observe that the interval estimate using Cornfield's interval (5.31) with continuity correction is wider than the other two estimates.

Note that, following Gart (1970), we can easily extend interval estimator (5.31) to accommodate the situation in which there is a series of 2×2 tables as discussed in Section 5.2 under cluster sampling. Note also that other relevant discussions on the use of the Mantel–Haenszel estimator of the OR under cluster sampling appear elsewhere (Donald and Donner, 1990).

5.4 ONE-TO-ONE MATCHED SAMPLING

To increase the efficiency or validity of our inference in cohort or case–control studies, we often employ matching design. Suppose that from the exposed population we take a random sample of n subjects, to each of whom we match a subject from the non-exposed population with respect to some matching variables. Following Ejigou and McHugh (1977), we assume that the combination of all matching variables consists of L distinct levels $V_1, V_2, \ldots, V_L$. Let D denote the random variable of disease status: $D = 1$ for a case, and $D = 0$ otherwise. Similarly, let E denote the random variable of exposure status: $E = 1$ denotes being exposed, and $E = 0$ otherwise. We assume that the distribution of the matching covariate level V_l in the exposed population is given by $p_l = \mathrm{P}(V_l|E = 1)$, where $p_l > 0$ and $\sum_l p_l = 1$. We assume further that, given the exposure status E and the matching level V_l, the conditional probability of being a case is $\mathrm{P}(D = 1|E, V_l) = \mathrm{e}^{\alpha_l+\beta E}/(1 + \mathrm{e}^{\alpha_l+\beta E})$, where the parameter β represents the effect of exposure on the disease. For simplicity, we assume that the conditional probabilities $\mathrm{P}(D|E = 1, V_l)$ and $\mathrm{P}(D|E = 0, V_l)$, given V_l fixed, are independent. Note that we assume the logistic regression to model these conditional probabilities here, which we will discuss in more detail in the next section.

Let $(D = d|E = e)$ denote the event that a subject has disease status d (1 for a case, or 0 for a non-case), given that a subject is exposed ($e = 1$) or non-exposed ($e = 0$). Furthermore, let π_{ij} denote the probabilities $\mathrm{P}((D = i|E = 1)\,\&\,(D = j|E = 0))$, where i and $j = 0, 1$. Then, we have $\pi_{ij} = \sum_l \mathrm{P}(D = i|E = 1, V_l)\mathrm{P}(D = j|E = 0, V_l)p_l$. Furthermore, note that, conditional upon each matched pair, the OR between the disease and the exposure is

$$\frac{\mathrm{P}(D = 1|E = 1, V_l)\,\mathrm{P}(D = 0|E = 0, V_l)}{\mathrm{P}(D = 0|E = 1, V_l)\,\mathrm{P}(D = 1|E = 0, V_l)} = \mathrm{e}^{\beta},$$

which is a constant. In fact, we can actually show that OR equals $\mathcal{O} = \pi_{10}/\pi_{01}$ (**Exercise 5.14**). For clarity, we use the following 2×2 table to summarize the data structure:

		Non-exposed		
		$D = 1$	$D = 0$	
Exposed	$D = 1$	π_{11}	π_{10}	$\pi_{1.}$
	$D = 0$	π_{01}	π_{00}	$\pi_{0.}$
		$\pi_{.1}$	$\pi_{.0}$	

where π_{ij} denotes the cell probability for i and $j = 0, 1$. Readers should not confuse with the cell probabilities π_{ij} defined here with those defined in the introduction to this chapter, which have a completely different meaning. Let N_{ij} denote the corresponding number of matched pairs among n matched pairs falling in the cell (i, j) with the cell probability π_{ij}. Then $(N_{11}, N_{10}, N_{01}, N_{00})'$ follows the multinomial distribution (2.25) with parameters n and $(\pi_{11}, \pi_{10}, \pi_{01}, \pi_{00})'$.

Note that the MLE of π_{ij} is $\hat{\pi}_{ij} = N_{ij}/n$ and hence the MLE of the OR is $\hat{\mathcal{O}} = \hat{\pi}_{10}/\hat{\pi}_{01}$, with asymptotic variance $(\pi_{10}/\pi_{01})^2[1/(n\pi_{10}) + 1/(n\pi_{01})]$ (**Exercise 5.15**). Because $\hat{\mathcal{O}}$ is a ratio of the two sample proportions, we do not recommend deriving the confidence interval for $\mathcal{O}$ directly based on $\hat{\mathcal{O}}$ unless n is large. To improve the normal approximation of $\hat{\mathcal{O}}$, we may apply the logarithmic transformation. We obtain an asymptotic $100(1-\alpha)$ percent confidence interval for $\mathcal{O}$ given by

$$[(\hat{\pi}_{10}/\hat{\pi}_{01})\exp(-Z_{\alpha/2}\sqrt{[1/(n\hat{\pi}_{10}) + 1/(n\hat{\pi}_{01})}),$$
$$(\hat{\pi}_{10}/\hat{\pi}_{01})\exp(Z_{\alpha/2}\sqrt{[1/(n\hat{\pi}_{10}) + 1/(n\hat{\pi}_{01})})]. \quad (5.34)$$

Define $Z = \hat{\pi}_{10} - \mathcal{O}\hat{\pi}_{01}$. We can easily see that this has expectation $E(Z) = 0$ and variance $\mathrm{Var}(Z) = [\pi_{10}(1-\pi_{10}) + \mathcal{O}^2\pi_{01}(1-\pi_{01}) + 2\mathcal{O}\pi_{10}\pi_{01}]/n$. Therefore, an asymptotic $100(1-\alpha)$ percent confidence interval for $\mathcal{O}$ is (**Exercise 5.16**)

$$[(B^{\ddagger} - \sqrt{B^{\ddagger} - A^{\ddagger}C^{\ddagger}})/A^{\ddagger},$$
$$(B^{\ddagger} + \sqrt{B^{\ddagger} - A^{\ddagger}C^{\ddagger}})/A^{\ddagger}], \quad (5.35)$$

where $A^{\ddagger} = \hat{\pi}_{01}^2 - Z_{\alpha/2}^2\hat{\pi}_{01}(1-\hat{\pi}_{01})/n$, $B^{\ddagger} = \hat{\pi}_{10}\hat{\pi}_{01} + Z_{\alpha/2}^2\hat{\pi}_{10}\hat{\pi}_{01}/\mathrm{n}$, and $C^{\ddagger} = \hat{\pi}_{10}^2 - Z_{\alpha/2}^2\hat{\pi}_{10}(1-\hat{\pi}_{10})/\mathrm{n}$.

Note that the conditional distribution of N_{10}, given a fixed total number of pairs with discordant outcomes (or disease status) $N_{10} + N_{01} = n_d$, is the binomial distribution with parameters n_d and $\pi_{10}/(\pi_{10} + \pi_{01})(= \mathcal{O}/(1+\mathcal{O}))$ (**Exercise 5.17**). Therefore, we can apply interval estimators (1.3), (1.5), and (1.6) to obtain a $100(1-\alpha)$ percent confidence interval for $\mathcal{O}/(1+\mathcal{O})$ first and then use the monotonic transformation $X/(1-X)$ to produce a $100(1-\alpha)$ percent confidence interval for $\mathcal{O}$. For example, when the number of pairs with discordant disease status n_d is small, we can use (1.6) to obtain an exact $100(1-\alpha)$ percent confidence interval for $\mathcal{O}$ given by

$$[L/(1-L), U/(1-U)], \quad (5.36)$$

where $L = N_{10}/(N_{10} + (N_{01}+1)F_{2(N_{01}+1), 2N_{10}, \alpha/2})$ and $U = ((N_{10}+1)F_{2(N_{10}+1), 2N_{01}, \alpha/2})/(N_{01} + (N_{10}+1)F_{2(N_{10}+1), 2N_{01}, \alpha/2})$. Note that when $N_{10} = 0$, we define the lower limit of (5.36) to be 0. Similarly, when $N_{01} = 0$, we define the upper limit of (5.36) to be ∞.

Example 5.5 Consider comparing two treatments for breast cancer: simple mastectomy and radical mastectomy. As reported elsewhere (Rosner, 1990, p. 384),

20 matched pairs of women, who were in the same decade of age and had the same clinical condition, were formed. For each pair, one patient was assigned to simple mastectomy and the other to radical mastectomy. Their 5-year survival was then monitored. There were (n_{11} =)10 pairs in which both women lived at least 5 years after receiving either form of surgery; (n_{10} =)8 pairs in which the woman receiving simple mastectomy lived at least 5 years but the woman receiving radical mastectomy died; (n_{01} =)1 pair in which the woman receiving simple mastectomy died but the woman receiving radical mastectomy survived at least 5 years; and (n_{00} =)1 pair in which both women died within 5 years. Using these data, the MLE $\hat{\mathcal{O}}$ of the OR of 5-year survival between women receiving simple mastectomy and radical mastectomy is 8. Because the number of discordant pairs in this example is so small ($n_{10} + n_{01} = 9$), we use interval estimator (5.36) and obtain a 95% confidence interval for $\mathcal{O}$ of [1.073, 354.981]. Since the lower limit excludes 1, we conclude that there is a significant evidence at the 5% level that patients receiving simple mastectomy tend to have a higher 5-year survival than patients receiving radical mastectomy. However, the confidence interval obtained is so wide that the data cannot be considered to provide us with a precise estimate of OR between 5-year survival and the treatments.

5.5 LOGISTIC MODELING

When the parameter of interest is the OR, logistic regression is probably the most common multivariate approach to modeling the binary response in epidemiology. This is because we can easily employ logistic regression to control confounders and effect modifiers simultaneously, and each parameter in the model has an easily understood practical interpretation. Let D denote the response random variable: 1 for a case, and 0 otherwise. Let $\mathbf{Z} = (Z_1, Z_2, \ldots, Z_K)'$ be a vector of K explanatory random variables, including the variable of primary interest, the confounding variables, such as gender, race, and socioeconomic class, and the interaction terms between these variables. Based on the logistic regression model, the conditional probability of being a case, given $\mathbf{Z} = \mathbf{z}$, is given by

$$\mathrm{P}(D = 1|\mathbf{z}) = \exp(\beta_0 + \boldsymbol{\beta}'\mathbf{z})/(1 + \exp(\beta_0 + \boldsymbol{\beta}'\mathbf{z})), \qquad (5.37)$$

where β_0 denotes the intercept term, and $\boldsymbol{\beta} = (\beta_1, \beta_2, \ldots, \beta_K)'$ denotes the coefficients corresponding to the vector $\mathbf{Z}$. On the basis of (5.37), the OR of being a case for a subject with $\mathbf{Z} = \mathbf{z}_a$ versus a subject with $\mathbf{Z} = \mathbf{z}_b$ is then equal to

$$\mathrm{OR} = \mathrm{P}(D = 1|\mathbf{z}_a)\mathrm{P}(D = 0|\mathbf{z}_b)/[\mathrm{P}(D = 0|\mathbf{z}_a)\mathrm{P}(D = 1|\mathbf{z}_b)] = \exp(\boldsymbol{\beta}'(\mathbf{z}_a - \mathbf{z}_b)). \qquad (5.38).$$

In particular, let Z_1 denote the dichotomous exposure variable, with value 1 for being exposed, and 0 otherwise. For simplicity, consider first the situation where there is no interaction between Z_1 and all the other covariates. The OR of being a case between exposure (i.e., $Z_1 = 1$) and non-exposure (i.e., $Z_1 = 0$), given all the other variables fixed, becomes

$$\mathrm{OR} = \exp(\beta_1). \tag{5.39}$$

In other words, the coefficient β_1 denotes the log-odds ratio of being a case between exposure and non-exposure after adjusting all the other covariates. From (5.39), we can easily see that OR $= 1$ if and only if $\beta_1 = 0$. Similarly, if the covariate Z_1 were on a continuous scale, as when representing the dosage level of a treatment, then the OR of being a case for $Z_1 = z_{1a}$ versus $Z_1 = z_{1b}$, adjusting all the other covariates, would equal

$$\mathrm{OR} = \exp(\beta_1(z_{1a} - z_{1b})). \tag{5.40}$$

The coefficient β_1 in (5.40) represents the log-odds ratio of being a case per unit increase in Z_1. If there were an interaction between Z_1 and other covariates, the OR of being a case for $Z_1 = z_{1a}$ versus $Z_1 = z_{1b}$ would depend on the levels of covariates interacting with Z_1 as well. To illustrate this point, suppose that $\mathbf{Z}$ contains only a single covariate, say Z_2, that interacts with Z_1. Let Z_k represent this interaction term $Z_1 Z_2$. Then, the OR of being a case for $Z_1 = z_{1a}$ versus $Z_1 = z_{1b}$, holding all the other covariates fixed, is then equal to

$$\mathrm{OR} = \exp(\beta_1(z_{1a} - z_{1b}) + \beta_k z_2(z_{1a} - z_{1b})), \tag{5.41}$$

which depends on the level of Z_2 (unless $\beta_k = 0$). Note that when $z_{1a} = 1$ and $z_{1b} = 0$ for dichotomous exposure, the OR in (5.41) simplifies to

$$\mathrm{OR} = \exp(\beta_1 + \beta_k z_2). \tag{5.42}$$

We can see that the above the OR in (5.42) is an increasing function of z_2 when $\beta_k > 0$ and a decreasing function of Z_2 when $\beta_k < 0$.

5.5.1 Estimation under multinomial or independent binomial sampling

Suppose that we follow a group of n disease-free subjects for a period of time and obtain the data $(D_i, \mathbf{z}_i)$, $i = 1, 2, \ldots, n$, where $D_i = 1$ if the ith subject is a case, and $D_i = 0$ otherwise, and $\mathbf{z}_i = (z_{i0}, z_{i1}, \ldots, z_{iK})'$ denotes the vector of covariate values on the ith subject. Based on the logistic regression model (5.37), the likelihood is

$$\prod_i \left(\frac{\exp(\beta_0 + \boldsymbol{\beta}'\mathbf{z}_i)}{1 + \exp(\beta_0 + \boldsymbol{\beta}'\mathbf{z}_i)}\right)^{D_i} \left(\frac{1}{1 + \exp(\beta_0 + \boldsymbol{\beta}'\mathbf{z}_i)}\right)^{1-D_i}. \tag{5.43}$$

The MLEs $\hat{\beta}_0, \hat{\beta}_1, \hat{\beta}_2, \ldots, \hat{\beta}_K$ and their estimated asymptotic covariance matrix can be obtained by solving a system of equations and using the inverse of the $(K+1) \times (K+1)$ observed information matrix $\mathbf{I}_o(\hat{\beta}_0, \hat{\boldsymbol{\beta}})$ (**Exercise 5.18**).

In practice, however, we can apply PROC LOGISTIC of SAS (1990) to obtain these estimates easily. For example, in the situation where the response variable Z_1 is dichotomous and does not interact with other covariates, an asymptotic $100(1-\alpha)$ percent confidence interval for the OR (5.39) based on Wald's statistic is given by

$$[\exp(\hat{\beta}_1 - Z_{\alpha/2}\sqrt{\widehat{\mathrm{Var}}(\hat{\beta}_1)}), \exp(\hat{\beta}_1 + Z_{\alpha/2}\sqrt{\widehat{\mathrm{Var}}(\hat{\beta}_1)})], \qquad (5.44)$$

where $\widehat{\mathrm{Var}}(\hat{\beta}_1)$ is the (2,2)th element of $\mathbf{I}_o(\hat{\beta}_0, \hat{\boldsymbol{\beta}})^{-1}$ and is available from the output of PROC LOGISTIC. Similarly, if Z_1 is on a continuous scale, an asymptotic $100(1-\alpha)$ percent for the OR (5.40) is given by

$$[\exp(\hat{\beta}_1(z_{1a} - z_{1b}) - Z_{\alpha/2}\sqrt{(z_{1a} - z_{1b})^2\widehat{\mathrm{Var}}(\hat{\beta}_1)}),$$
$$\exp(\hat{\beta}_1(z_{1a} - z_{1b}) + Z_{\alpha/2}\sqrt{(z_{1a} - z_{1b})^2\widehat{\mathrm{Var}}(\hat{\beta}_1)})]. \qquad (5.45)$$

When there is an interaction between Z_1 and other covariates, we need to account for the levels of other covariates when calculating the confidence interval of OR. For example, suppose that we are interested in obtaining a confidence interval for $\mathrm{OR} = \exp(\beta_1 + \beta_k z_2)$ (5.42). Because the asymptotic variance $\widehat{\mathrm{Var}}(\hat{\beta}_1 + \hat{\beta}_k z_2) = \widehat{\mathrm{Var}}(\hat{\beta}_1) + \widehat{\mathrm{Var}}(\hat{\beta}_k)z_2^2 + 2z_2\widehat{\mathrm{Cov}}(\hat{\beta}_1, \hat{\beta}_k)$, where $\widehat{\mathrm{Var}}(\hat{\beta}_1)$, $\widehat{\mathrm{Var}}(\hat{\beta}_k)$, and $\widehat{\mathrm{Cov}}(\hat{\beta}_1, \hat{\beta}_k)$ are the (2,2)th, $(k+1, k+1)$th, and $(2, k+1)$th elements of $\mathbf{I}_o(\hat{\beta}_0, \hat{\boldsymbol{\beta}})^{-1}$, an asymptotic $100(1-\alpha)$ percent for the OR (5.42) is given by

$$[\exp(\hat{\beta}_1 + \hat{\beta}_k z_2 - Z_{\alpha/2}\sqrt{\widehat{\mathrm{Var}}(\hat{\beta}_1 + \hat{\beta}_k z_2)}),$$
$$\exp(\hat{\beta}_1 + \hat{\beta}_k z_2 + Z_{\alpha/2}\sqrt{\widehat{\mathrm{Var}}(\hat{\beta}_1 + \hat{\beta}_k z_2)})]. \qquad (5.46)$$

Recall that one of the advantages of using the OR to measure the extent of association between disease and exposure is that the OR is estimable from a retrospective case–control study. The above likelihood (5.43), assuming the logistic regression model, is for a cohort or a cross-sectional study. However, Farewell (1979) and Prentice and Pyke (1979) show that the vector $\boldsymbol{\beta}$ would be estimable if the sampling scheme used to collect cases and controls were independent of **Z**. For clarity, we briefly outline the arguments in exercises and present the main result in the following.

Suppose that we independently sample n_1 subjects from the case $(j = 1)$ population and n_0 from the control $(j = 0)$ population, respectively. We define the indicator random variable $\mathcal{I}$ as 1 for a sampled subject, and as 0 otherwise. Then the likelihood is given by

$$\prod_{i=1}^{n_1} \mathrm{P}(\mathbf{z}_i | D_i = 1, \mathcal{I}_i = 1) \prod_{i=1}^{n_0} \mathrm{P}(\mathbf{z}_i | D_i = 0, \mathcal{I}_i = 1). \qquad (5.47)$$

Under the logistic regression model (5.37), we can show that if the sampling scheme used to sample the cases and controls is independent of the random vector $\mathbf{Z}_i$, then the likelihood (5.47) will actually be equivalent to (Hosmer and Lemeshow, 1989; **Exercise 5.19**)

$$\prod_{i=1}^{n}\left(\frac{\exp(\beta_0^* + \boldsymbol{\beta}'\mathbf{z}_i)}{1+\exp(\beta_0^* + \boldsymbol{\beta}'\mathbf{z}_i)}\right)^{D_i}\left(\frac{1}{1+\exp(\beta_0^* + \boldsymbol{\beta}'\mathbf{z}_i)}\right)^{1-D_i}\prod_{i=1}^{n}K(\mathbf{Z}_i, D_i), \tag{5.48}$$

where $n = n_1 + n_0$, $\beta_0^* = \beta_0 + \log(\tau_1/\tau_0)$, $\tau_j = \mathrm{P}(\mathcal{I}_i = 1|D_i = j)$, which is the sampling fraction for the case $(j = 1)$ or the control $(j = 0)$ population, and $K(\mathbf{Z}_i, D_i) = \mathrm{P}(\mathbf{Z}_i)/\mathrm{P}(D_i|\mathcal{I}_i = 1)$. When we assume that $\mathrm{P}(\mathbf{Z}_i)$ contains no information on the coefficients in $\boldsymbol{\beta}$, the MLEs of β_0^* and $\boldsymbol{\beta}$ on the basis of the likelihood (5.48) are algebraically equivalent to the MLEs based on the first part of this likelihood (Anderson, 1972; Farewell, 1979; Hosmer and Lemeshow, 1989). In other words, we can obtain the MLEs of $\boldsymbol{\beta}$ using the data obtained from a case–control study by proceeding exactly the same way as for data obtained from a cohort study. Nevertheless, inferences on the intercept parameter β_0 are not feasible unless we can estimate the sampling fractions τ_j.

Example 5.6 Consider the data (Table 5.1) given in Hosmer and Lemeshow (1989, p. 72). We cross-classified mothers by smoker status during pregnancy (1 = Yes, 0 = No), low birthweight (1 for less than 2500 g; =0 for more than 2500 g), and race (of the mother). Since there are three levels (white, black, and other) for the race variable, we create two design variables R1 and R2: R1 = 0, R2 = 0, for white; R1 = 1, R2 = 0 for Black; and R1 = 0, R2 = 1 for Other. First, we test if there is an interaction between the variables Smoke and Race by using the asymptotic likelihood ratio test: $-2\log(L_1 - L_2)$, where L_1 is the maximum likelihood under model I, $P(D = 1|\mathbf{Z}) = 1/(1+\exp(-(\beta_0 + \beta_1\mathrm{Smoke} + \beta_2\mathrm{R1} + \beta_3\mathrm{R2}))$, and L_2 is the maximum likelihood under model II, $P(D = 1|\mathbf{Z}) = 1/(1+\exp(-(\beta_0 + \beta_1\mathrm{Smoke} + \beta_2\mathrm{R1} + \beta_3\mathrm{R2} + \beta_4\mathrm{R1Smoke} + \beta_5\mathrm{R2Smoke}))$. We employ PROC LOGISTIC in SAS (1990) and obtain $-2\log(L_1 - L_2) = 3.157$. This gives a p-value of 0.21 based on the χ^2 distribution with two degrees of freedom. Therefore, it is reasonable to make inferences based on model I. We obtain an MLE $\hat{\beta}_1 = 1.116$ and an estimated standard error $\widehat{\mathrm{SD}}(\hat{\beta}_1) = 0.369$. Thus, the MLE of the OR is $3.05 (\doteq \exp(1.116))$ with a 95% confidence interval (5.44) given by [1.481, 6.292]. These results are essentially similar to those obtained previously when we did not assume any model for analyzing the data (see Example 5.3).

5.5.2 Estimation in the case of paired-sample data

In Section 5.4, we focus discussion on the situation where matching covariates are the only confounders. In many situations, we may have other confounders or even effect modifiers in addition to those matching covariates. Thus, we generalize the discussion of Section 5.4 to the case where there are confounders besides matching variables. Define the vector $\mathbf{Z} = (\mathbf{Z}_R', \mathbf{Z}_M')'$, where $\mathbf{Z}_M$ and $\mathbf{Z}_R$ are respectively vectors of matching variables and the remaining variables in $\mathbf{Z}$. Suppose that we form n matched pairs. Among these, suppose further that we obtain n_d pairs with discordant disease status, in which $\mathbf{Z}_{1i} = (\mathbf{Z}_{1Ri}', \mathbf{Z}_{1Mi}')'$ and $\mathbf{Z}_{0i} = (\mathbf{Z}_{0Ri}', \mathbf{Z}_{0Mi}')'(i = 1, 2, \ldots, n_d)$ denote the vector of explanatory variables for the case and the non-case, respectively. Conditional upon each matched pair i with discordant disease status, the probability of the observed data is given by

$$L_i(\boldsymbol{\beta}_R) = \frac{\exp(\boldsymbol{\beta}_R'\mathbf{z}_{1Ri})}{\exp(\boldsymbol{\beta}_R'\mathbf{z}_{1Ri}) + \exp(\boldsymbol{\beta}_R'\mathbf{z}_{0Ri})}, \tag{5.49}$$

where $\boldsymbol{\beta}_R$ denotes the subvector of $\boldsymbol{\beta}$ corresponding to $\mathbf{Z}_R$. Note that this conditional probability (5.49) does not depend on the nuisance parameters β_0 and β_M. On the basis of (5.49), the conditional likelihood based on n_d matched pairs with discordant disease status is

$$L(\boldsymbol{\beta}_R) = \prod_{i=1}^{n_d} \frac{\exp(\boldsymbol{\beta}_R'\mathbf{z}_{1Ri})}{\exp(\boldsymbol{\beta}_R'\mathbf{z}_{1Ri}) + \exp(\boldsymbol{\beta}_R'\mathbf{z}_{0Ri})}. \tag{5.50}$$

Note that we can rewrite $L(\boldsymbol{\beta}_R)$ (5.50) as

$$= \prod_i \frac{\exp(\boldsymbol{\beta}_R'(\mathbf{z}_{1Ri} - \mathbf{z}_{0Ri}))}{1 + \exp(\boldsymbol{\beta}_R'(\mathbf{z}_{1Ri} - \mathbf{z}_{0Ri}))}, \tag{5.51}$$

which is identical to the unconditional likelihood based on the logistic regression model with no intercept term for the data $\{(D_i, \mathbf{z}_i^*)|D_i = 1$ for all $i, i = 1, \ldots, n_d$ and $\mathbf{z}_i^* = \mathbf{z}_{1Ri} - \mathbf{z}_{0Ri}\}$. Thus, we can apply PROC LOGISTIC in SAS (1990) to obtain the MLE of $\boldsymbol{\beta}_R$ and the estimated covariance matrix as before. The confidence interval for the OR can also be similarly derived as in Section 5.5.1. When using logistic regression model to analyze paired-sample data for a case–control study, we may apply Bayes' theorem to show that the conditional likelihood (5.51) is still applicable (**Exercise 5.21**).

5.6 INDEPENDENT INVERSE SAMPLING

Consider a case–control study in which we employ independent inverse sampling (Haldane, 1945a, 1945b; Singh and Aggarwal, 1991) to collect cases and controls. Suppose that we continue independently sampling subjects from the case ($j = 1$) and the control ($j = 0$) populations until we obtain a predetermined number x_j of subjects with exposure, respectively. Let $\pi_{1|j}$ denote the probability of exposure in population j and Y_j denote the corresponding number of subjects without exposure before obtaining the first x_j subjects with exposure. The OR of exposure between the cases and the controls is then

$\mathcal{O} = \pi_{1|1}(1 - \pi_{1|0})/[\pi_{1|0}(1 - \pi_{1|1})]$. Given these assumptions, the number Y_j of subjects without exposure collected before accumulating the first x_j subjects with exposure then follows the negative binomial distribution (1.13) with parameters x_j and $\pi_{1|j}$; its mean is $x_j(1 - \pi_{1|j})/\pi_{1|j}$ and its variance $x_j(1 - \pi_{1|j})/\pi_{1|j}^2$.

Define $\overline{Y}_j = Y_j/x_j$. When the number of subjects with exposure x_j is large, the random variable $\overline{Y}_j$ asymptotically has the normal distribution with mean $(1 - \pi_{1|j})/\pi_{1|j}$ and variance $(1 - \pi_{1|j})/(x_j\pi_{1|j}^2)$. Therefore, using the delta method (Bishop *et al.*, 1975; Agresti, 1990), we can show that the random variable $\log(\overline{Y}_j)$ asymptotically has the normal distribution with mean $\log((1 - \pi_{1|j})/\pi_{1|j})$ and variance $1/(x_j(1 - \pi_{1|j}))$ (**Exercise 5.22**). These results suggest that an asymptotic $100(1 - \alpha)$ percent confidence interval for $\mathcal{O}$ is given by

$$[(\overline{Y}_0/\overline{Y}_1)\exp(-Z_{\alpha/2}\sqrt{1/(x_1(1 - \hat{\pi}^*_{1|1})) + 1/(x_0(1 - \hat{\pi}^*_{1|0}))}),$$
$$(\overline{Y}_0/\overline{Y}_1)\exp(Z_{\alpha/2}\sqrt{1/(x_1(1 - \hat{\pi}^*_{1|1})) + 1/(x_0(1 - \hat{\pi}^*_{1|0}))})], \qquad (5.52)$$

where $\hat{\pi}^*_{1|j}$ can be either the MLE $\hat{\pi}_{1|j}(= x_j/(x_j + Y_j))$ or the UMVUE $\hat{\pi}^{(u)}_{1|j}(= (x_j - 1)/(x_j + Y_j - 1))$.

Define $\overline{Z} = \overline{Y}_0 - \mathcal{O}\overline{Y}_1$. If the numbers of subjects x_i with exposure in both samples were large, the random variable $\overline{Z}$ would approximately follow a normal distribution with mean 0 and variance $\mathrm{Var}(\overline{Z}) = \mathrm{Var}(\overline{Y}_0) + \mathcal{O}^2\mathrm{Var}(\overline{Y}_1)$, where $\mathrm{Var}(\overline{Y}_j) = (1 - \pi_{1|j})/(x_j\pi_{1|j}^2)$. This suggests that the probability $P(\overline{Z}^2/\mathrm{Var}(\overline{Z}) \leq Z^2_{\alpha/2}) \doteq 1 - \alpha$, as the predetermined number of subjects with exposure x_i in both samples is large. Note that the inequality $\overline{Z}^2/\mathrm{Var}(\overline{Z}) \leq Z^2_{\alpha/2}$ is equivalent to (**Exercise 5.23**)

$$\mathcal{A}\mathcal{O}^2 - 2\mathcal{B}\mathcal{O} + \mathcal{C} \leq 0, \qquad (5.53)$$

where $\mathcal{A} = \overline{Y}_1^2 - Z^2_{\alpha/2}(1 - \hat{\pi}^*_{1|1})/[x_1(\hat{\pi}^*_{1|1})^2]$, $\mathcal{B} = \overline{Y}_1\overline{Y}_2$, and $\mathcal{C} = \overline{Y}_0^2 - Z^2_{\alpha/2}(1 - \hat{\pi}^*_{1|0})/[x_0(\hat{\pi}^*_{1|0})^2]$, and where $\hat{\pi}^*_{1|j}$ can be either the MLE or the UMVUE of $\pi_{1|j}$. Therefore, if $\mathcal{A} > 0$ and $\mathcal{B}^2 - \mathcal{A}\mathcal{C} > 0$, then an asymptotic $100(1 - \alpha)$ percent confidence interval for $\mathcal{O}$ is given by

$$[(\mathcal{B} - \sqrt{\mathcal{B}^2 - \mathcal{A}\mathcal{C}})/\mathcal{A}, \quad (\mathcal{B} + \sqrt{\mathcal{B}^2 - \mathcal{A}\mathcal{C}})/\mathcal{A}]. \qquad (5.54)$$

When both the probabilities of exposure $\pi_{1|j}$ are close to 0, note that the odds ratio $\mathcal{O} = \pi_{1|1}(1 - \pi_{1|0})/(\pi_{1|0}(1 - \pi_{1|1})) \doteq \pi_{1|1}/\pi_{1|0}$, similar in form to the RR. Therefore, we may also apply the interval estimator originally developed for the RR under inverse sampling (Bennett, 1981; Lui, 1995). As shown in **Exercise 1.20**, the random variable $2(x_j + Y_j)\pi_{1|j}$ approximately has a χ^2 distribution with $2x_j$ degrees of freedom when $\pi_{1|j}$ is small. Therefore, the random variable $\{[2(x_1 + Y_1)\pi_{1|1}]/(2x_1)\}/\{[2(x_0 + Y_0)\pi_{1|0}]/(2x_0)\}$ has an approximate F distribution with degrees of freedom given by $2x_1$ and $2x_0$, respectively. These results suggest that when both $\pi_{1|j}$ are small, an approximate $100(1 - \alpha)$ percent confidence interval for OR ($\doteq \pi_{1|1}/\pi_{1|0}$) should be

$$\left[\frac{1+\overline{Y}_0}{1+\overline{Y}_1}F_{2x_1,2x_0,1-\alpha/2},\ \frac{1+\overline{Y}_0}{1+\overline{Y}_1}F_{2x_1,2x_0,\alpha/2}\right], \tag{5.55}$$

where $F_{f_1,f_2,\alpha}$ is the upper 100αth percentile of the central F distribution with degrees of freedom equal to f_1 and f_2, respectively. Note that the validity of (5.55) depends on the assumption that both the probabilities $\pi_{1|j}$ are small, while the validity of (5.52) and (5.54) depends on the assumption that both the x_j are large.

Lui (1996a) compares the performance of the three interval estimators (5.52), (5.54), and (5.55) and notes that (5.52) is generally preferable to the other two. This is because (5.54) is conservative if the number of subjects with exposure in both comparison groups is small ($x_i \le 20$), while (5.55) may have coverage probability substantially less than the nominal level of 95% if the underlying $\pi_{1|1}$ is moderate or large and $\mathcal{O}$ is much different from 1.

Example 5.7 In a case–control study, suppose that we decide to continue independently sampling subjects until we obtain exactly $x_j = 20$ subjects with exposure from the case ($j = 1$) and the control ($j = 0$) populations, respectively. Suppose further that we obtain $y_1 = 80$ and $y_0 = 180$. Applying (5.52), (5.54), and (5.55) with substitution of the UMVUE $(x_j - 1)/(x_j + y_j - 1)$ for $\hat{\pi}^*_{1|j}$, we obtain 95% confidence intervals for $\mathcal{O}$ of [1.150, 4.401], [1.036, 5.073], and [1.067, 3.750]. Since neither x_j in this example is large, the interval estimate obtained from (5.54) is longer than the other two. Furthermore, because neither the MLE $x_j/(x_j + y_j)$ nor the UMVUE $(x_j - 1)/(x_j + y_j - 1)$ is small, inference based on (5.55) can be misleading, even though the interval obtained from it is the shortest.

5.7 NEGATIVE MULTINOMIAL SAMPLING FOR PAIRED-SAMPLE DATA

When the underlying risk factor is strongly associated with the disease of interest (e.g., oral contraceptives and thrombosis) and the sample size n is not large, there is a non-negligible probability under matched-pair sampling that there will be no matched pairs with case unexposed and control exposed and that the MLE $\hat{\mathcal{O}}$ is infinite (discussed in Section 5.4). In fact, the MLE $\hat{\mathcal{O}}$ under the multinomial distribution (2.25) has infinitely large bias and no exact finite variance. To reduce the bias of this estimator, Jewell (1984) develops an estimator and compares its small-sample performance with that of several other estimators. None of these estimators is, however, unbiased under one-to-one matched sampling.

When the underlying disease is rare, the cases are likely to arrive sequentially. Therefore, in this situation it is quite natural to employ inverse sampling (George and Elston, 1993; Haldane, 1945a, 1945b), in which we continue to sample cases to form matched pairs until we obtain a predetermined (positive) number of index pairs with certain attributes. Furthermore, when regarding the pair with case unexposed and control exposed as the index pair under inverse sampling, one can avoid the above theoretical concern of obtaining infinite MLE $\hat{\mathcal{O}}$.

Suppose that we continue sampling cases to form matched pairs until we obtain the predetermined number $n_{01}(\geq 1)$ of pairs in which the case is not exposed, and the control is exposed. For clarity, we use the following table to summarize the resulting data:

		Control	
		Exposed	Non-exposed
Case	Exposed	N_{11}	N_{10}
	Non-exposed	n_{01}	N_{00}

where n_{01} is a positive integer and is determined in advance, and random variables N_{11}, N_{10}, and N_{00} denote the number of matched pairs falling in the corresponding categories before obtaining exactly n_{01} matched pairs as defined above. The joint distribution of N_{11}, N_{10}, and N_{00} is then given by

$$f_{N_{11},N_{10},N_{00}}(n_{11}, n_{10}, n_{00}) = \frac{(n_{11} + n_{10} + n_{01} + n_{00} - 1)!}{n_{11}!n_{10}!(n_{01} - 1)!n_{00}!} \times \pi_{01}^{n_{01}} \pi_{11}^{n_{11}} \pi_{10}^{n_{10}} \pi_{00}^{n_{00}}, \qquad (5.56)$$

where $0 < \pi_{ii'} < 1$, $\sum_i \sum_{i'} \pi_{ii'} = 1$, i and $i' = 0, 1$, and where n_{11}, n_{10}, and $n_{00} = 0, 1, 2, \ldots$. This is a negative multinomial distribution with parameters $n_{01}, \pi_{11}, \pi_{10}$, and π_{00} (Ratnaparkhi, 1985). Under (5.56), the MLE of the OR is simply $\hat{\mathcal{O}} = N_{10}/n_{01}$, which is of the same form as that for traditional one-to-one matched sampling discussed in Section 5.4. Note also that we can easily show that the marginal distribution of N_{10} follows the negative binomial distribution with parameters n_{01} and $\mathcal{P} = \pi_{01}/(\pi_{10} + \pi_{01}) = 1/(1 + \mathcal{O})$ (**Exercise 5.25**):

$$f_{N_{10}}(n_{10}) = \frac{(n_{10} + n_{01} - 1)!}{n_{10}!(n_{01} - 1)!} \mathcal{P}^{n_{01}} (1 - \mathcal{P})^{n_{10}}, \qquad n_{10} = 0, 1, 2, \ldots. \qquad (5.57)$$

The expectation of N_{10} is simply $E(N_{10}) = n_{01}(1 - \mathcal{P})/\mathcal{P} = n_{01}\mathcal{O}$, and the variance $\mathrm{Var}(N_{10}) = n_{01}(1 - \mathcal{P})/\mathcal{P}^2 = n_{01}\mathcal{O}(\mathcal{O} + 1)$. Thus, on the basis of distribution (5.57), the estimator N_{10}/n_{01} has expectation $E(N_{10}/n_{01}) = \mathcal{O}$ and variance $\mathrm{Var}(N_{10}/n_{01}) = \mathcal{O}(\mathcal{O} + 1)/n_{01}$. Furthermore, because distribution (5.56) belongs to the exponential family (Lehmann, 1983, p. 46), the random vector (N_{11}, N_{10}, N_{00}) is a complete sufficient statistic. Following the Rao–Blackwell and Lehmann–Scheffé theorems (Graybill, 1976), we can conclude that N_{10}/n_{01} is, in fact, the UMVUE of the OR.

Note that the probability for the cumulative distribution $P(N_{10} \leq n_{10}|n_{01}, \mathcal{P})$, where N_{10} follows the negative binomial distribution (5.57) with parameters n_{01} and $\mathcal{P}$, equals the probability $P(X \geq n_{01}|n, \mathcal{P})$, where X follows the binomial distribution with parameters $n = n_{10} + n_{01}$ and $\mathcal{P}$ (**Exercise 1.14**). Therefore, for a given observed value n_{10}, an exact $100(1-\alpha)$ percent confidence interval for $\mathcal{P}$ (**Exercise 1.15**) is given by $[\mathcal{L}(n_{10}, n_{01}, \alpha/2), \mathcal{U}(n_{10}, n_{01}, \alpha/2)]$, where $\mathcal{L}(n_{10}, n_{01}, \alpha/2) = n_{01}/[n_{01} + (n_{10}+1)F_{2(n_{10}+1), 2n_{01}, \alpha/2}$, $\mathcal{U}(n_{10}, n_{01}, \alpha/2) = n_{01} F_{2n_{01}, 2n_{10}, \alpha/2}/[n_{01}F_{2n_{01}, 2n_{10}, \alpha/2} + n_{10}]$. Note that if n_{10} were 0, we would define the upper limit $\mathcal{U}(0, n_{01}, \alpha/2) = 1$ by convention. Furthermore, because $\mathcal{O} = (1-\mathcal{P})/\mathcal{P}$ and the transformation $f(x) = (1-x)/x$ is a strictly decreasing function, an exact $100(1-\alpha)$ percent confidence interval for OR is then given by

$$[(1-\mathcal{U}(n_{10}, n_{01}, \alpha/2))/\mathcal{U}(n_{10}, n_{01}, \alpha/2),$$
$$(1-\mathcal{L}(n_{10}, n_{01}, \alpha/2))/\mathcal{L}(n_{10}, n_{01}, \alpha/2)]. \qquad (5.58)$$

Applying interval estimator (5.58) can ensure that the coverage probability is always larger than or equal to the desired confidence level. Note that other relevant discussions, such as an exact test procedure for testing if the underlying OR equals any specified value, an asymptotic test similar to McNemar's test (**Exercise 5.27**), as well as sample size calculation under (5.56), appear elsewhere (Lui, 1996b).

Example 5.8 Consider the data (Schlesselman, 1982, p. 209) for a case–control study of oral conjugated estrogens and endometrial cancer, in which cases and controls are matched on age, race, hospital, and date of admission. For illustration purposes only, we assume that these data are collected by means of inverse sampling, and that there are $n_{01} = 7$ index pairs in which the case is unexposed but the control is exposed; the numbers of matched pairs falling in the other categories are $n_{11} = 12$, $n_{10} = 43$, and $n_{00} = 121$. Given these data, the UMVUE of the underlying OR between use of oral conjugated estrogens and endometrial cancer is $\hat{\mathcal{O}} = 6.14 (= n_{10}/n_{01})$ with estimated variance $6.27 (= \hat{\mathcal{O}}(\hat{\mathcal{O}}+1)/n_{01})$. The exact 95% confidence interval for the OR is [3.04, 16.18]. Because the resulting confidence interval excludes 1, we may conclude that there is a significant association at the 5% level between taking oral conjugated estrogens and developing endometrial cancer.

EXERCISES

5.1. Show that the odds ratio OR is invariant when we multiply a row and/or a column by a positive constant.

5.2. Let E and D denote the random variables for the status of exposure and disease, respectively. Define $E = 1$ for exposure, and $E = 0$ otherwise. Similarly, define $D = 1$ for a case, and $D = 0$ otherwise. Show that when the underlying disease is rare, OR $(= \pi_{11}\pi_{00}/(\pi_{10}\pi_{01}))$ approximately equals relative risk (RR $= (\pi_{11}/\pi_{1.})/(\pi_{01}/\pi_{0.})$), where $\pi_{ij} = \mathrm{P}(E = i, D = j)$, i and $j = 1, 0$.

5.3. Show that under independent binomial sampling in Section 5.1, the estimated asymptotic variance of $\log(\hat{\mathcal{O}})$ is $\widehat{\mathrm{Var}}(\log(\hat{\mathcal{O}})) = 1/X_1 + 1/X_0 + 1/(n_1 - X_1) + 1/(n_0 - X_0)$, where $\hat{\mathcal{O}} = X_1(n_0 - X_0)/[(n_1 - X_1)X_0]$.

5.4. When n is large, we may expect that $X_1(n_0 - X_0)/[X_0(n_1 - X_1)] = \mathcal{O}$. Therefore, given $\mathcal{O} = \mathcal{O}_0$ and $x_{.}$, show that we can calculate the expected frequency of X_1 from the quadratic equation given by (5.2).

5.5. Show that, in Section 5.1.1:
(a) $[\mathcal{O}_0(n_1 + x_{.}) + (n_0 - x_{.})]^2 - 4(\mathcal{O}_0 - 1)\mathcal{O}_0 n_1 x_{.} \geq 0$;
(b) when $\mathcal{O}_0 > 1$, $X_2(\mathcal{O}_0) > \min\{x_{.}, n_1\}$ and when $\mathcal{O}_0 < 1$, $X_2(\mathcal{O}_0) < \max\{0, x_{.} - n_0\}$, where

$$X_2(\mathcal{O}_0) = \frac{\mathcal{O}_0(n_1 + x_{.}) + (n_0 - x_{.}) + \sqrt{[\mathcal{O}_0(n_1 + x_{.}) + (n_0 - x_{.})]^2 - 4(\mathcal{O}_0 - 1)\mathcal{O}_0 n_1 x_{.}}}{2(\mathcal{O}_0 - 1)};$$

(c) $\max\{0, x_{.} - n_0\} < X_1(\mathcal{O}_0) < \min\{x_{.}, n_1\}$, where $X_1(\mathcal{O}_0)$ is given in (5.3).

5.6. In Section 5.1.1, show that the asymptotic variance of X_1 can be approximated by $\mathrm{Var}(X_1|X_1 + X_0 = x_{.}, \mathcal{O} = \mathcal{O}_0) = [1/X_1(\mathcal{O}_0) + 1/(x_{.} - X_1(\mathcal{O}_0)) + 1/(n_1 - X_1(\mathcal{O}_0)) + 1/(n_0 - x_{.} + X_1(\mathcal{O}_0))]^{-1}$, as given in (5.4).

5.7. Show that under multinomial sampling with parameters n and $(\pi_{11}, \pi_{10}, \pi_{01}, \pi_{00})'$ in cross-sectional studies, the estimated asymptotic variance $\widehat{\mathrm{Var}}(\log(\mathcal{O}))$ is given by $1/N_{11} + 1/N_{10} + 1/N_{01} + 1/N_{00}$, where N_{ij} denotes the observed number of subjects falling into the cell with probability π_{ij}, and $\hat{\mathcal{O}} = N_{11}N_{00}/(N_{10}N_{01})$.

5.8. When the underlying OR equals 1, we may claim that $(\log(\hat{\mathcal{O}}))^2/\mathrm{Var}(\log(\hat{\mathcal{O}})) \doteq \chi^2$, where χ^2 is the chi-squared test value without the use of continuity correction (Miettinen, 1976). Therefore, $\mathrm{Var}(\log(\hat{\mathcal{O}})) \doteq (\log(\hat{\mathcal{O}}))^2/\chi^2$. On the basis of this result, show that an approximate $100(1-\alpha)$ percent confidence interval for OR is given by $[\hat{\mathcal{O}}^{(1-Z_{\alpha/2}/\chi)}, \hat{\mathcal{O}}^{(1+Z_{\alpha/2}/\chi)}]$.

5.9. In Section 5.1.2, show that the conditional probability mass function of $X_1 = x_1$, given $X_1 + X_0 = x_{.}$, is given by $P(X_1 = x_1|X_1 + X_0 = x_{.}, \mathcal{O}) = \binom{n_1}{x_1}\binom{n_0}{x_{.} - x_1}\mathcal{O}^{x_1}/\sum_x \binom{n_1}{x}\binom{n_0}{x_{.} - x}\mathcal{O}^x$, where X_j follows the binomial distribution with parameters n_j and $\pi_{1|j}$, and $\mathcal{O} = \pi_{1|1}(1 - \pi_{1|0})/[(1 - \pi_{1|1})\pi_{1|0}]$. Therefore, when $\mathcal{O} = 1$ under the assumption of no association between the disease and the risk factor, this probability mass function simplifies to the hypergeometric distribution. Note that $\sum_x \binom{n_1}{x}\binom{n_0}{x_{.} - x} = \binom{n_1 + n_0}{x_{.}}$.

5.10. In Section 5.1.2, show that the conditional MLE of $\mathcal{O}$ on the basis of (5.8), when the observed value $X_1 = x_1$, is the solution $\mathcal{O}$ of the equation $x_1 = E(X_1|x_{.}, \mathcal{O})$.

5.11. Show that the cumulative probability distribution $P(X_1 \leq x_1|x_{.}, \mathcal{O}) = \sum_{x \leq x_1} P(X_1 = x|X_1 + X_0 = x_{.}, \mathcal{O})$, where $P(X_1 = x|X_1 + X_0 = x_{.}, \mathcal{O})$ is given by (5.8), is a decreasing function of $\mathcal{O}$.

5.12. Consider the hypothetical data given by Fleiss (1979). There are 20 out of 50 cases with exposure, while there are only 10 out of 50 controls with exposure. What is the conditional MLE $\hat{\mathcal{O}}_{\text{cond}}$ of the OR? What is the exact 95% confidence interval (5.10) for the OR?

5.13. Consider the all-cause mortality data in Table 2.1.
(a) What are the p-values when we apply statistics (5.25) and (5.26) to test the homogeneity of the OR for the first five trials ($S = 1, 2, \ldots, 5$)?
(b) What are the summary point estimates of $\mathcal{O}_c$ using $\sum_s \hat{W}_s^* \hat{\mathcal{O}}_s / \sum_s \hat{W}_s^*$ and $\hat{\mathcal{O}}_{\text{MH}}$ for these five trials?
(c) What are the 95% confidence intervals for the OR using (5.12), (5.15), and (5.18) for the first five trials?
(d) What is the p-value for the test of the homogeneity of the OR across all six trials?

5.14. In Section 5.4, show that under the logistic model assumptions the ratio of the cell probabilities $\pi_{10}/\pi_{01} = e^{\beta}$, which is equal to the OR,

$$\mathcal{O} = \frac{P(D = 1|E = 1, V_l)P(D = 0|E = 0, V_l)}{P(D = 0|E = 1, V_l)P(D = 1|E = 0, V_l)}.$$

5.15. In Section 5.4, show that the asymptotic variance of $\hat{\pi}_{10}/\hat{\pi}_{01}$ is given by $(\pi_{10}/\pi_{01})^2[1/(n\pi_{10}) + 1/(n\pi_{01})]$.

5.16. In Section 5.4, show that when the number of matched pairs n is large, an asymptotic $100(1 - \alpha)$ percent confidence interval for the OR is given by $[(B^{\ddagger} - \sqrt{B^{\ddagger} - A^{\ddagger}C^{\ddagger}})/A^{\ddagger}, (B^{\ddagger} + \sqrt{B^{\ddagger} - A^{\ddagger}C^{\ddagger}})/A^{\ddagger}]$, where $A^{\ddagger} = \hat{\pi}_{01}^2 - Z_{\alpha/2}^2\hat{\pi}_{01}(1 - \hat{\pi}_{01})/n$, $B^{\ddagger} = \hat{\pi}_{10}\hat{\pi}_{01} + Z_{\alpha/2}^2\hat{\pi}_{10}\hat{\pi}_{01}/n$, and $C^{\ddagger} = \hat{\pi}_{10}^2 - Z_{\alpha/2}^2\hat{\pi}_{10}(1 - \hat{\pi}_{10})/n$.

5.17. In Section 5.4, show that the conditional probability mass function of n_{10}, given $n_{10} + n_{01} = n_d$, is the binomial distribution with parameters n_d and $\pi_{10}/(\pi_{10} + \pi_{01}) (= \mathcal{O}/(1 + \mathcal{O}))$.

5.18. Under the likelihood (5.43), show that: (a) the MLEs $\hat{\beta}_0, \hat{\beta}_1, \hat{\beta}_2, \ldots, \hat{\beta}_K$ can be obtained by solving the equations $\sum_i z_{ik}(D_i - P(D_i = 1|\mathbf{z}_i)) = 0$ for $k = 0, 1, 2, \ldots, K$ (we define $z_{i0} = 1$ for all i); (b) the observed information matrix $\mathbf{I}_o(\hat{\beta}_0, \hat{\boldsymbol{\beta}})$ is the $(K + 1) \times (K + 1)$ matrix with the diagonal element $(k + 1, k + 1)$ equal to $\sum_i z_{ik}^2 \hat{P}(D_i = 1|\mathbf{z}_i)(1 - \hat{P}(D_i = 1|\mathbf{z}_i))$ and off-diagonal element (k, l) equal to $\sum_i z_{ik}z_{il}\hat{P}(D_i = 1|\mathbf{z}_i)(1 - \hat{P}(D_i = 1|\mathbf{z}_i))$, where $\hat{P}(D_i = 1|\mathbf{z}_i)$ denotes $P(D_i = 1|\mathbf{z}_i)$ (5.37) in which parameters β_0 and $\boldsymbol{\beta}$ are replaced by their corresponding MLEs, respectively. By large-sample theory, the estimated covariance matrix of $(\hat{\beta}_0, \hat{\boldsymbol{\beta}}')'$ can be given by the inverse of the above observed information matrix.

5.19. Assume that $P(D = 1|\mathbf{Z} = \mathbf{z}) = \exp(\beta_0 + \boldsymbol{\beta}'\mathbf{z})/(1 + \exp(\beta_0 + \boldsymbol{\beta}'\mathbf{z}))$ and that the sampling scheme used to collect the cases and controls is independent of the random vector $\mathbf{Z}$. Show that the likelihood (5.47) is proportional to (5.48). (Hint: First, show that the conditional probability

$P(\mathbf{Z}|D, \mathcal{I}=1)$ of $\mathbf{Z}$, given D and $\mathcal{I}=1$, is equal to $P(D|\mathbf{Z}, \mathcal{I}=1)P(\mathbf{Z}|\mathcal{I}=1)/P(D|\mathcal{I}=1)$. Secondly, show that the conditional probability $P(D=1|\mathbf{Z}, \mathcal{I}=1)$ is equal to $P(\mathcal{I}=1|\mathbf{Z}, D=1)P(D=1|\mathbf{Z})/[P(\mathcal{I}=1|\mathbf{Z}, D=1)P(D=1|\mathbf{Z}) + P(\mathcal{I}=1|\mathbf{Z}, D=0)P(D=0|\mathbf{Z})]$. Note that, by assumption, $P(\mathcal{I}=1|\mathbf{Z}, D=j) = P(\mathcal{I}=1|D=j) = \tau_j$. Then show that $P(D=1|\mathbf{Z}, \mathcal{I}=1) = \exp(\beta_0^* + \boldsymbol{\beta}'\mathbf{z})/(1+\exp(\beta_0^* + \boldsymbol{\beta}'\mathbf{z}))$, where $\beta_0^* = \beta_0 + \log(\tau_1/\tau_0)$. Note that, by assumption, $P(\mathbf{Z}|\mathcal{I}=1) = P(\mathbf{Z})$, which is the joint probability distribution of $\mathbf{Z}$.)

5.20. Consider the all-cause mortality data in Table 2.1 from the six trials. Assume a logistic regression model.
(a) What is the p-value when we apply the likelihood ratio test to detect if there is an interaction between aspirin and centers for the first five trials?
(b) Assuming that there is no interaction between taking aspirin and centers, what is the point estimate $\widehat{\text{OR}}(=\exp(\hat{\beta}_1))$ and the 95% confidence interval for the OR between aspirin ($Z_1 = 1$) and placebo ($Z_1 = 0$) for the five trials?
(c) What is the p-value when we apply the likelihood ratio test to detect if there is an interaction between aspirin and centers for the data from all six trials?

5.21. Under the assumption of a logistic regression model (5.37), show that the likelihood (5.51) is also appropriate for paired-sample data in a case–control study.

5.22. Show that the asymptotic variance of the random variable $\log(\overline{Y}_j)$ is given by $1/(x_j(1-\pi_{1|j}))$ under the negative binomial distribution with parameters x_j and $\pi_{1|j}$ (1.13).

5.23. In Section 5.6, show that the inequality $\overline{Z}^2/\text{Var}(\overline{Z}) \le Z_{\alpha/2}^2$ is equivalent to $\mathcal{A}\mathcal{O}^2 - 2\mathcal{B}\mathcal{O} + \mathcal{C} \le 0$, where $\mathcal{A} = \overline{Y}_1^2 - Z_{\alpha/2}^2(1-\hat{\pi}_{1|1}^*)/[x_1(\hat{\pi}_{1|1}^*)^2]$, $\mathcal{B} = \overline{Y}_1\overline{Y}_2$, and $\mathcal{C} = \overline{Y}_0^2 - Z_{\alpha/2}^2(1-\hat{\pi}_{1|0}^*)/[x_0(\hat{\pi}_{1|0}^*)^2]$.

5.24. Suppose that we employ independent inverse sampling to collect the cases and the controls in a case–control study. Suppose further that we have collected $Y_1 = 270$ subjects with non-exposure before obtaining the predetermined $x_1 = 30$ subjects with exposure from the case population, while we have collected $Y_0 = 550$ controls with non-exposure before obtaining the predetermined 50 subjects with exposure from the control population. What are the 95% confidence intervals for the OR using (5.52), (5.54), and (5.55)?

5.25. In Section 5.7, show that the marginal distribution of N_{10} follows the negative binomial distribution (5.57) with parameters n_{01} and $\mathcal{P} = \pi_{01}/(\pi_{10} + \pi_{01}) = 1/(1+\text{OR})$ where $\text{OR} = \pi_{10}/\pi_{01}$.

5.26. Consider the data taken from a case–control study of tonsillectomy and Hodgkin's disease (Vianna *et al.*, 1971). As given elsewhere (Mausner and Bahn, 1974, p. 317), we have 67 cases with prior tonsillectomy out of 109 cases, while we have only 43 subjects with prior tonsillectomy out of 109 controls. What is the MLE $\hat{\mathcal{O}}$? What are the 95% confidence intervals for the OR when we use (5.1) and (5.5)?

5.27. Show that under distribution (5.57), if $H_0 : \mathcal{O} = 1$, then the statistic $(N_{10} - n_{01})^2/(2n_{01})$ asymptotically follows a χ^2 distribution with one degree of freedom. By contrast, McNemar's test without continuity correction under one-to-one matched sampling is $(N_{10} - N_{01})^2/(N_{10} + N_{01})$. When H_0 is true, we expect N_{10} to be approximately equal to n_{01}, and hence the values of these two test statistics are expected to be similar.

5.28. Consider a case–control study of the association between the diabetic ketoacidosis (DKA) in patients before and after the onset of pump therapy (Mecklenburg *et al.*, 1984). As given elsewhere, we have 128 patients with no DKA before and after pump therapy, 7 patients with DKA before pump therapy but with no DKA after pump therapy, 19 patients with no DKA before pump therapy but with DKA after pump therapy, and 7 patients with DKA before and after pump therapy. What is the MLE of the OR of possessing DKA for patients after pump therapy versus before pump therapy? What are the 95% confidence intervals for the OR using various estimators?

REFERENCES

Anderson, J. A. (1972) Separate sample logistic discrimination. *Biometrika*, **59**, 19–35.

Agresti, A. (1990) *Categorical Data Analysis*. Wiley, New York.

Agresti, A. (1999) On logit confidence intervals for the odds ratio with small samples. *Biometrics*, **55**, 597–602.

Bennett, B. M. (1981) On the use of the negative binomial in epidemiology. *Biometrical Journal*, **23**, 69–72.

Bishop, Y. M. M., Fienberg, S. E. and Holland, P. W. (1975) *Discrete Multivariate Analysis: Theory and Practice*. MIT Press, Cambridge, MA.

Breslow, N. E. and Day, N. E. (1980) *Statistical Methods in Cancer Research, Volume 1. The Analysis of Case–Control Studies*. International Agency for Research on Cancer, Lyon.

Brown, C. C. (1981) The validity of approximation methods for interval estimation of the odds ratio. *American Journal of Epidemiology*, **113**, 474–480.

Casella, G. and Berger, R. L. (1990) *Statistical Inference*. Duxbury Press, Belmont, CA.

Cornfield, J. (1951) A method of estimating comparative rates from clinical data: Applications to cancer of the lung, breast and cervix. *Journal of the National Cancer Institute*, **11**, 1269–1275.

Cornfield, J. (1956) A statistical problem arising from retrospective studies. In J. Neyman (ed.), *Proceedings of the Third Berkeley Symposium on Mathematical Statistics and Probability*, Vol. 4. University of California Press, Berkeley, pp. 136–148.

Donald, A. and Donner, A. (1990) A simulation study of the analysis of sets of 2 × 2 contingency tables under cluster sampling: estimation of a common odds ratio. *Journal of the American Statistical Association*, **85**, 537–543.

Ejigou, A. and McHugh, R. (1977) Estimation of relative risk from matched pairs in epidemiologic research. *Biometrics*, **33**, 552–556.

Ejigou, A. and McHugh, R. (1984) Testing the homogeneity of the relative risk under multiple matching. *Biometrika*, **71**, 408–411.

Elston, R. C. (1977) Response to query: estimating 'inheritability' of a dichotomous trait. *Biometrics*, **33**, 232–233.

Farewell, V. T. (1979) Some results on the estimation of logistic models based on retrospective data. *Biometrika*, **66**, 27–32.

Fleiss, J. L. (1979) Confidence intervals for the odds ratio in case–control studies: The state of the art. *Journal of Chronic Diseases*, **32**, 69–77.
Fleiss, J. L. (1981) *Statistical Methods for Rates and Proportions*. Wiley, New York.
Fleiss, J. L. (1986) *The Design and Analysis of Clinical Experiments*. Wiley, New York.
Gart, J. J. (1970) Point and interval estimation of the common odds ratio in the combination of 2×2 tables with fixed marginals. *Biometrika*, **57**, 471–475.
Gart, J. J. and Thomas, D. G. (1972) Numerical results on approximate confidence limits for the odds ratio. *Journal of Royal Statistical Society B*, **34**, 441–447.
Gart, J. J. and Thomas, D. G. (1982) The performance of three approximate confidence limit methods for the odds ratio. *American Journal of Epidemiology*, **115**, 453–470.
Gart, J. J. and Zweifel, J. R. (1967) On the bias of the logit and its variance with application to quantal bioassay. *Biometrika*, **54**, 181–187.
George, V. T. and Elston, R. C. (1993) Confidence limits based on the first occurrence of an event. *Statistics in Medicine*, **12**, 685–690.
Graham, S., Dayal, H., Rohrer, T., *et al.* (1977) Dentition, diet, tobacco, and alcohol in the epidemiology of oral cancer. *Journal of the National Cancer Institute*, **59**, 1611–1616.
Graybill, F. A. (1976) *Theory and Application of the Linear Model*. Duxbury Press, North Scituate, MA.
Haldane, J. B. S. (1945a) A labour-saving method of sampling. *Nature*, **155**, 49–50.
Haldane, J. B. S. (1945b) On a method of estimating frequencies. *Biometrika*, **33**, 222–225.
Hauck, W. W. (1989) Odds ratio inference from stratified samples. *Communication in Statistics*, A18, 767–800.
Hosmer, D. W. and Lemeshow, S. (1989) *Applied Logistic Regression*. Wiley, New York.
Jewell, N. P. (1984) Small-sample bias of point estimators of the odds ratio from matched sets. *Biometrics*, **40**, 421–435.
Johnson, N. L. and Kotz, S. (1970) *Distributions in Statistics: Continuous Univariate Distributions 2*. Wiley, New York.
Lehmann, E. L. (1983) *Theory of Point Estimation*. Wiley, New York.
Liang, K. -Y. and Self, S. G. (1985) Tests for homogeneity of odds ratio when the data are sparse. *Biometrika*, **72**, 353–358.
Lui, K. -J. (1991) Sample size for repeated measurements in dichotomous data. *Statistics in Medicine*, **10**, 463–472.
Lui, K. -J. (1995) Confidence intervals for the risk ratio in cohort studies under inverse sampling. *Biometrical Journal*, **37**, 965–971.
Lui, K. -J. (1996a) Notes on confidence limits for the odds ratio in case–control studies under inverse sampling. *Biometrical Journal*, **38**, 221–229.
Lui, K. -J. (1996b) Notes in case–control studies with matched pairs under inverse sampling. *Biometrical Journal*, **38**, 681–693.
Lui, K. -J. and Lin, C. D. (2003) A revisit on comparing the asymptotic interval estimators of odds ratio in a single 2×2 table. *Biometrical Journal*, **45**, 226–237.
Lui, K. -J., Cumberland, W. G. and Kuo, L. (1996) An interval estimate for the intraclass correlation in beta-binomial sampling. *Biometrics*, **52**, 412–425.
Mantel, N. and Haenszel, W. (1959) Statistical aspects of the analysis of data from retrospective studies of disease. *Journal of the National Cancer Institute*, **22**, 719–748.
Mausner, J. and Bahn, A. K. (1974) *Epidemiology, An Introductory Text*. W. B. Saunders, Philadelphia.
Mayer, J., Slymen, D. J., Eckhardt, L., *et al.* (1997) Reducing ultraviolet radiation exposure in children. *Preventive Medicine*, **26**, 516–522.
Mecklenburg, R. S., Benson, E. A., Benson, J. W., *et al.* (1984) Acute complications associated with insulin pump therapy: Report of experience with 161 patients. *Journal of the American Medical Association*, **252**, 3265–3269.
Miettinen, O. S. (1976) Estimability and estimation in case-referent studies. *American Journal of Epidemiology*, **103**, 226–235.

Prentice, R. L. and Pyke, R. (1979) Logistic disease incidence models and case–control studies. *Biometrika*, **66**, 403–411.

Ratnaparkhi, M. V. (1985) Negative multinomial distribution. In S. Kotz and N. L. Johnson (eds), *Encyclopedia of Statistical Sciences*, Vol. 5 John Wiley & Sons, Inc., New York, pp. 662–665.

Robins, J., Breslow, N. and Greenland, S. (1986) Estimators of the Mantel–Haenszel variance consistent in both sparse data and large-strata limiting models. *Biometrics*, **42**, 311–323.

Rosner, B. (1990) *Fundamentals of Biostatistics*, 3rd edition. PWS-Kent, Boston.

SAS Institute, Inc. (1990) *SAS/STAT User's Guide, Volume 2, Version 6*, 4th edition. SAS Institute, Inc., Cary, NC.

Schlesselman, J. J. (with contributions by Stolley, P. D.) (1982) *Case–Control Studies: Design, Conduct, Analysis*. Oxford University Press, New York.

Singh, P. and Aggarwal, A. R. (1991) Inverse sampling in case control studies. *Environmetrics*, **2**, 293–299.

Skarin, A. T., Pinkus, G. S., Myerowitz, R. L., *et al.* (1973) Combination chemotherapy of advanced lymphocytic lymphoma: importance of histologic classification in evaluating response. *Cancer*, **34**, 1023–1029.

Stevens, W. L. (1951) Mean and variance of an entry in a contingency table. *Biometrika*, **38**, 468–470.

Vianna, N. J., Greenwald, P. and Davies, J. N. P. (1971) Tonsillectomy and Hodgkin's disease: the lymphoid tissue barrier. *Lancet*, **1**, 431–432.

Woolf, B. (1955) On estimating the relation between blood group and disease. *Annals of Human Genetics*, **19**, 251–253.

6

Generalized Odds Ratio

To measure the strength of association between a dichotomous risk factor and a dichotomous response, the odds ratio (OR) discussed in Chapter 5 is probably the most frequently used epidemiological index. However, we commonly encounter the situation in which the outcomes are on an ordinal scale with more than two categories. For example, consider the study of tonsil size in carriers of *Streptococcus pyogenes* (Clayton, 1974). Because tonsil size is measured on an ordinal scale (tonsils present but not enlarged; tonsils enlarged; and tonsils greatly enlarged), the OR cannot directly be used without arbitrarily grouping multiple levels of response into two categories. Furthermore, the collapsing of data may cause a loss of efficiency. The generalized odds ratio (GOR) (Agresti, 1980) is useful for summarizing the difference between two stochastically ordered distributions of an ordinal categorical variable without the need to assume any specific parametric models (Hosmer and Lemeshow, 1989). Readers may find some discussions on other generalized summary measures as well (Edwardes and Baltzan, 2000; Lui, 2002b).

In this chapter, we first discuss estimation of the GOR under independent multinomial sampling for randomized trials. We then extend this discussion to accommodate situations where we take repeated measurements per subject or we employ cluster sampling to collect data. We also discuss estimation of the GOR for an ordinal exposure variable in case–control studies with matched pairs. Finally, to alleviate the bias of our GOR estimator under multinomial sampling, we discuss estimation of the GOR under a mixed negative multinomial and multinomial sampling procedure.

6.1 INDEPENDENT MULTINOMIAL SAMPLING

Suppose that we wish to quantify the extent of association between the distributions of ordinal responses over J categories for two comparison groups. Define $\boldsymbol{\pi}_i' = (\pi_{1|i}, \pi_{2|i}, \pi_{3|i}, \ldots, \pi_{J|i})$, where $\pi_{j|i}$ (satisfying $0 < \pi_{j|i} < 1$ and $\sum_j \pi_{j|i} = 1$) denotes the probability of the outcome for a randomly selected subject from group

Statistical Estimation of Epidemiological Risk K-J. Lui
 ISBN: 0-470-85071-X (HB)

$i(i = 1, 2)$ falling into category $j(j = 1, 2, \ldots, J)$. Without loss of generality, we assume that the response in category j' is more severe than the response in category j when $j < j'$. The GOR is defined as $\mathcal{G} = \Pi_c/\Pi_d$ (Agresti, 1980), where $\Pi_c = \sum_{r=1}^{J-1}\sum_{s=r+1}^{J} \pi_{r|1}\pi_{s|2}$ and $\Pi_d = \sum_{r=2}^{J}\sum_{s=1}^{r-1} \pi_{r|1}\pi_{s|2}$. Note that Π_c denotes the probability that the response of a randomly selected subject from group 2 is severer than the response of a randomly selected subject from group 1. Similarly, Π_d denotes the probability that the response for a randomly selected subject from group 1 is severer than the response for a randomly selected subject from group 2. When $J = 2$, the GOR reduces to the OR for a single 2×2 table: $\pi_{1|1}\pi_{2|2}/(\pi_{2|1}\pi_{1|2})$. By definition, the range of the GOR is $0 < \mathcal{G} < \infty$. When the underlying distributions of ordinal responses are identical between two comparison groups (i.e., $\boldsymbol{\pi}_1 = \boldsymbol{\pi}_2$), $\mathcal{G} = 1$. Note that the index $(\mathcal{G} - 1)/(\mathcal{G} + 1)$ is actually the γ proposed by Goodman and Kruskal (1954) to measure the association between two ordinal variables. The GOR considered here is, in fact, Agresti's α (Agresti, 1980).

Assume that from group $i(i = 1, 2)$, we take a random sample of size n_i. Assume further that these two samples are independent. Let X_{ij} denote the number of subjects out of n_i subjects falling into category j. Then, the random vector $(X_{i1}, X_{i2}, \ldots, X_{iJ})$ follows the multinomial distribution with parameters n_i and $\boldsymbol{\pi}_i' = (\pi_{1|i}, \pi_{2|i}, \pi_{3|i}, \ldots, \pi_{J|i})$. The MLE of $\pi_{j|i}$ is simply $\hat{\pi}_{j|i} = X_{ij}/n_i$. As n_i is large, by the central limit theorem (Casella and Berger, 1990), we claim that $\sqrt{n_i}(\hat{\boldsymbol{\pi}}_i - \boldsymbol{\pi}_i)$, where $\hat{\boldsymbol{\pi}}_i' = (\hat{\pi}_{1|i}, \hat{\pi}_{2|i}, \hat{\pi}_{3|i}, \ldots, \hat{\pi}_{J|i})$, asymptotically follows the multivariate normal distribution with mean vector $\mathbf{0}$ and covariance matrix with the diagonal terms equal to $\pi_{j|i}(1 - \pi_{j|i})$, and off-diagonal terms equal to $-\pi_{j|i}\pi_{j'|i}$ for $j \neq j'$. To estimate $\mathcal{G}$, we may use the MLE $\hat{\mathcal{G}} = \hat{\Pi}_c/\hat{\Pi}_d$, where $\hat{\Pi}_c = \sum_{r=1}^{J-1}\sum_{s=r+1}^{J} \hat{\pi}_{r|1}\hat{\pi}_{s|2}$ and $\hat{\Pi}_d = \sum_{r=2}^{J}\sum_{s=1}^{r-1} \hat{\pi}_{r|1}\hat{\pi}_{s|2}$. Furthermore, using the delta method (see the Appendix), we can show that the asymptotic variance of $\hat{\mathcal{G}}$ is (**Exercise 6.1**; Agresti, 1980)

$$\mathrm{Var}(\hat{\mathcal{G}}) = \frac{\sum_{r=1}^{J}\left[\sum_{s=r+1}^{J}\pi_{s|2} - \mathcal{G}\sum_{s=1}^{r-1}\pi_{s|2}\right]^2 \pi_{r|1}}{n_1\Pi_d^2} + \frac{\sum_{s=1}^{J}\left[\sum_{r=1}^{s-1}\pi_{r|1} - \mathcal{G}\sum_{r=s+1}^{J}\pi_{r|1}\right]^2 \pi_{s|2}}{n_2\Pi_d^2}. \quad (6.1)$$

Note that by convention we define $\sum_{j=J+1}^{J} \pi_{j|i} = 0$ and $\sum_{j=1}^{0} \pi_{j|i} = 0$ for $i = 1, 2$ in (6.1). Note also that the variance $\mathrm{Var}(\hat{\mathcal{G}})$, which is a function of unknown parameters, cannot be used in practice. We can substitute $\hat{\pi}_{j|i}$ for $\pi_{j|i}$, $\hat{\mathcal{G}}$ for $\mathcal{G}$, and $\hat{\Pi}_d$ for Π_d to estimate $\mathrm{Var}(\hat{\mathcal{G}})$, and we denote the resulting estimator by $\widehat{\mathrm{Var}}(\hat{\mathcal{G}})$. Thus, as both n_i are large, we obtain an asymptotic $100(1 - \alpha)$ percent confidence interval for the GOR given by

$$[\max\{\hat{\mathcal{G}} - Z_{\alpha/2}\sqrt{\widehat{\mathrm{Var}}(\hat{\mathcal{G}})}, 0\}, \hat{\mathcal{G}} + Z_{\alpha/2}\sqrt{\widehat{\mathrm{Var}}(\hat{\mathcal{G}})}], \quad (6.2)$$

where Z_α is the upper 100αth percentile of the standard normal distribution. When neither n_i is large, the sampling distribution $\hat{\mathcal{G}}$ can be skewed and hence (6.2) may not perform well. To improve the normal approximation of the statistic $\hat{\mathcal{G}}$, we may consider use of the logarithmic transformation (Agresti, 1980; Katz *et al.*, 1978). This leads to a $100(1-\alpha)$ percent asymptotic confidence interval for the GOR given by (**Exercise 6.2**).

$$[\hat{\mathcal{G}}\exp(-Z_{\alpha/2}\sqrt{\widehat{\text{Var}}(\hat{\mathcal{G}})}/\hat{\mathcal{G}}),\ \hat{\mathcal{G}}\exp(Z_{\alpha/2}\sqrt{\widehat{\text{Var}}(\hat{\mathcal{G}})}/\hat{\mathcal{G}})]. \tag{6.3}$$

Similarly, to avoid having to base inference on the possibly skewed sampling distribution of $\widehat{\text{GOR}}$ when the n_i are not large, we may also consider a method analogous to the idea of using Fieller's theorem (Casella and Berger, 1990). Define $Z = \hat{\Pi}_c - \mathcal{G}\hat{\Pi}_d$. We can easily see that the expectation of $E(Z) = 0$ (**Exercise 6.3**). Furthermore, if both n_i were large, we would have the probability $P((\hat{\Pi}_c - \mathcal{G}\hat{\Pi}_d)^2/\text{Var}(\hat{\Pi}_c - \mathcal{G}\hat{\Pi}_d) \leq Z^2_{\alpha/2}) \doteq 1-\alpha$. This leads to the following quadratic equation in $\mathcal{G}$ (**Exercise 6.4**):

$$A\mathcal{G}^2 - 2B\mathcal{G} + C \leq 0, \tag{6.4}$$

where

$$A = \hat{\Pi}_d^2 - Z^2_{\alpha/2}\left[\sum_{r=1}^{J}\left(\sum_{s=1}^{r-1}\hat{\pi}_{s|2}\right)^2\hat{\pi}_{r|1}/n_1 + \sum_{s=1}^{J}\left(\sum_{r=s+1}^{J}\hat{\pi}_{r|1}\right)^2\hat{\pi}_{s|2}/n_2\right],$$

$$B = \hat{\Pi}_c\hat{\Pi}_d - Z^2_{\alpha/2}\left[\sum_{r=1}^{J}\left(\sum_{s=1}^{r-1}\hat{\pi}_{s|2}\right)\left(\sum_{s=r+1}^{J}\hat{\pi}_{s|2}\right)\hat{\pi}_{r|1}/n_1\right.$$
$$\left.+\sum_{s=1}^{J}\left(\sum_{r=s+1}^{J}\hat{\pi}_{r|1}\right)\left(\sum_{r=1}^{s-1}\hat{\pi}_{r|1}\right)\hat{\pi}_{s|2}/n_2\right],$$

$$C = \hat{\Pi}_c^2 - Z^2_{\alpha/2}\left[\sum_{r=1}^{J}\left(\sum_{s=r+1}^{J}\hat{\pi}_{s|2}\right)^2\hat{\pi}_{r|1}/n_1 + \sum_{s=1}^{J}\left(\sum_{r=1}^{s-1}\hat{\pi}_{r|1}\right)^2\hat{\pi}_{s|2}/n_2\right].$$

Therefore, if $A > 0$ and $B^2 - AC > 0$, then an asymptotic $100(1-\alpha)$ percent confidence interval for GOR would be

$$[\max\{(B - \sqrt{B^2 - AC})/A, 0\}, (B + \sqrt{B^2 - AC})/A]. \tag{6.5}$$

Note that following similar arguments to those in Section 5.2, it is straightforward to extend the above results on estimation of the GOR to accommodate a series of independent $2 \times J$ tables (Agresti, 1980).

Example 6.1 Consider the data (Holmes and Williams, 1954) on the tonsil size for carriers and non-carriers of *S. pyogenes* analyzed by both Clayton (1974) and Agresti (1980). In the group of 1326 non-carriers, there were 497 subjects with tonsils present but not enlarged, 560 subjects with tonsils enlarged, and 269 subjects with tonsils greatly enlarged. In the group of 72 carriers, there were 19 subjects with tonsils present but not enlarged, 29 subjects with tonsils enlarged, and 24 subjects with tonsils greatly enlarged. The MLE $\hat{\mathcal{G}}$ of the GOR is 1.689. This suggests that there are 1.69 times as many carrier–non-carrier pairs in the sample for which the carrier has the larger tonsils as there are pairs for which the non-carrier has the larger tonsils. Applying interval estimators (6.2), (6.3), and (6.5), we obtain 95% confidence intervals for the GOR of [1.006, 2.372], [1.127, 2.531], and [1.121, 2.613], respectively. We can see that the lower limits are all above 1, and we may conclude that there is a significant association at the 5% level between tonsil size and *S. pyogenes*. Furthermore, we can see that the interval estimate based on (6.2) tends to be shifted to the left compared to the other two estimates.

6.2 DATA WITH REPEATED MEASUREMENTS (OR UNDER CLUSTER SAMPLING)

In many biomedical or epidemiological studies, we may have more than one measurement per subject. For example, consider the lung transplant study (Song *et al.*, 1997) in which the investigators were interested in assessing whether one method of surgery resulted in better healing after transplant than the other. Healing was measured by the degree of inflammation (recorded as: no, some, much, and very much inflammation) on two sites of surgery for each rabbit. The data are summarized in Table 6.1. We wish to compare the distributions of inflammation grade for the two methods of surgery. Because measurements taken from the same rabbit are likely correlated, interval estimators of the GOR for independent multinomial sampling discussed in Section 6.1 for mutually independent responses are inadequate for use in the situations discussed here. We discuss how to extend the discussion on interval estimation of GOR to accommodate the data with repeated measurements per subject in this section.

Consider a trial in which we randomly assign n_1 and n_2 subjects to receive treatments 1 and 2, respectively. Suppose that on subject $k (k = 1, 2, \ldots, n_i)$ in treatment $i (i = 1, 2)$, we take m_{ik} repeated measurements. Suppose further that the measurements are on an ordinal scale with J ordered categories. Define $X_{ijkl} = 1$ if the lth $(l = 1, 2, \ldots, m_{ik})$ measurement on the kth subject in treatment $i (i = 1, 2)$ falls into category j, and $X_{ijkl} = 0$ otherwise. We assume that the probability $\mathrm{P}(X_{ijkl} = 1) = p_{jk|i}$ and $\mathrm{P}(X_{ijkl} = 0) = 1 - p_{jk|i}$, where $\sum_{j=1}^{J} p_{jk|i} = 1$. Let $X_{ijk.} = \sum_l X_{ijkl}$. Given $\mathbf{p}'_{ik} = (p_{1k|i}, p_{2k|i}, p_{3k|i}, \ldots, p_{Jk|i})$,

Table 6.1 Inflammation grade (0 = no; 1 = some; 2 = much; and 3 = serious) from repeated measurements on rabbits receiving two types of surgery.

Surgery	
A	(0, 0), (0, 0), (0, 0), (1, 1), (1, 1), (1, 1), (2, 1), (2, 1), (2, 2), (2, 2), (3, 2), (3, 2)
B	(1, 1), (1, 1), (1, 1), (1, 1), (1, 2), (1, 2), (1, 2), (1, 2), (2, 2), (2, 2), (2, 2), (2, 2), (2, 2), (3, 2)

Source: Jung and Kang (2001).

the random vector $\mathbf{X}'_{ik} = (X_{i1k.}, X_{i2k.}, X_{i3k.}, \ldots, X_{iJk.})$ follows the multinomial distribution with parameters m_{ik} and $\mathbf{p}'_{ik}$. To account for the intraclass correlation between repeated measurements within subjects, we further assume that $\mathbf{p}'_{ik} = (p_{1k|i}, p_{2k|i}, p_{3k|i}, \ldots, p_{Jk|i})$ independently identically follow the multivariate version of the beta distribution, known as:

$$\frac{\Gamma(T_i)}{\prod_j \Gamma(\alpha_{ij})} \prod_j (p_{jk|i})^{\alpha_{ij}-1}, \tag{6.6}$$

where $T_i = \sum_{j=1}^{J} \alpha_{ij}$, $\alpha_{ij} > 0$ for $i = 1, 2$, and $j = 1, 2, \ldots, J$ (Johnson and Kotz, 1970; Lui, 1991). A discussion of some of the good statistical properties of the Dirichlet-multinomial distribution appears elsewhere (Lui, 2000). Under the above model assumptions, we can easily show that the expectation $\mathrm{E}(X_{ijkl}) = \mathrm{E}(\mathrm{E}(X_{ijkl}|P_{jk|i})) = \pi_{j|i}$, where $\pi_{j|i} = \alpha_{ij}/T_i$. We can further show that the intraclass correlation ρ_i between repeated measurements X_{ijkl} and $X_{ijkl'}$ for $l \neq l'$ is simply $1/(T_i + 1)$ (**Exercise 1.7**). Define $\boldsymbol{\pi}'_i = (\pi_{1|i}, \pi_{2|i}, \pi_{3|i}, \ldots, \pi_{J|i})$, the vector of the mean response distribution over the J ordered categories for treatment i, where $i = 1, 2$. As defined previously, the GOR is simply $\mathcal{G} = \Pi_c/\Pi_d$, where $\Pi_c = \sum_{r=1}^{J-1} \sum_{s=r+1}^{J} \pi_{r|1}\pi_{s|2}$ and $\Pi_d = \sum_{r=2}^{J} \sum_{s=1}^{r-1} \pi_{r|1}\pi_{s|2}$.

First, note that $\hat{\pi}_{j|i} = X_{ij..}/m_{i.}$, where $X_{ij..} = \sum_k X_{ijk.}$ and $m_{i.} = \sum_k m_{ik}$, is an unbiased estimator of $\pi_{j|i}$. When $m_{i.}$ goes to ∞, $\sqrt{m_{i.}}(\hat{\boldsymbol{\pi}}_i - \boldsymbol{\pi}_i)$, where $\hat{\boldsymbol{\pi}}'_i = (\hat{\pi}_{1|i}, \hat{\pi}_{2|i}, \hat{\pi}_{3|i}, .., \hat{\pi}_{J|i})$, asymptotically follows the multivariate normal distribution with mean vector $\mathbf{0}$ and covariance matrix with diagonal terms equal to $\pi_{j|i}(1 - \pi_{j|i})f(\mathbf{m}_i, \rho_i)$, and off-diagonal terms equal to $-\pi_{j|i}\pi_{j'|i}f(\mathbf{m}_i, \rho_i)$, where $f(\mathbf{m}_i, \rho_i) = \sum_k m_{ik}[1 + (m_{ik} - 1)\rho_i]/m_{i.}$ is the variance inflation factor due to the intraclass correlation ρ_i, and $\mathbf{m}'_i = (m_{i1}, m_{i2}, \ldots, m_{in_i})$. As noted in previous chapters, the component $f(\mathbf{m}_i, \rho_i)$ is always at least 1. When $\rho_i = 0$ or $m_{ik} = 1$ for k, $f(\mathbf{m}_i, \rho_i)$ reduces to 1. A consistent estimator of $\mathcal{G}$ under cluster sampling is $\hat{\mathcal{G}} = \hat{\Pi}_c/\hat{\Pi}_d$, where $\hat{\Pi}_c = \sum_{r=1}^{J-1} \sum_{s=r+1}^{J} \hat{\pi}_{r|1}\hat{\pi}_{s|2}$ and $\hat{\Pi}_d = \sum_{r=2}^{J} \sum_{s=1}^{r-1} \hat{\pi}_{r|1}\hat{\pi}_{s|2}$. Furthermore, using the delta method, we can show that an estimated asymptotic variance of $\hat{\mathcal{G}}$ is

$$\widehat{\mathrm{Var}}(\hat{\mathcal{G}}) = f(\mathbf{m}_1, \hat{\rho}_1)\frac{\sum_{r=1}^{J}\left[\sum_{s=r+1}^{J} \hat{\pi}_{s|2} - \hat{\mathcal{G}}\sum_{s=1}^{r-1} \hat{\pi}_{s|2}\right]^2 \hat{\pi}_{r|1}}{m_{1.}\hat{\Pi}_d^2} + f(\mathbf{m}_2, \hat{\rho}_2)\frac{\sum_{s=1}^{J}\left[\sum_{r=1}^{s-1} \hat{\pi}_{r|1} - \hat{\mathcal{G}}\sum_{r=s+1}^{J} \hat{\pi}_{r|1}\right]^2 \hat{\pi}_{s|2}}{m_{2.}\hat{\Pi}_d^2}, \tag{6.7}$$

where $\hat{\rho}_i = (\mathrm{BMS}_i - \mathrm{WMS}_i)/[\mathrm{BMS}_i + (m_i^* - 1)\mathrm{WMS}_i]$, $\mathrm{WMS}_i = \sum_k [m_{ik} - \sum_j (X_{ijk.}^2/m_{ik})]/\sum_k (m_{ik} - 1)$, $\mathrm{BMS}_i = \left[\sum_k \sum_j (X_{ijk.}^2/m_{ik}) - \left(\sum_j X_{ij..}^2/m_{i.}\right)\right]/(n_i - 1)$, and $m_i^* = \left(m_{i.}^2 - \sum_k m_{ik}^2\right)/[(n_i - 1)m_{i.}]$. When $m_{ik} = 1$ for all i and k, the estimated variance $\widehat{\mathrm{Var}}(\hat{\mathcal{G}})$ in (6.7) simplifies to that for a single measurement per subject. A discussion of interval estimation of the intraclass correlation under the Dirichlet-multinomial distribution appears elsewhere (Lui *et al.*, 1999). Based on (6.7), we obtain an asymptotic $100(1-\alpha)$ percent confidence interval for the GOR given by

$$[\max\{\hat{\mathcal{G}} - Z_{\alpha/2}\sqrt{\widehat{\mathrm{Var}}(\hat{\mathcal{G}})}, 0\}, \hat{\mathcal{G}} + Z_{\alpha/2}\sqrt{\widehat{\mathrm{Var}}(\hat{\mathcal{G}})}]. \tag{6.8}$$

As noted before, if both n_i are not large, the sampling distribution of $\hat{\mathcal{G}}$ may be skewed and hence interval estimator (6.8) may not perform well. To improve the normal approximation of the statistic $\hat{\mathcal{G}}$, we may again consider using the logarithmic transformation. Using the delta method, we obtain an $100(1-\alpha)$ percent asymptotic confidence interval for GOR given by

$$[\hat{\mathcal{G}}\exp(-Z_{\alpha/2}\sqrt{\widehat{\mathrm{Var}}(\hat{\mathcal{G}})}/\hat{\mathcal{G}}), \hat{\mathcal{G}}\exp(Z_{\alpha/2}\sqrt{\widehat{\mathrm{Var}}(\hat{\mathcal{G}})}/\hat{\mathcal{G}})]. \tag{6.9}$$

Similarly, we may look to Fieller's theorem as for deriving interval estimator (6.5). Define $Z = \hat{\Pi}_c - \mathcal{G}\hat{\Pi}_d$. Because $E(X_{ijkl}) = E(P_{jk|i}) = \pi_{j|i}$, we can easily see that $E(Z) = 0$. Furthermore, if both n_i are large, we have the probability $P((\hat{\Pi}_c - \mathcal{G}\hat{\Pi}_d)^2/\mathrm{Var}(\hat{\Pi}_c - \mathcal{G}\hat{\Pi}_d) \leq Z^2_{\alpha/2}) \doteq 1-\alpha$. This leads to the following quadratic equation in $\mathcal{G}$:

$$A^\dagger(\mathcal{G})^2 - 2B^\dagger(\mathcal{G}) + C^\dagger \leq 0, \tag{6.10}$$

where

$$A^\dagger = \hat{\Pi}_d^2 - Z^2_{\alpha/2}\left[f(\mathbf{m}_1, \hat{\rho}_1)\sum_{r=1}^{J}\left(\sum_{s=1}^{r-1}\hat{\pi}_{s|2}\right)^2 \hat{\pi}_{r|1}/m_1.\right.$$
$$\left. + f(\mathbf{m}_2, \hat{\rho}_2)\sum_{s=1}^{J}\left(\sum_{r=s+1}^{J}\hat{\pi}_{r|1}\right)^2 \hat{\pi}_{s|2}/m_2.\right],$$

$$B^\dagger = \hat{\Pi}_c\hat{\Pi}_d - Z^2_{\alpha/2}\left[f(\mathbf{m}_1, \hat{\rho}_1)\sum_{r=1}^{J}\left(\sum_{s=1}^{r-1}\hat{\pi}_{s|2}\right)\left(\sum_{s=r+1}^{J}\hat{\pi}_{s|2}\right)\hat{\pi}_{r|1}/m_1.\right.$$
$$\left. + f(\mathbf{m}_2, \hat{\rho}_2)\sum_{s=1}^{J}\left(\sum_{r=s+1}^{J}\hat{\pi}_{r|1}\right)\left(\sum_{r=1}^{s-1}\hat{\pi}_{r|1}\right)\hat{\pi}_{s|2}/m_2.\right],$$

$$C^\dagger = \hat{\Pi}_c^2 - Z^2_{\alpha/2}\left[f(\mathbf{m}_1, \hat{\rho}_1)\sum_{r=1}^{J}\left(\sum_{s=r+1}^{J}\hat{\pi}_{s|2}\right)^2 \hat{\pi}_{r|1}/m_1.\right.$$
$$\left. + f(\mathbf{m}_2, \hat{\rho}_2)\sum_{s=1}^{J}\left(\sum_{r=1}^{s-1}\hat{\pi}_{r|1}\right)^2 \hat{\pi}_{s|2}/m_2.\right].$$

Therefore, if $A^\dagger > 0$ and $B^{\dagger 2} - A^\dagger C^\dagger > 0$, then an asymptotic $100(1-\alpha)$ percent confidence interval for the GOR would be

$$[\max\{(B^\dagger - \sqrt{B^{\dagger 2} - A^\dagger C^\dagger})/A^\dagger, 0\}, (B^\dagger + \sqrt{B^{\dagger 2} - A^\dagger C^\dagger})/A^\dagger]. \qquad (6.11)$$

Note that when $m_{ik} = 1$ for all i and k, interval estimators (6.8), (6.9), (6.11) reduce to (6.2), (6.3), and (6.5), respectively. Using Monte Carlo simulation, Lui (2002a) compares and evaluates the performance of interval estimators (6.8), (6.9), and (6.11). Lui notes that when the number of subjects n_0 per group is not large, using (6.8) tends to produce an interval estimate with coverage probability less than the desired confidence level, while using either (6.9) or (6.11) can produce an interval estimate with coverage probability approximately equal to or larger than the desired confidence level even when n_0 is not large. Furthermore, Lui finds that using (6.11) can cause loss of efficiency as compared with using (6.9), especially when n_0 is not large and ρ_0 is large. When n_0 is large (at least 100), all three interval estimators (6.8), (6.9), and (6.11) are all essentially equivalent with respect to coverage probability and average length. Lui also evaluates the performance of (6.8), (6.9), and (6.11) under some alternatives to the Dirichlet distribution, and finds good robustness for all three estimators in a variety of situations.

Example 6.2 Consider comparing the distribution of inflammation grade after transplant on 12 rabbits receiving surgery A with that on 14 rabbits receiving surgery B (Jung and Kang, 2001). For each animal, we take a pair ($m_0 = 2$) of measurements on the inflammation grade, divided into four ($J = 4$) categories: 0 = no; 1 = some; 2 = much; and 3 = serious (Table 6.1). Given these data, we obtain the estimates $\hat{\mathcal{G}} = 2.027$ and $\hat{\rho}_0 = 0.45 (= (m_{1.}\hat{\rho}_1 + m_{2.}\hat{\rho}_2)/(m_{1.} + m_{2.}))$. Applying interval estimators (6.8), (6.9), and (6.11), we obtain 95% confidence intervals for GOR of [0.00, 4.33], [0.65, 6.33], and [0.43, 82.19]. We can see that interval estimate (6.8) using Wald's statistic is shifted to the left compared to the other two interval estimates. We can also see that using (6.11) can leads to substantial loss of efficiency as compared with using (6.9) in the particular situation considered in this example. Note that the estimate $\hat{\mathcal{G}} = 2.027$ suggests the probability for 'the inflammation grade of a randomly selected animal from surgery A less than the inflammation grade of a randomly selected animal from surgery B' is approximately twice of the probability for "the inflammation grade of a randomly selected animal receiving surgery B less than the inflammation grade of a randomly selected animal from surgery A". However, because the numbers of subjects in both comparison groups are small, all the above interval estimates are wide and cover 1. Thus, there is no significant evidence at the 5% level against the null hypothesis that the distributions of inflammation grade between interventions A and B are equal.

6.3 PAIRED-SAMPLE DATA

In epidemiologic studies or clinical trials it is common to employ a matching design to increase efficiency. For example, consider the paired-sample data (Breslow, 1982) from a study of the number of beverages drunk at 'burning hot' temperature and esophageal cancer. Because the number of beverages ranges from 0 to 3, the statistics commonly used for dichotomous exposure in paired-sample data are not directly applicable here. To account for multiple levels of exposure in studies with matched pairs, we may apply the conditional logistic regression model as proposed elsewhere (Breslow, 1982). In this section, we concentrate our attention on estimates of the GOR that may be used without the need to assume a parametric model.

Consider a data set consisting of n matched pairs of subjects with ordinal responses in J categories. Within each pair, suppose that exactly one subject is taken from one of the two comparison groups. Let $\pi_{rs}(r, s = 1, 2, \ldots, J)$ denote the probability of obtaining a pair where the subject from group 1 has response in category r and the subject from group 2 has response in category s. Define $\boldsymbol{\pi}' = (\pi_{11}, \ldots, \pi_{1J}, \pi_{21}, \ldots \pi_{2J}, \ldots, \pi_{J1}, \pi_{J2}, \ldots, \pi_{JJ})$. Let N_{rs} (r and $s = 1, 2, \ldots, J$) denote the observed frequency of pairs falling in cell (r, s). Then the random vector $\mathbf{N}' = (N_{11}, N_{12}, \ldots, N_{1J}, N_{21}, \ldots, N_{2J}, \ldots, N_{J1}, N_{J2}, \ldots, N_{JJ})$ follows the multinomial distribution with parameters n $\left(= \sum_r \sum_s N_{rs}\right)$ and $\boldsymbol{\pi}$. Without loss of generality, we assume that exposure level 1 represents non-exposure. We further assume that the higher the level of exposure, the more severe is the extent of exposure. Define $\pi_c = \sum_{r=1}^{J-1} \sum_{s=r+1}^{J} \pi_{rs}$, the probability of a randomly selected matched pair in which the subject from group 2 has a higher level of exposure than the subject from group 1. Similarly, define $\pi_d = \sum_{r=2}^{J} \sum_{s=1}^{r-1} \pi_{rs}$, the probability of a randomly selected matched pair in which the subject from group 1 has a higher level than the subject from group 2. Thus, we may define the GOR for paired-sample data as $\mathfrak{G} = \pi_c/\pi_d$. Note that the GOR defined here for paired-sample data is actually Agresti's α' (Agresti, 1980). Note also that when $J = 2$, the GOR reduces to the common definition of OR for dichotomous exposure in paired-sample data, π_{12}/π_{21}.

Under the multinomial distribution, the MLE of π_{rs} is $\hat{\pi}_{rs} = N_{rs}/n$. Thus, the MLE of the GOR is $\hat{\mathfrak{G}} = \hat{\pi}_c/\hat{\pi}_d$, where $\hat{\pi}_c = \sum_{r=1}^{J-1} \sum_{s=r+1}^{J} \hat{\pi}_{rs}$ and $\hat{\pi}_d = \sum_{r=2}^{J} \sum_{s=1}^{r-1} \hat{\pi}_{rs}$. Note that the MLE $\hat{\mathfrak{G}}$ depends on the random vector $\mathbf{N}$ only through the two random subtotals: $N_c = \sum_{r=1}^{J-1} \sum_{s=r+1}^{J} N_{rs}$ and $N_d = \sum_{r=2}^{J} \sum_{s=1}^{r-1} N_{rs}$. Thus, when studying the sampling distribution of $\hat{\mathfrak{G}}$, we can consider the joint marginal probability mass function of N_c and N_d:

$$f_{N_c, N_d}(n_c, n_d) = \frac{n!}{n_c! n_d! (n - n_c - n_d)!} \pi_c^{n_c} \pi_d^{n_d} (1 - \pi_c - \pi_d)^{(n - n_c - n_d)}, \qquad (6.12)$$

which is simply a trinomial distribution. Applying the delta method (Agresti, 1990), we obtain that the estimated asymptotic variance of $\hat{\mathfrak{G}}$ is given by $\widehat{\mathrm{Var}}(\hat{\mathfrak{G}}) =$

$\hat{\pi}_c(\hat{\pi}_c + \hat{\pi}_d)/(n\hat{\pi}_d^3)$ (**Exercise 6.5**). Therefore, an asymptotic $100(1-\alpha)$ percent confidence interval for the GOR using Wald's test statistic is

$$[\max\{\hat{\mathfrak{G}} - Z_{\alpha/2}\sqrt{\widehat{\mathrm{Var}}(\hat{\mathfrak{G}})}, 0\}, \hat{\mathfrak{G}} + Z_{\alpha/2}\sqrt{\widehat{\mathrm{Var}}(\hat{\mathfrak{G}})}]. \tag{6.13}$$

Note that $\hat{\mathfrak{G}}(=\hat{\pi}_c/\hat{\pi}_d)$ is a ratio of two sample proportions, hence the sampling distribution of $\hat{\mathfrak{G}}$ is probably skewed when n is not large. To improve the normal approximation of the statistic $\hat{\mathfrak{G}}$, we may use the logarithmic transformation (Katz *et al.*, 1978). This leads to an asymptotic $100(1-\alpha)$ percent confidence interval for $\mathfrak{G}$ given by

$$[\hat{\mathfrak{G}}\exp(-Z_{\alpha/2}\sqrt{\widehat{\mathrm{Var}}(\log(\hat{\mathfrak{G}}))}), \hat{\mathfrak{G}}\exp(Z_{\alpha/2}\sqrt{\widehat{\mathrm{Var}}(\log(\hat{\mathfrak{G}}))})], \tag{6.14}$$

where $\widehat{\mathrm{Var}}(\log(\hat{\mathfrak{G}})) = (\hat{\pi}_c + \hat{\pi}_d)/(n\hat{\pi}_c\hat{\pi}_d)$.

On the basis of the trinomial distribution (6.12), we can easily show that the conditional probability mass function of $N_c = n_c$, given $N_c + N_d = n_t$ follows the binomial distribution (**Exercise 5.14**):

$$f(N_c = n_c|n_t) = \frac{n_t!}{n_c!(n_t - n_c)!}\left(\frac{\mathfrak{G}}{\mathfrak{G}+1}\right)^{n_c}\left(\frac{1}{\mathfrak{G}+1}\right)^{n_t-n_c}. \tag{6.15}$$

Define $Q = \mathfrak{G}/(\mathfrak{G}+1)$. From (6.15), if n_t is large, the probability $P((\hat{Q} - Q)^2/[Q(1-Q)/n_t] \le Z^2_{\alpha/2}|n_t) \doteq 1-\alpha$, where $\hat{Q} = N_c/n_t$. Therefore, we may obtain an asymptotic $100(1-\alpha)$ percent confidence interval $[Q_l, Q_u]$ for Q, where Q_l and Q_u are the two distinct roots of $A^\ddagger Q^2 - 2B^\ddagger Q + C^\ddagger = 0$, where $A^\ddagger = (1 + Z^2_{\alpha/2}/n_t)$, $B^\ddagger = \hat{Q} + Z^2_{\alpha/2}/(2n_t)$, and $C^\ddagger = \hat{Q}^2$ (see (1.5)). Note that $f(Q) = Q/(1-Q) = \mathfrak{G}$, which is a monotonically increasing function of Q. These lead to an asymptotic $100(1-\alpha)$ percent confidence interval for the GOR given by

$$\left[\frac{Q_l}{1-Q_l}, \frac{Q_u}{1-Q_u}\right]. \tag{6.16}$$

On the other hand, when n_t is not large, we may employ the exact $100(1-\alpha)$ percent confidence limits (Clopper and Pearson, 1934; see (1.6)) for Q and then apply the monotonic transformation $f(Q)$ again. Thus, we obtain an exact $100(1-\alpha)$ percent confidence interval for $\mathfrak{G}$,

$$\left[\frac{N_c}{(N_d+1)F_{2(N_d+1),2N_c,\alpha/2}}, \frac{(N_c+1)F_{2(N_c+1),2N_d,\alpha/2}}{N_d}\right], \tag{6.17}$$

where $F_{f_1,f_2,\alpha}$ is the upper 100αth percentile of the central F distribution with f_1 and f_2 degrees of freedom. Note that when $N_c = 0$, the upper percentile $F_{2(n_t-N_c+1),2N_c,\alpha/2}$ in (6.17) is not defined. In this case, we define the lower limit of (6.17) to be 0. Similarly, when $N_d = 0$, we define the upper limit of (6.17) to be ∞. Note that applying interval estimator (6.17) can always ensure that the coverage probability is greater than or equal to the desired confidence level $1-\alpha$.

Example 6.3 Consider Breslow's (1982) data consisting of 80 matched pairs taken from a case–control study of esophageal cancer. The response is the number of beverages (ranging from 0 to 3) reported drunk at 'burning hot' temperatures. Suppose that we want to estimate the GOR of the number of beverages drunk at 'burning hot' temperature and the risk of having esophageal cancer. Here, the random number $N_{rs}(r, s = 1, 2, \ldots, 4)$ denotes the observed frequency of pairs in which the numbers of beverages reported drunk at 'burning hot' temperature are $r - 1$ and $s - 1$ for case and control, respectively. Given the data (Table 6.2), we obtain the number of matched pairs in which the case has a higher level of exposure than the control, $n_c = 35$, and the number of pairs in which the control has a higher level of exposure than the case, $n_d = 11$. The MLE of the GOR is $\hat{\mathfrak{G}} = 3.18(= 35/11)$. Applying interval estimators (6.13), (6.14), (6.16), and (6.17), we obtain 95% confidence intervals for GOR of [1.026, 5.337], [1.616, 6.265], [1.636, 6.188], and [1.580, 6.945], respectively because the lower limits of these intervals are all greater than 1, there is evidence at the 5% level of an association between the number of beverages drunk at 'burning hot' temperature and esophageal cancer. Note that using (6.13) based on Wald's test statistic seems to produce an interval estimate that is shifted to the left compared to all the other interval estimates. Note also that since using (6.17) can always ensure that the coverage probability is equal to or greater than the desired confidence level, it is not surprising to see that the interval derived from (6.17) is longer than the other intervals considered here.

Recall that exposure level 1 represents non-exposure. If we collapsed all the categories with exposure levels 2 or greater in one category, then we would obtain a 2×2 table with cell probabilities $\pi^*_{11} = \pi_{11}, \pi^*_{12} = \sum_{s=2}^{J} \pi_{1s}, \pi^*_{21} = \sum_{r=2}^{J} \pi_{r1}$, and $\pi^*_{22} = \sum_{r\geq 2}^{J} \sum_{s\geq 2}^{J} \pi_{rs}$. Note that π^*_{11} represents the probability of a randomly selected pair with both case and control unexposed; π^*_{12} represents the probability of a randomly selected pair with case exposed and control unexposed; π^*_{21} represents the probability of a randomly selected pair with control exposed and

Table 6.2 Observed frequency of matched pairs cross-classified by the number of beverages drunk at 'burning hot' temperatures between esophageal cancer patients and their matched controls.

	Case			
Control	0	1	2	3
0	31	12	14	6
1	5	1	1	1
2	5	0	2	1
3	0	0	1	0

Source: Breslow (1982).

case unexposed; and π^*_{22} represents the probability of a randomly selected pair with both case and control exposed. The OR of the probability of exposure (ever versus never) for cases versus controls is then equal to π^*_{12}/π^*_{21} $\left(= \sum_{s=2}^{J} \pi_{1s}/\sum_{r=2}^{J} \pi_{r1}\right)$, which is different from the GOR defined here. Under the multinomial distribution, the MLE of the OR is $\hat{\mathcal{O}} = \sum_{s=2}^{J} \hat{\pi}_{1s}/\sum_{r=2}^{J} \hat{\pi}_{r1}$, which completely ignores the information on $\sum_{r=2}^{J-1}\sum_{s=r+1}^{J} n_{rs}$ and $\sum_{r=3}^{J}\sum_{s=2}^{r-1} n_{rs}$. Thus, using the statistic related to $\hat{\mathcal{O}}$ is expected to be less efficient than using the statistic related to $\hat{\mathfrak{G}}$. To illustrate the point here, consider the matched-pair case–control study of esophageal cancer (Example 6.3). The MLE of the OR $(= \pi^*_{12}/\pi^*_{21})$ is $\hat{\mathcal{O}} = 3.2$. Applying interval estimators similar to (6.13), (6.14), (6.16), and (6.17) to estimate OR, we obtain [0.928, 5.472], [1.573, 6.509], [1.596, 6.418], and [1.535, 7.298], respectively. A comparison of these interval estimates with those obtained previously for the GOR reveals that every interval estimate in the former is longer than the corresponding one in the latter. The ratio of the GOR interval length to the corresponding OR interval length ranges from 0.93 to 0.95. In other words, using the test statistic related to $\hat{\mathfrak{G}}$ is likely to be more efficient than using the test statistic related to $\hat{\mathcal{O}}$ in detecting the association between the number of beverages drunk at 'burning hot' temperature and esophageal cancer. In fact, Lui (2002b) provides algebraic arguments to explain why this is generally true with respect to the coefficient of variation.

6.4 MIXED NEGATIVE MULTINOMIAL AND MULTINOMIAL SAMPLING

Recall that under the multinomial distribution in the previous section on paired-sample data, the MLE of π_{rs} is $\hat{\pi}_{rs} = N_{rs}/n$. Thus, the MLE of the GOR is $\hat{\mathfrak{G}} = \hat{\pi}_c/\hat{\pi}_d$, where $\hat{\pi}_c = \sum_{r=1}^{J-1}\sum_{s=r+1}^{J} \hat{\pi}_{rs}$ and $\hat{\pi}_d = \sum_{r=2}^{J}\sum_{s=1}^{r-1} \hat{\pi}_{rs}$. Because there is a positive probability that the estimator $\hat{\pi}_d$ equals 0, the MLE $\hat{\mathfrak{G}}$ has neither finite expectation nor finite variance. Although we can always apply the *ad hoc* procedure of adding 0.50 to each cell whenever this occurs, it is difficult to claim any optimal statistical property for this procedure. However, we can avoid the need for this by employing inverse sampling (Haldane, 1945a, 1945b), in which we continue to collect subjects until we obtain a predetermined number n_d of subjects falling into the set of cells (r, s), $r = 2, \ldots, J$, $s = 1, \ldots, r-1$. The joint probability mass function for the random vector $\mathbf{N} = \mathbf{n}$ is then

$$f_{\mathbf{N}}(\mathbf{n}) = \frac{n_d!}{\prod_{r=2}^{J}\prod_{s=1}^{r-1} n_{rs}!} \prod_{r=2}^{J}\prod_{s=1}^{r-1} \left(\frac{\pi_{rs}}{\pi_d}\right)^{n_{rs}} \frac{(n_d + n_c + n_0 - 1)!}{(n_d - 1)!\prod_{u=1}^{J}\prod_{v=u}^{J} n_{uv}!} \pi_d^{n_d} \prod_{u=1}^{J}\prod_{v=u}^{J} \pi_{uv}^{n_{uv}}, \quad (6.18)$$

where $n_d = \sum_{r=2}^{J}\sum_{s=1}^{r-1} n_{rs}$, $n_c = \sum_{r=1}^{J-1}\sum_{s=r+1}^{J} n_{rs}$, $n_0 = \sum_{r=1}^{J} n_{rr}$, and $n_{uv} = 0, 1, 2, \ldots$, for $(u, v) \in \{(r, s) | r = 1, 2, \ldots, J, s = r, \ldots, J\}$. This is a mixed negative multinomial and multinomial distribution. Note that n_d is fixed, while n_c and n_0

are random in (6.18). Define $n = n_c + n_0 + n_d$. Note also that the likelihood of (6.18) is actually proportional to the likelihood of the corresponding multinomial distribution with the same realization of random vector $\mathbf{n}$. Therefore, the MLE of π_{rs}, given $\mathbf{N} = \mathbf{n}$, is simply $\hat{\pi}_{rs} = n_{rs}/n$, which is of the same form as that under the multinomial sampling. This also suggests that the MLE of the GOR under (6.18) is $\hat{\mathfrak{G}} = N_c/n_d$. Note that the MLE $\hat{\mathfrak{G}}$ depends on the sample vector $\mathbf{n}$ only through the marginal random subtotal n_c. Thus, we may study the sampling distribution of $\hat{\mathfrak{G}}$ on the basis of the marginal probability mass function of $N_c = n_c$ (**Exercise 5.25**):

$$f_{N_c}(n_c) = \frac{(n_d + n_c - 1)!}{(n_d - 1)!\, n_c!} \left(\frac{\pi_d}{\pi_d + \pi_c}\right)^{n_d} \left(\frac{\pi_c}{\pi_d + \pi_c}\right)^{n_c} \tag{6.19}$$

which is, in fact, the negative binomial distribution with parameters n_d and $\mathcal{Q} = \pi_d/(\pi_d + \pi_c)$. Under (6.19), we can easily show that the random variable N_c has expectation $E(N_c) = n_d\mathfrak{G}$ and variance $n_d\mathfrak{G}(\mathfrak{G} + 1)$, where $\mathfrak{G} = \pi_c/\pi_d$. This implies that the estimator $\hat{\mathfrak{G}}(= N_c/n_d)$ is an unbiased estimator of $\mathfrak{G}$ with variance $\mathfrak{G}(\mathfrak{G} + 1)/n_d$. In fact, following the same arguments as in Section 5.7, we can claim that $\hat{\mathfrak{G}}$ is the UMVUE of $\mathfrak{G}$.

Note that the probability for the cumulative distribution $P(N_c \leq n_c | n_d, \mathcal{Q})$, where N_c follows the negative binomial distribution (6.19) with parameters n_d and $\mathcal{Q}$, equals the probability $P(X \geq n_d | n_t, \mathcal{Q})$, where X follows the binomial distribution with parameters $n_t = n_c + n_d$ and $\mathcal{Q}$ (**Exercise 1.14**). Therefore, for a given observed value n_c, an exact $100(1 - \alpha)$ percent confidence interval for $\mathcal{O}$ (**Exercise 1.15**) is given by $[\mathcal{L}(N_c, n_d, \alpha/2), \mathcal{U}(N_c, n_d, \alpha/2)]$, where $\mathcal{L}(N_c, n_d, \alpha/2) = n_d/[n_d + (N_c + 1)F_{2(N_c+1),2n_d,\alpha/2}$, $\mathcal{U}(N_c, n_d, \alpha/2) = n_d F_{2n_d,2N_c,\alpha/2}/[n_d F_{2n_d,2N_c,\alpha/2} + N_c]$. Note that if N_c were 0, we would define the upper limit $\mathcal{U}(0, n_d, \alpha/2) = 1$ by convention. Furthermore, because $\mathcal{O} = (1 - \mathcal{P})/\mathcal{P}$ and the transformation $f(x) = (1 - x)/x$ is a strictly decreasing function, an exact $100(1 - \alpha)$ percent confidence interval for $\mathfrak{G}$ is given by

$$[(1 - \mathcal{U}(N_c, n_d, \alpha/2))/\mathcal{U}(N_c, n_d, \alpha/2), (1 - \mathcal{L}(N_c, n_d, \alpha/2))/\mathcal{L}(N_c, n_d, \alpha/2)]. \tag{6.20}$$

Applying interval estimator (6.20) can ensure that the coverage probability is always larger than or equal to the desired confidence level. Note that an exact test procedure for testing equality of the underlying OR to any specified value, an asymptotic test similar to McNemar's test, as well as calculation of sample size under (6.19) can be found elsewhere Lui (1996). A discussion on interval estimation of the generalized risk difference in paired-sample ordinal data is also given elsewhere (Lui, 2002b).

EXERCISE

6.1. Using the delta method, show that in Section 6.1, the asymptotic $\mathrm{Var}(\hat{\mathcal{G}})$ of $\hat{\mathcal{G}}$ is (6.1).

6.2. Show that when we use the logarithmic transformation, an asymptotic $100(1-\alpha)$ percent confidence interval for $\mathcal{G}$ is given by (6.3).

6.3. Show that in Section 6.1 the expectation E(Z) equals 0, where $Z = \hat{\Pi}_c - \mathcal{G}\hat{\Pi}_d$, $\hat{\Pi}_c = \sum_{r=1}^{J-1}\sum_{s=r+1}^{J}\hat{\pi}_{r|1}\hat{\pi}_{s|2}$ and $\hat{\Pi}_d = \sum_{r=2}^{J}\sum_{s=1}^{r-1}\hat{\pi}_{r|1}\hat{\pi}_{s|2}$.

6.4. From the result that $P((\hat{\Pi}_c - \mathcal{G}\hat{\Pi}_d)^2/\text{Var}(\hat{\Pi}_c - \mathcal{G}\hat{\Pi}_d) \le Z^2_{\alpha/2}) \doteq 1-\alpha$, derive the quadratic equation given in (6.4).

6.5. Consider the following data (Hedlund, 1978) on the political ideology of Democrats and Republicans (Agresti, 1990, p. 273). The numbers of Democrats falling into the categories liberal, moderate, and conservative were 143, 156, and 100. The corresponding numbers of Republicans were 15, 72, and 127. What is the MLE $\hat{\mathcal{G}}$? What are the 95% confidence intervals for the GOR using (6.2), (6.3), and (6.5)?

6.6. Consider the data (Smith, 1976) on education and attitudes toward abortion taken from the 1972 General Social Survey (Agresti, 1990, p. 197). We combine the data for subjects who have not completed high school and those who have completed high school into one category and compare the distribution of attitudes toward abortion for this combined category with that for subjects educated beyond high school level. In the former category, 360 generally disapprove of abortion, 227 take a middle position, and 663 generally approve; in the latter category, the corresponding numbers are 16, 21, and 138, respectively. What is the MLE $\hat{\mathcal{G}}$? What are the 95% confidence intervals for the GOR using (6.2), (6.3), and (6.5)?

REFERENCES

Agresti, A. (1980) Generalized odds ratios for ordinal data. *Biometrics*, **36**, 59–67.
Agresti, A. (1990) *Categorical Data Analysis*. Wiley, New York.
Breslow, N. (1982) Covariance adjustment of relative risk estimates in matched studies. *Biometrics*, **38**, 661–672.
Casella, G. and Berger, R. L. (1990) *Statistical Inference*. Wadsworth, Belmont, CA.
Clayton, D. G. (1974) Some odds ratio statistics for the analysis of ordered categorical data. *Biometrika*, **61**, 525–531.
Clopper, C. J. and Pearson, E. S. (1934) The use of confidence or fiducial limits illustrated in the case of the binomial. *Biometrika*, **26**, 404–413.
Edwardes, M. D. and Baltzan, M. (2000) The generalization of the odds ratio, risk ratio and risk difference to $r \times k$ tables. *Statistics in Medicine*, **19**, 1901–1914.
Goodman, L. A. and Kruskal, W. H. (1954) Measure of association for cross classification. *Journal of the American Statistical Association*, **49**, 732–764.
Haldane, J. B. S. (1945a) A labour-saving method of sampling. *Nature*, **155**, 49–50.
Haldane, J. B. S. (1945b) On a method of estimating frequencies. *Biometrika*, **33**, 222–225.
Hedlund, R. D. (1978) Cross-over voting in a 1976 presidential primary. *Public Opinion Quarterly*, **41**, 498–514.
Holmes, M. C. and Williams, R. E. O. (1954) The distribution of carriers of *Streptococcus pyogenes* among 2413 healthy children. *Journal of Hygiene (Cambridge)*, **52**, 165–179.
Hosmer, D. W. and Lemeshow, S. (1989) *Applied Logistic Regression*. Wiley, New York.

Johnson, N. L. and Kotz, S. (1970) *Distributions in Statistics: Continuous Univariate Distributions 2*. Wiley, New York.

Jung, S.-H. and Kang, S.-H. (2001) Tests for $2 \times K$ contingency tables with clustered ordered categorical data. *Statistics in Medicine*, **20**, 785–794.

Katz, D., Baptista, J., Azen, S. P. and Pike, M. C. (1978) Obtaining confidence intervals for risk ratio in cohort studies. *Biometrics*, **34**, 469–474.

Lui, K.-J. (1991) Sample size for repeated measurements in dichotomous data. *Statistics in Medicine*, **10**, 463–472.

Lui, K.-J. (1996) Notes in case–control studies with matched pairs under inverse sampling. *Biometrical Journal*, **38**, 681–693.

Lui, K.-J. (2000) Notes on life table analysis in correlated observations. *Biometrical Journal*, **42**, 93–110.

Lui, K.-J. (2002a) Interval estimation of generalized odds ratio in data with repeated measurements. *Statistics in Medicine*, **21**, 3107–3117.

Lui, K.-J. (2002b) Notes on estimation of the general odds ratio and the general risk difference for paired-sample data. *Biometrical Journal*, **44**, 957–968.

Lui, K.-J., Cumberland, W. G., Mayer, J. A. and Eckhardt, L. (1999) A note on interval estimation for the intraclass correlation in Dirichlet-multinomial data. *Psychometrika*, **64**, 355–369.

Smith, K. W. (1976) Table standardization and table shrinking: Aids in the traditional analysis of contingency tables. *Social Forces*, **54**, 669–693.

Song, W. Y., Lee, Y. J., Hwang, S. W., Kim, H. Y., Yoo, B. H. and Kwon, O. J. (1997) Comparative study of tracheal anastomotic techniques. *Korean Journal of Thoracic and Cardiovascular Surgery*, **30**, 1–7.

7
Attributable Risk

The attributable risk (AR) of a disease due to a risk factor represents the proportion of cases that are preventable in a population if this particular risk factor is completely eliminated (Levin, 1953). It is well known that a risk factor that has a strong association with the underlying disease of interest need not be a top priority for control if its prevalence is rare in the general population. Because it can reflect the strength of the association between a risk factor and a disease, as well as the prevalence of the risk factor, the AR is one of the most widely used epidemiologic indices for public health administrators to rank the importance of risk factors for intervention. Note that the AR has also been referred to as the attributable fraction (Greenland and Robins, 1988), the population attributable risk percent (Cole and MacMahon, 1971), the etiologic fraction (Miettinen, 1974); many other terms have also been used (Gefeller, 1990, 1992a).

To enable readers to more easily appreciate the practical interpretation and usefulness of the AR, we consider the case of no confounders. Let D and $\overline{D}$ denote the events of being a case and being a non-case, respectively. Similarly, let E and $\overline{E}$ denote the events of being exposed and not exposed to a risk factor of interest. Further, let $P(D)$, $P(\overline{D})$, $P(E)$, and $P(\overline{E})$ denote the probabilities of these events. If we completely eliminated this risk factor from the population, then the risk $P(D|E)$ of possessing the disease of interest for a subject with exposure would reduce to $P(D|\overline{E})$, the risk for a subject with non-exposure. Thus, when the risk factor is eliminated from a population of size N, we can expect to reduce the number $NP(E)[P(D|E) - P(D|\overline{E})]$ out of the total number $NP(D)$ of cases. The AR is simply equal to the ratio of these two parameters: $NP(E)[P(D|E) - P(D|\overline{E})]/[NP(D)]$. We can re-express the AR as $[P(D) - P(D|\overline{E})]/P(D)$ (**Exercise 7.1**) or $P(E)(RR - 1)/[P(E)(RR - 1) + 1]$ (**Exercise 7.2**), where $RR = P(D|E)/P(D|\overline{E})$ is the risk ratio between the exposed and non-exposed populations (see Chapter 4). Note that the AR lies in the range $-\infty < AR < 1$ (Greenland and Drescher, 1993). When $AR < 0$ (or equivalently, $RR < 1$), it is typical to consider the preventable fraction (PR), defined as $P(\overline{E})(RR^* - 1)/[P(\overline{E})(RR^* - 1)+1]$, where $RR^* = 1/RR$ (Last, 1983, p. 82). Note that the PR is obtained by switching the labels between the exposure and the non-exposure groups in the definition of the AR and is equal

Statistical Estimation of Epidemiological Risk K-J. Lui
 ISBN: 0-470-85071-X (HB)

to $[P(D) - P(D|E)]/P(D)$. Thus, all interval estimators for the AR can be easily modified to estimate the PR.

In this chapter, we begin by discussing estimation of the AR for both cross-sectional and case–control studies with no confounders. We then consider the situations where one can apply stratified analysis to control the effects of confounders for both these study designs. We further discuss estimation of the AR in case–control studies with matched pairs. We also include a discussion on estimation of the AR for case–control studies when the underlying exposure risk factor is polychotomous rather than dichotomous. When considering a multivariate model-based approach to control confounders, we discuss the use of the logistic regression model for case–control studies. Finally, we provide a brief discussion on estimation of the AR for case–control studies under inverse sampling.

7.1 STUDY DESIGNS WITH NO CONFOUNDERS

Note that the AR depends on the exposure prevalence of the underlying risk factor. We can estimate the exposure prevalence in a cross-sectional (or unstratified cohort) study or in a case–control study for a rare disease. We first consider the simplest case in which there are no confounders.

7.1.1 Cross-sectional sampling

Consider a cross-sectional study in which we take a random sample of n subjects and classify each subject by the presence or absence of a disease and a suspected risk factor. For clarity, we use the following table to summarize the data structure:

		Status of disease		
		Yes	No	
Exposure to risk Factor	Yes	π_{11}	π_{10}	$\pi_{1.}$
	No	π_{01}	π_{00}	$\pi_{0.}$
		$\pi_{.1}$	$\pi_{.0}$	1

where $0 < \pi_{ij} < 1$ denotes the cell probability, $\pi_{i.} = \pi_{i1} + \pi_{i0}$, and $\pi_{.j} = \pi_{1j} + \pi_{0j}$ for i and $j = 1, 0$. Let N_{ij} (where $\sum_i \sum_j N_{ij} = n$) denote the random frequency falling into cell (i, j) with probability π_{ij}. Define $N_{i.} = N_{i1} + N_{i0}$ and $N_{.j} = N_{1j} + N_{0j}$. The random vector $\mathbf{N}' = (N_{11}, N_{10}, N_{01}, N_{00})$ then follows the multinomial distribution (2.25) with parameters n and $\boldsymbol{\pi}' = (\pi_{11}, \pi_{10}, \pi_{01}, \pi_{00})$. Note that the maximum likelihood estimator (MLE) of π_{ij} is simply the sample proportion $\hat{\pi}_{ij} = N_{ij}/n$, and hence the MLEs for $\pi_{i.}$ and $\pi_{.j}$ are $\hat{\pi}_{i.} = N_{i.}/n$ and $\hat{\pi}_{.j} = N_{.j}/n$, respectively.

In terms of parameters π_{ij}, we have $P(D) = \pi_{.1}$ and $P(D|\overline{E}) = \pi_{01}/\pi_{0.}$. Thus, we express the AR as $1 - \phi$, where $\phi = \pi_{01}/(\pi_{0.}\pi_{.1})$ (**Exercise 7.1**). Note that the MLE of ϕ is $\hat{\phi} = \hat{\pi}_{01}/(\hat{\pi}_{0.}\hat{\pi}_{.1})$. Because $\phi > 0$, we can reasonably restrict the estimate $\hat{\phi}$ to the same range by applying the adjustment procedure of adding 0.5 to all cells whenever any of the observed cell frequencies n_{ij} equals 0. Thus, the statistic $\log(\hat{\phi})$ is always defined. By the delta method (Anderson, 1958; see Appendix), $\sqrt{n}(\log(\hat{\phi}) - \log(\phi))$ has an asymptotic normal distribution with mean 0 and variance $(1 - \pi_{01})/\pi_{01} - (\pi_{0.} + \pi_{.1} - 2\pi_{01})/(\pi_{0.}\pi_{.1})$ (**Exercise 7.4**). This leads to an asymptotic $100(1 - \alpha)$ percent confidence interval for the AR (Fleiss, 1979, 1981) given by

$$[1 - \hat{\phi}\exp(Z_{\alpha/2}\sqrt{\widehat{\mathrm{Var}}(\log(\hat{\phi}))}), 1 - \hat{\phi}\exp(-Z_{\alpha/2}\sqrt{\widehat{\mathrm{Var}}(\log(\hat{\phi}))})], \qquad (7.1)$$

where Z_α is the upper 100αth percentile of the standard normal distribution, and $\widehat{\mathrm{Var}}(\log(\hat{\phi})) = (1 - \hat{\pi}_{01})/(n\hat{\pi}_{01}) - (\hat{\pi}_{0.} + \hat{\pi}_{.1} - 2\hat{\pi}_{01})/(n\hat{\pi}_{0.}\hat{\pi}_{.1})$.

Using the delta method again, we can show that the asymptotic variance of $\hat{\phi}$ is $\mathrm{Var}(\hat{\phi}) = \phi^2\mathrm{Var}(\log(\phi))$. Thus, an asymptotic $100(1 - \alpha)$ percent confidence interval for the AR directly based on $\hat{\phi}$ is

$$[1 - \hat{\phi} - Z_{\alpha/2}\sqrt{\widehat{\mathrm{Var}}(\hat{\phi})}, 1 - \max\{\hat{\phi} - Z_{\alpha/2}\sqrt{\widehat{\mathrm{Var}}(\hat{\phi})}, 0\}], \qquad (7.2)$$

where $\widehat{\mathrm{Var}}(\hat{\phi}) = \hat{\phi}^2\widehat{\mathrm{Var}}(\log(\hat{\phi}))$. In fact, (7.2) is essentially equivalent to the interval estimator using Wald's test statistic proposed by Walter (1976). Lui (2001a) derives another asymptotic interval estimator from a quadratic equation using an idea analogous to that of Fieller's theorem. The details can be found in **Exercise 7.5**.

Note that the MLE of the AR is simply $\widehat{\mathrm{AR}} = 1 - \hat{\phi}$. Leung and Kupper (1981) propose using the logit transformation $\mathrm{logit}(\widehat{\mathrm{AR}}/(1 - \widehat{\mathrm{AR}})) = \log(\hat{\pi}_{1.}(\widehat{\mathrm{RR}} - 1))$, where $\widehat{\mathrm{RR}} = \hat{\pi}_{11}\hat{\pi}_{0.}/(\hat{\pi}_{01}\hat{\pi}_{1.})$. Thus, we obtain an asymptotic $100(1 - \alpha)$ percent confidence interval for the AR given by

$$[\mathrm{LT_l}/(\mathrm{LT_l} + 1), \mathrm{LT_u}/(\mathrm{LT_u} + 1)], \qquad (7.3)$$

where

$$\mathrm{LT_l} = \exp\{\log(\hat{\pi}_{1.}(\widehat{\mathrm{RR}} - 1)) - Z_{\alpha/2}\sqrt{\widehat{\mathrm{Var}}(\log(\hat{\pi}_{1.}(\widehat{\mathrm{RR}} - 1)))}\},$$

$$\mathrm{LT_u} = \exp\{\log(\hat{\pi}_{1.}(\widehat{\mathrm{RR}} - 1)) + Z_{\alpha/2}\sqrt{\widehat{\mathrm{Var}}(\log(\hat{\pi}_{1.}(\widehat{\mathrm{RR}} - 1)))}\},$$

and

$$\widehat{\mathrm{Var}}[\log(\hat{\pi}_{1.}(\widehat{\mathrm{RR}} - 1))] = \left\{\frac{\hat{\pi}_{.1}\hat{\pi}_{0.}}{\hat{\pi}_{11}\hat{\pi}_{00} - \hat{\pi}_{10}\hat{\pi}_{01}}\right\}^2 \left\{\frac{\hat{\pi}_{11}\hat{\pi}_{00}(1 - \hat{\pi}_{01}) + \hat{\pi}_{10}\hat{\pi}_{01}^2}{n\hat{\pi}_{01}\hat{\pi}_{.1}\hat{\pi}_{0.}}\right\}.$$

The logarithmic function $\log(x)$ is defined only for $x > 0$. When the underlying RR is close to 1, the probability of interval estimator (7.3) being inapplicable due to $\widehat{\mathrm{RR}} \leq 1$ can be substantial. Leung and Kupper (1981) suggest using (7.2) whenever (7.3) is inapplicable. However, use of this *ad hoc* strategy does not produce an accurate interval estimator when AR $\doteq$ 1 (Lui, 2001a).

Lui (2001a) compares the performance of interval estimators (7.1)–(7.3) with respect to the coverage probability and the average length. Lui finds that when the probability of exposure $\pi_{1.}$ is moderate or large (0.20 or more), the coverage probability of (7.1) will generally agree well with the desired confidence level as long as the sample size is adequate or large (at least 100). However, using this estimator generally leads to loss of efficiency with respect to the average length of the confidence interval. When the underlying RR is close to 1, using (7.3) may produce inaccurate interval estimates; the coverage probability can be much less than the desired confidence level. On the other hand, when RR is large (at least 4) and the probability of exposure $\pi_{1.}$ is not small (0.05 or greater), interval estimator (7.3) is preferable to (7.1) or (7.2). When the RR is large (at least 4) and $\pi_{1.}$ is very small (say, 0.005), interval estimator (7.2) is recommended.

Example 7.1 Consider the data (Fleiss, 1981, p. 10) consisting of 2784 subjects classified by the presence or absence of a respiratory disease and a locomotor disease. We have $n_{11} = 17$, $n_{10} = 207$, $n_{01} = 184$, and $n_{00} = 2376$. Suppose that we are interested in estimation of the AR of the locomotor disease due to the respiratory disease. Applying interval estimators (7.1)–(7.3), we obtain asymptotic 95% confidence intervals for the AR of [−0.037, 0.044], [−0.036, 0.045], and [0.000, 0.974], respectively. The intervals obtained from (7.1) and (7.2) are similar to each other, while that from (7.3) is shifted to the right. Because $\widehat{\text{RR}} = 1.06$, which is close to 1, (7.3) is an unsuitable estimator in this instance.

Example 7.2 In a sample of 1329 subjects from the Framingham study of heart disease (Leung and Kupper, 1981), 72 subjects developed coronary heart disease (CHD) after 6 years out of 756 subjects with initial serum cholesterol (mg%) of 220 or greater, while only 20 subjects developed CHD after 6 years out of 573 subjects with initial serum cholesterol less than 220. Given these data, we obtain estimates $\widehat{\text{RR}} = 2.73$, $\hat{\pi}_{1.} = 0.569$, $\hat{\pi}_{11} = 0.054$, and $\widehat{\text{AR}} = 0.496$. Applying interval estimators (7.1)–(7.3), we obtain 95% confidence intervals for AR of [0.263, 0.655], [0.305, 0.687], and [0.314, 0.679], respectively. We can see that the interval obtained from (7.1) is slightly less efficient than the other two, while the interval from (7.3) is the shortest.

7.1.2 Case–control studies

Note that the AR can be expressed as $\text{P}(E|D)(\text{RR} - 1)/\text{RR}$, where $\text{P}(E|D)$ denotes the exposure prevalence in the case population, and $\text{RR} = \text{P}(D|E)/\text{P}(D|\bar{E})$ (Exercise 7.6). When the underlying disease is rare, the RR can be approximated by the odds ratio (OR) (Exercise 5.2). Note that both OR and $\text{P}(E|D)$ are estimable from a case–control study, as is AR via the equation AR $\doteq$ P($E|D$)(OR − 1)/OR. In the following discussion, we shall assume that the underlying disease is so rare that the difference between AR and P($E|D$)(OR − 1)/OR is negligible.

Consider a case–control study in which we take an independent random sample of n_j subjects from the case ($j = 1$) and control groups ($j = 0$), respectively. We then retrospectively classify each subject by whether he/she is exposed ($i = 1$) or unexposed ($i = 0$) to the risk factor of interest. For clarity, we use the following table to summarize the data structure:

		Status of disease	
		Case	Control
Exposure to risk factor	Yes	$\pi_{1\|1}$	$\pi_{1\|0}$
	No	$\pi_{0\|1}$	$\pi_{0\|0}$
		1	1

where $0 < \pi_{i|j} < 1$ denotes the conditional probability that a subject has exposure i (1 denoting exposed, and 0 unexposed), given the subject is from group j (1 for the case group, and 0 for the control). Let X_j denote the number of subjects who are exposed in group j. Then X_j follows the binomial distribution with parameters n_j and $\pi_{1|j}$. The sample proportion $\hat{\pi}_{1|j} = X_j/n_j$ is the MLE of $\pi_{1|j}$ for $j = 1, 0$. The conditional probabilities P($E|D$) and P($E|\bar{D}$) are simply $\pi_{1|1}$ and $\pi_{1|0}$, respectively. We can show that P($E|D$)(OR − 1)/OR $= (\pi_{1|1} - \pi_{1|0})/(1 - \pi_{1|0}) = 1 - \phi^*$, where $\phi^* = (1 - \pi_{1|1})/(1 - \pi_{1|0})$ (**Exercise 7.7**). Note that the estimator $\widehat{\text{AR}}$ in this case is of the same form as the relative difference (3.1) discussed in Chapter 3. Therefore, using ideas for deriving interval estimators of the relative difference, we can derive interval estimators of AR.

First, note that because the MLE of the AR is $\widehat{\text{AR}} = 1 - \hat{\phi}^*$, where $\hat{\phi}^* = (1 - \hat{\pi}_{1|1})/(1 - \hat{\pi}_{1|0})$, we may apply the delta method to obtain an estimator of the asymptotic variance Var($\widehat{\text{AR}}$) (**Exercise 7.8**), namely,

$$\widehat{\text{Var}}(\widehat{\text{AR}}) = (\hat{\phi}^*)^2\{\hat{\pi}_{1|1}/[n_1(1 - \hat{\pi}_{1|1})] + \hat{\pi}_{1|0}/[n_0(1 - \hat{\pi}_{1|0})]\}. \quad (7.4)$$

Thus, an asymptotic $100(1 - \alpha)$ percent confidence interval for the AR is

$$[\widehat{\text{AR}} - Z_{\alpha/2}\sqrt{\widehat{\text{Var}}(\widehat{\text{AR}})}, \min\{\widehat{\text{AR}} + Z_{\alpha/2}\sqrt{\widehat{\text{Var}}(\widehat{\text{AR}})}, 1\}]. \quad (7.5)$$

Note that the maximum value of AR is 1, and therefore we define the upper limit in (7.5) as the minimum of $\widehat{\text{AR}} + Z_{\alpha/2}\sqrt{\widehat{\text{Var}}(\widehat{\text{AR}})}$ and 1 to ensure that this limit is valid. As noted for (3.3), interval estimator (7.5) is not likely to perform well unless both n_i are large. To improve the normal approximation of $\widehat{\text{AR}}$, we may use the logarithmic transformation on $\hat{\phi}^*$. This leads to an asymptotic $100(1 - \alpha)$ percent confidence interval for the AR given by

$$[1 - \hat{\phi}^* \exp(Z_{\alpha/2}\sqrt{\widehat{\text{Var}}(\log(\hat{\phi}^*))}), 1 - \hat{\phi}^* \exp(-Z_{\alpha/2}\sqrt{\widehat{\text{Var}}(\log(\hat{\phi}^*))})], \quad (7.6)$$

where $\widehat{\text{Var}}(\log(\hat{\phi}^*)) = \hat{\pi}_{1|1}/[n_1(1 - \hat{\pi}_{1|1})] + \hat{\pi}_{1|0}/[n_0(1 - \hat{\pi}_{1|0})]$. Note that, to use (7.5) and (7.6), whenever $\hat{\pi}_{i|j} = 0$ for some i and j, we may add 0.50 to each cell frequency so that the cell probability estimate $0 < \hat{\pi}_{i|j} < 1$ always holds for all i and j.

Example 7.3 Consider the data taken from a case–control study (Cole *et al.*, 1971; Miettinen, 1976; Schlesselman, 1982) that identifies all newly diagnosed cases of bladder cancer over an 18-month period ending June 30, 1968 among residents of the Boston and Brockton Standard Metropolitan Statistical Areas in eastern Massachusetts. Exposure is defined as whether a subject has smoked at least 100 cigarettes during his/her lifetime. For the purpose of illustration, we consider subjects between the ages of 75 and 79 years only. Given the data $X_1 = 39$, $n_1 - X_1 = 14$, $X_0 = 32$, and $n_0 - X_0 = 20$, we obtain $\widehat{\text{AR}} = 0.313$. This suggests that 31.3% of bladder cancers for individuals aged between 75 to 79 can be prevented if we reduce the number of cigarettes smoked during the lifetime to below 100. Applying interval estimators (7.5) and (7.6), we obtain $[-0.075, 0.702]$ and $[-0.209, 0.610]$, respectively. Because both interval estimates cover 0, there is no significant evidence at the 5% level that the AR due to smoking in this particular age category exceeds 0.

7.2 STUDY DESIGNS WITH CONFOUNDERS

When there are confounders, estimation of the AR can be biased if we do not account for the confounding effects. In this section, we focus discussion on estimation of the AR by employing post-stratified analysis to control confounders for cross-sectional and case–control studies. Benichou (1991) and Coughlin *et al.* (1994) both provide excellent reviews on various aspects of AR estimation in case–control studies. The approach presented in this section is model-free. The use of logistic regression to control confounders for estimation of the AR in case–control studies (Deubner *et al.*, 1980; Bruzzi *et al.*, 1985; Benichou and Gail, 1990; Drescher and Schill, 1991; Greenland and Drescher, 1993) will be discussed in a separate section.

7.2.1 Cross-sectional sampling

Suppose that we take a random sample of n subjects and determine the status of each subject by the presence or the absence of a disease and a suspected risk factor. Suppose further that the combination of all confounders forms S levels, denoted by C_s, $s = 1, 2, \ldots, S$. We post-stratify the n sampled subjects according to the level of C_s to control the effects resulting from these confounders. For clarity, we use the following table to summarize the data structure at the confounder level

$C_s (s = 1, 2, \ldots, S)$:

			Level of confounders C_s Status of disease		
			D Yes	$\bar{D}$ No	
Exposure to risk Factor	E	Yes	π_{11s}	π_{10s}	$\pi_{1.s}$
	$\bar{\text{E}}$	No	π_{01s}	π_{00s}	$\pi_{0.s}$
			$\pi_{.1s}$	$\pi_{.0s}$	

where $0 < \pi_{ijs} < 1, i = 1, 0, j = 1, 0$, and $s = 1, 2, \ldots, S$, and $\sum_i \sum_j \sum_s \pi_{ijs} = 1$. Let N_{ijs} denote the random frequency with the corresponding cell probability π_{ijs}, $N_{i.s} = N_{i1s} + N_{i0s}$, $N_{.js} = N_{1js} + N_{0js}$, $\sum_i \sum_j N_{ijs} = N_{..s}$, and $\sum_s N_{..s} = n$. The random vector $\mathbf{N}' = (N_{111}, N_{101}, N_{011}, N_{001}, N_{112}, N_{102}, N_{012}, N_{002}, \ldots, N_{11S}, N_{10S}, N_{01S}, N_{00S})$ then follows the multinomial distribution with parameters n and $\boldsymbol{\pi}' = (\pi_{111}, \pi_{101}, \pi_{011}, \pi_{001}, \pi_{112}, \pi_{102}, \pi_{012}, \pi_{002}, \ldots, \pi_{11S}, \pi_{10S}, \pi_{01S}, \pi_{00S})$. Let $\hat{\pi}_{ijs} = N_{ijs}/n$, which is simply the sample proportion estimate of π_{ijs}. By the central limit theorem, the random vector $\sqrt{n}(\hat{\boldsymbol{\pi}} - \boldsymbol{\pi})$ asymptotically has the normal distribution with mean vector $\mathbf{0}$ and covariance matrix $\mathbf{D}(\boldsymbol{\pi}) - \boldsymbol{\pi}\boldsymbol{\pi}'$, where $\hat{\boldsymbol{\pi}} = (\hat{\pi}_{111}, \hat{\pi}_{101}, \hat{\pi}_{011}, \hat{\pi}_{001}, \hat{\pi}_{112}, \hat{\pi}_{102}, \hat{\pi}_{012}, \hat{\pi}_{002}, \ldots, \hat{\pi}_{11S}, \hat{\pi}_{10S}, \hat{\pi}_{01S}, \hat{\pi}_{00S})'$, $\mathbf{0} = (0, 0, \ldots, 0)'$ is a $1 \times 4S$ vector with all terms equal to 0, and $\mathbf{D}(\boldsymbol{\pi})$ is a $4S \times 4S$ diagonal matrix with diagonal terms given by π_{ijs}. Define $\pi_{i.s} = \pi_{i1s} + \pi_{i0s}$ and $\pi_{.js} = \pi_{1js} + \pi_{0js}$. Similarly, when a subscript on the estimated sample proportion $\hat{\pi}_{ijs}$ (or the random frequency N_{ijs}) is replaced by a dot, we mean summation of $\hat{\pi}_{ijs}$ (or N_{ijs}) over that subscript; for example, $\hat{\pi}_{i..} = \sum_s \hat{\pi}_{i.s} = \sum_j \sum_s \hat{\pi}_{ijs}$ and $\hat{\pi}_{.j.} = \sum_s \hat{\pi}_{.js} = \sum_i \sum_s \hat{\pi}_{ijs}$, for i and $j = 1, 0$. When the underlying risk factor is eliminated, the AR, denoting the proportional reduction of disease, is defined as $\sum_s \mathrm{P}(C_s)[\mathrm{P}(D|C_s) - \mathrm{P}(D|\bar{\mathrm{E}}, C_s)]/\mathrm{P}(D) = 1 - \sum_s \mathrm{P}(C_s)\mathrm{P}(D|\bar{\mathrm{E}}, C_s)/\mathrm{P}(D)$, where $\mathrm{P}(C_s)$ denotes the prevalence of the confounder level C_s, $\mathrm{P}(D|\bar{\mathrm{E}}, C_s)$ denotes the conditional probability of the disease given a subject from the non-exposed group at the level C_s, and $\mathrm{P}(D)$ denotes the overall rate of the disease in the general population (Gefeller, 1992a; Whittemore, 1982, 1983; Basu and Landis, 1995). In this notation, the AR can be expressed as $1 - \sum_s (\pi_{01s}\pi_{..s}/(\pi_{0.s}\pi_{.1.}))$. Due to the functional invariance property (Casella and Berger, 1990; see the Appendix) of the MLE, the corresponding MLE of the AR is simply $\widehat{\mathrm{AR}} = 1 - \sum_s (\hat{\pi}_{01s}\hat{\pi}_{..s}/(\hat{\pi}_{0.s}\hat{\pi}_{.1.}))$.

Define $\phi_{\text{conf}} = 1 - \mathrm{AR} = \sum_s (\pi_{01s}\pi_{..s}/(\pi_{0.s}\pi_{.1.}))$. The MLE of ϕ_{conf} is $\hat{\phi}_{\text{conf}} = \sum_s (\hat{\pi}_{01s}\hat{\pi}_{..s}/(\hat{\pi}_{0.s}\hat{\pi}_{.1.}))$. Using the delta method (Anderson, 1958), we can easily see that $\sqrt{n}(\hat{\phi}_{\text{conf}} - \phi_{\text{conf}})$ has the asymptotic normal distribution with mean 0 and variance

$$
\begin{aligned}
n\mathrm{Var}(\hat{\phi}_{\mathrm{conf}}) = \frac{1}{\pi_{.1.}^2} \Bigg\{ & \mathrm{AR}^2 \pi_{.1.} + 2\mathrm{AR} \sum_s \frac{\pi_{00s}(\pi_{10s}\pi_{01s} - \pi_{11s}\pi_{00s})}{\pi_{0.s}^2} \\
& + \sum_s \left[\frac{\pi_{00s}^2 \pi_{11s} + \pi_{10s}\pi_{01s}^2}{\pi_{0.s}^2} + \frac{\pi_{00s}\pi_{1.s}^2 \pi_{01s}}{\pi_{0.s}^3} \right] \\
& - \left[\mathrm{AR}\pi_{.1.} + \sum_s \frac{\pi_{10s}\pi_{01s} - \pi_{11s}\pi_{00s}}{\pi_{0.s}} \right]^2 \Bigg\}.
\end{aligned}
$$

The estimated variance $\widehat{\mathrm{Var}}(\hat{\phi}_{\mathrm{conf}})$ can then be obtained by simply substituting $\widehat{\mathrm{AR}}$ for AR and the corresponding $\hat{\pi}_{ijs}$ for π_{ijs}. Note that because the estimated asymptotic variance $\widehat{\mathrm{Var}}(\widehat{\mathrm{AR}})(= \widehat{\mathrm{Var}}(1 - \hat{\phi}_{\mathrm{conf}}))$ is equal to $\widehat{\mathrm{Var}}(\hat{\phi}_{\mathrm{conf}})$, an asymptotic $100(1 - \alpha)$ percent confidence interval for the AR is

$$
[\widehat{\mathrm{AR}} - Z_{\alpha/2}\sqrt{\widehat{\mathrm{Var}}(\widehat{\mathrm{AR}})}, \min\{\widehat{\mathrm{AR}} + Z_{\alpha/2}\sqrt{\widehat{\mathrm{Var}}(\widehat{\mathrm{AR}})}, 1\}]. \tag{7.7}
$$

To improve the normal approximation of $\widehat{\mathrm{AR}}$, following Leung and Kupper (1981), we consider the logit transformation $\log(\widehat{\mathrm{AR}}/(1 - \widehat{\mathrm{AR}}))$. We thus obtain an asymptotic $100(1 - \alpha)$ percent confidence interval for the AR given by (Whittemore, 1982)

$$
\begin{aligned}
\Bigg[& \left\{ 1 + \frac{1 - \widehat{\mathrm{AR}}}{\widehat{\mathrm{AR}}} \exp\left(\frac{Z_{\alpha/2}\sqrt{\widehat{\mathrm{Var}}(\widehat{\mathrm{AR}})}}{\widehat{\mathrm{AR}}(1 - \widehat{\mathrm{AR}})} \right) \right\}^{-1}, \\
& \left\{ 1 + \frac{1 - \widehat{\mathrm{AR}}}{\widehat{\mathrm{AR}}} \exp\left(\frac{-Z_{\alpha/2}\sqrt{\widehat{\mathrm{Var}}(\widehat{\mathrm{AR}})}}{\widehat{\mathrm{AR}}(1 - \widehat{\mathrm{AR}})} \right) \right\}^{-1} \Bigg].
\end{aligned} \tag{7.8}
$$

Note that the logit transformation $\log(\widehat{\mathrm{AR}}/(1 - \widehat{\mathrm{AR}}))$ is defined only for $\widehat{\mathrm{AR}} > 0$. When the underlying $\mathrm{RR}_s(= P(D|E, C_s)/P(D|\bar{E}, C_s))$ in all strata is less than or equal to 1, the probability that $\widehat{\mathrm{AR}}$ is negative or zero may be substantial. In this case, interval estimator (7.8) is no longer applicable. However, applying the logarithmic transformation directly to $\hat{\phi}_{\mathrm{conf}}$ escapes this limitation. This is because $\phi_{\mathrm{conf}} > 0$, so that we can reasonably restrict the estimate $\hat{\phi}_{\mathrm{conf}}$ to be positive. Thus, we may consider use of $\log(\hat{\phi}_{\mathrm{conf}})$ here. This leads to an asymptotic $100(1 - \alpha)$ percent confidence interval for the AR given by

$$
[1 - \hat{\phi}_{\mathrm{conf}} \exp(Z_{\alpha/2}\sqrt{\widehat{\mathrm{Var}}(\hat{\phi}_{\mathrm{conf}})/\hat{\phi}_{\mathrm{conf}}^2}), 1 - \hat{\phi}_{\mathrm{conf}} \exp(-Z_{\alpha/2}\sqrt{\widehat{\mathrm{Var}}(\hat{\phi}_{\mathrm{conf}})/\hat{\phi}_{\mathrm{conf}}^2})]. \tag{7.9}
$$

Finally, if the total number of subjects n is large, we have $P((\widehat{\mathrm{AR}} - \mathrm{AR})^2/\mathrm{Var}(\widehat{\mathrm{AR}})) \doteq 1 - \alpha$, where $\mathrm{Var}(\widehat{\mathrm{AR}})$ is a function of AR. This leads us to consider the following quadratic equation in AR:

$$
\mathfrak{A}(\mathrm{AR})^2 - 2\mathfrak{B}(\mathrm{AR}) + \mathfrak{C} \le 0, \tag{7.10}
$$

where

$$\mathfrak{A} = 1 - Z_{\alpha/2}^2 \hat{\pi}_{.1.}(1 - \hat{\pi}_{.1.})/(n\hat{\pi}_{.1.}^2),$$

$$\mathfrak{B} = \widehat{\text{AR}} + Z_{\alpha/2}^2 \left[\sum \frac{\hat{\pi}_{00s}(\hat{\pi}_{10s}\hat{\pi}_{01s} - \hat{\pi}_{11s}\hat{\pi}_{00s})}{\hat{\pi}_{0.s}^2} \right.$$
$$\left. - \hat{\pi}_{.1.} \sum \frac{\hat{\pi}_{10s}\hat{\pi}_{01s} - \hat{\pi}_{11s}\hat{\pi}_{00s}}{\hat{\pi}_{0.s}} \right] \Big/ (n\hat{\pi}_{.1.}^2),$$

$$\mathfrak{C} = \widehat{\text{AR}}^2 - Z_{\alpha/2}^2 \left[\sum \left(\frac{\hat{\pi}_{00s}^2\hat{\pi}_{11s} + \hat{\pi}_{10s}\hat{\pi}_{01s}^2}{\hat{\pi}_{0.s}^2} + \frac{\hat{\pi}_{00s}\hat{\pi}_{1.s}^2\hat{\pi}_{01s}}{\hat{\pi}_{0.s}^3} \right) \right.$$
$$\left. - \left(\sum \frac{\hat{\pi}_{10s}\hat{\pi}_{01s} - \hat{\pi}_{11s}\hat{\pi}_{00s}}{\hat{\pi}_{0.s}} \right)^2 \right] \Big/ (n\hat{\pi}_{.1.}^2).$$

If both $\mathfrak{A} > 0$ and $\mathfrak{B}^2 - \mathfrak{A}\mathfrak{C} > 0$, then an asymptotic $100(1 - \alpha)$ percent confidence interval for the AR would be

$$\left[\left(\mathfrak{B} - \sqrt{\mathfrak{B}^2 - \mathfrak{A}\mathfrak{C}}\right) \Big/ \mathfrak{A}, \min\left\{ \left(\mathfrak{B} + \sqrt{\mathfrak{B}^2 - \mathfrak{A}\mathfrak{C}}\right) \Big/ \mathfrak{A}, 1 \right\} \right]. \tag{7.11}$$

Example 7.4 Consider the prospective observational study (Table 7.1) of pregnancy and child development in Germany reported by Wermuth (1976). A total of $(n =)6751$ subjects are cross-classified by the number of cigarettes smoked per day ($< 5, \geq 5$), the length of gestation (LG) in days ($\leq 260, > 260$), and the age of mother (AM) in years ($< 30, \geq 30$). Suppose that we want to estimate the AR for 'smoking 5 or more cigarettes per day' versus 'smoking less than 5 cigarettes per day' on the perinatal death, while controlling the two confounders: LG and AM. We define the level of C_1 by (LG ≤ 260 days, AM < 30 years), C_2 by

Table 7.1 Contingency table for perinatal mortality by the number of cigarettes smoked per day during the prenatal period stratified by the length of gestation in days and the age category of mother.

			Perinatal survival	
Age	Gestation	Smoking	No	Yes
<30	≤260	5+	9	40
		<5	50	315
	>260	5+	6	459
		<5	24	4012
30+	≤260	5+	4	11
		<5	41	147
	>260	5+	1	124
		<5	14	1494

Source: Wermuth (1976).

(LG > 260 days, AM < 30 years), C_3 by (LG ≤ 260 days, AM > 30 years), and C_4(LG > 260 days, AM > 30 years). From the data, we obtain MLEs of $\mathrm{RR}_s(= \mathrm{P(D|E}, C_s)/\mathrm{P(D|\bar{E}}, C_s))$ for $s = 1, 2, 3, 4$ of 1.341, 2.170, 1.223, and 0.862, respectively. We further obtain the MLE for the AR to be 0.041. Applying (7.7), (7.8), (7.9), and (7.11) gives 95% confidence intervals for AR of [−0.016, 0.098], [0.010, 0.153], [−0.017, 0.096], and [−0.016, 0.098]. Except for the interval obtained from (7.8), these estimates are similar to one another. Because the lower limits of the intervals obtained from (7.7), (7.9), and (7.11) are all less than 0, these results suggest that there is no significant evidence at the 5% level that the proportional reduction of perinatal deaths would be greater than 0 if the mothers reduced the number of cigarettes smoked per day to 5 or less. Note that the estimated $\widehat{\mathrm{RR}}_s$ for the three levels C_1, C_3, and C_4 are not much different from 1. Thus, the estimated $\widehat{\mathrm{AR}}$ of perinatal mortality due to the reduction of the number of cigarettes smoked per day to below 5 is small ($\widehat{\mathrm{AR}} = 0.041$). Lui (2001b) notes that (7.8) may in this case not perform well.

When there are no confounders, Lui (2001b) notes that the interval estimator using the logarithmic transformation as suggested by Fleiss (1979) may improve the performance of the interval estimator using Wald's test statistic (Walter, 1976). By contrast, the latter outperforms the former when there are confounders requiring adjustment in estimation of the AR. Note also that both interval estimators (7.7) and (7.11) can consistently perform well with respect to coverage probability. If we had no prior knowledge about the possible range of the underlying RR in a study, we would recommend these two interval estimators for general use. However, if we know that the underlying RR_s are likely to be large (at least 4) for all k, then we will recommend use of interval estimator (7.8). This is because using (7.8) when RR ≥ 4 can generally gain efficiency. Finally, note that when the underlying RR_s is constant over strata (i.e., $\mathrm{RR}_s = \mathrm{RR}_0$), as shown in **Exercise 7.10**, the AR reduces to $\mathrm{P}(E|D)(\mathrm{RR}_0 - 1)/\mathrm{RR}_0$. Thus, we can estimate AR by $(N_{11.}/N_{.1.})(\widehat{\mathrm{RR}}_0 - 1)/\widehat{\mathrm{RR}}_0$, where $\widehat{\mathrm{RR}}_0 = \left(\sum_s N_{11s}N_{0.s}/N_{..s}\right) / \left(\sum_s N_{01s}N_{1.s}/N_{..s}\right)$ (Gefeller, 1992b; Tarone, 1981; see also (4.10)). Gefeller (1992b) notes that this point estimator performs quite well under the homogeneity risk model, but is subject to remarkable bias irrespective of the sample size under the heterogeneity risk model.

7.2.2 Case–control studies

Suppose that we take an independent random sample of n_j subjects from the case ($j = 1$) and the control ($j = 0$) populations, respectively. For each sampled subject, we retrospectively determine the exposure i (1 for being exposed, and 0 for being non-exposed). Suppose further that the combination of all confounders forms S levels, denoted by C_s, $s = 1, 2, \ldots, S$. To control the effect due to these confounders, we post-stratify the n_j sampled subjects according to the level of C_s.

For clarity, we use the following table to summarize the data structure:

			Level of confounder C_s Status of disease D Yes	$\overline{D}$ No
Exposure to risk factor	E	Yes	$\pi_{1s\mid 1}$	$\pi_{1s\mid 0}$
	E	No	$\pi_{0s\mid 1}$	$\pi_{0s\mid 0}$
			$\pi_{.s\mid 1}$	$\pi_{.s\mid 0}$

where $\pi_{is|j}$ denotes the conditional probability that a subject has exposure i and confounder level C_s, given this subject is from group j. Let $N_{is|j}$ denote the corresponding random frequency with cell probability $0 < \pi_{is|j} < 1$. Define $N_{i.|j} = \sum_s N_{is|j}$, $N_{.s|j} = N_{1s|j} + N_{0s|j}$, and $\sum_i \sum_s N_{is|j} = n_j$. The random vector $\mathbf{N}_j' = (N_{11|j}, N_{01|j}, N_{12|j}, N_{02|j}, \ldots, N_{1S|j}, N_{0S|j})$ follows the multinomial distribution with parameters n_j and $\boldsymbol{\pi}_j' = (\pi_{11|j}, \pi_{01|j}, \pi_{12|j}, \pi_{02|j}, \ldots, \pi_{1S|j}, \pi_{0S|j})$. Let $\hat{\pi}_{is|j} = N_{is|j}/n_j$. By the central limit theorem, the random vector $\sqrt{n_j}(\hat{\boldsymbol{\pi}}_j - \boldsymbol{\pi}_j)$ asymptotically follows the normal distribution with mean vector $\mathbf{0}$ and covariance matrix $\mathbf{D}_j(\boldsymbol{\pi}_j) - \boldsymbol{\pi}_j\boldsymbol{\pi}_j'$, where $\hat{\boldsymbol{\pi}}_j' = (\hat{\pi}_{11|j}, \hat{\pi}_{01|j}, \hat{\pi}_{12|j}, \hat{\pi}_{02|j}, \ldots, \hat{\pi}_{1S|j}, \hat{\pi}_{0S|j})$, $\mathbf{0}' = (0, 0, \ldots, 0)$ is a $1 \times 2S$ vector with all terms equal to 0, and $\mathbf{D}_j(\boldsymbol{\pi}_j)$ is a $2S \times 2S$ diagonal matrix with diagonal entries given by $\pi_{is|j}$. As noted in the previous section, when a subscript of the estimated sample proportion $\hat{\pi}_{is|j}$ is replaced by a dot, we mean summation of $\hat{\pi}_{is|j}$ over that subscript. For example, $\hat{\pi}_{i.|j} = \sum_s \hat{\pi}_{is|j}$. Note that the AR in the presence of confounders C_s can be expressed as $\sum_s \mathrm{P}(C_s|D)(1 - \mathrm{P}(D|\overline{E}, C_s)/\mathrm{P}(D|C_s))$ (**Exercise 7.10**). When the underlying disease is rare, the AR can then be approximated by (**Exercise 7.12**)

$$\mathrm{AR}_{\mathrm{apx}} = 1 - \sum_s \mathrm{P}(C_s|D)\mathrm{P}(\overline{E}|C_s, D)/\mathrm{P}(\overline{E}|C_s, \overline{D}), \tag{7.12}$$

which is a function of parameters estimable from a case–control study. In the following discussion we shall assume that the disease is so rare that the difference between $\mathrm{AR}_{\mathrm{apx}}$ and AR is negligible. Because $\mathrm{P}(\overline{E}|C_s, D) = \mathrm{P}(\overline{E}, C_s|D)/\mathrm{P}(C_s|D)$ and $\mathrm{P}(\overline{E}|C_s, \overline{D}) = \mathrm{P}(\overline{E}, C_s|\overline{D})/\mathrm{P}(C_s|\overline{D})$, we can write $\mathrm{AR}_{\mathrm{apx}} = 1 - \phi^*_{\mathrm{conf}}$, where $\phi^*_{\mathrm{conf}} = \sum_s \pi_{0s|1}\pi_{.s|0}/\pi_{0s|0}$, a weighted average of the ratio $\pi_{0s|1}/\pi_{0s|0}$ with weight given by $\pi_{.s|0}$. Thus, the MLE of $\mathrm{AR}_{\mathrm{apx}}$ is $\widehat{\mathrm{AR}}_{\mathrm{apx}} = 1 - \hat{\phi}^*_{\mathrm{conf}}$, where $\hat{\phi}^*_{\mathrm{conf}} = \sum_s \hat{\pi}_{0s|1}\hat{\pi}_{.s|0}/\hat{\pi}_{0s|0}$. By the delta method, we can show that the estimated asymptotic variance (Whittemore, 1982) of $\hat{\phi}^*_{\mathrm{conf}}$ is given by (**Exercise 7.13**)

$$\widehat{\mathrm{Var}}(\hat{\phi}^*_{\mathrm{conf}}) = \frac{1}{n_1}\left\{\sum_s \left(\frac{\hat{\pi}^2_{0s|1}\hat{\pi}^2_{.s|0}}{\hat{\pi}^2_{0s|0}}\right)\left(\frac{1}{\hat{\pi}_{0s|1}} + \frac{n_1\hat{\pi}_{1s|0}}{n_0\hat{\pi}_{.s|0}\hat{\pi}_{0s|0}}\right) - (\hat{\phi}^*_{\mathrm{conf}})^2\right\}. \tag{7.13}$$

Therefore, an asymptotic $100(1-\alpha)$ percent confidence interval for AR is given by

$$[1-\hat{\phi}^*_{\text{conf}}-Z_{\alpha/2}\sqrt{\widehat{\text{Var}}(\hat{\phi}^*_{\text{conf}})},\ \min\{1-\hat{\phi}^*_{\text{conf}}+Z_{\alpha/2}\sqrt{\widehat{\text{Var}}(\hat{\phi}^*_{\text{conf}})},\ 1\}]. \quad (7.14)$$

In order to improve the normal approximation of $\hat{\phi}^*_{\text{conf}}$, we may consider using the logarithmic transformation. It is easy to see that the estimated asymptotic variance $\widehat{\text{Var}}(\log(\hat{\phi}^*_{\text{conf}}))$ equals $\widehat{\text{Var}}(\hat{\phi}^*_{\text{conf}})/(\hat{\phi}^*_{\text{conf}})^2$. Thus, an asymptotic $100(1-\alpha)$ percent confidence interval for the AR using the logarithmic transformation is given by

$$[1-\hat{\phi}^*_{\text{conf}}\exp(Z_{\alpha/2}\sqrt{\widehat{\text{Var}}(\hat{\phi}^*_{\text{conf}})}/\hat{\phi}^*_{\text{conf}}),\ 1-\hat{\phi}^*_{\text{conf}}\exp(-Z_{\alpha/2}\sqrt{\widehat{\text{Var}}(\hat{\phi}^*_{\text{conf}})}/\hat{\phi}^*_{\text{conf}})]. \quad (7.15)$$

Again, following Leung and Kupper (1981), we may consider use of the logit transformation $\log(\widehat{\text{AR}}/(1-\widehat{\text{AR}}))$. Because the estimated asymptotic variance $\widehat{\text{Var}}(\log(\widehat{\text{AR}}/(1-\widehat{\text{AR}}))) = \widehat{\text{Var}}(\hat{\phi}^*_{\text{conf}})/(\widehat{\text{AR}}\hat{\phi}^*_{\text{conf}})^2$, an asymptotic $100(1-\alpha)$ percent confidence interval for the AR using the logit transformation is given by

$$[\{1+(\hat{\phi}^*_{\text{conf}}/\widehat{\text{AR}})\exp(Z_{\alpha/2}\sqrt{\widehat{\text{Var}}(\hat{\phi}^*_{\text{conf}})}/(\widehat{\text{AR}}\hat{\phi}^*_{\text{conf}}))\}^{-1},$$
$$\{1+(\hat{\phi}^*_{\text{conf}}/\widehat{\text{AR}})\exp(-Z_{\alpha/2}\sqrt{\widehat{\text{Var}}(\hat{\phi}^*_{\text{conf}})}/(\widehat{\text{AR}}\hat{\phi}^*_{\text{conf}}))\}^{-1}]. \quad (7.16)$$

The interval estimators (7.14), (7.15), and (7.16), derived on the basis of a weighted average of stratum-specific estimates of AR_s, were first developed by Whittemore (1982, 1983). They are valid even when the underlying OR varies between strata. However, their use does require sufficient subjects from each stratum. When the data are sparse, readers may refer to the approach proposed by Kuritz and Landis (1988a). When the underlying disease is rare and OR is constant, because AR can be approximated by $P(E|D)(\text{OR}-1)/\text{OR}$, we may use the Mantel–Haenszel summary odds-ratio estimator (5.13) to estimate the common OR and $N_{1.|1}/n_1$ to estimate $P(E|D)$ (Kleinbaum *et al.*, 1982). Using the delta method, Greenland (1987) derives an asymptotic variance for this estimator.

Example 7.5 Consider the lung cancer mortality study of white US uranium miners which included data on their cigarette smoking and radiation exposure status (Whittemore, 1982). The 776 miners in the study were matched to the 194 lung cancer decedents on year of birth. Suppose we are interested in estimation of the AR due to cumulative radon-daughter exposure, measured in working-level months (WLM); the cumulative smoking variable, measured in pack-years, is a confounder. Following Whittemore (1982), we dichotomize the radiation exposure status by using 120 WLM as the cutoff point; we also dichotomize cumulative smoking, with 20 pack-years as the cutoff point. For simplicity, we consider only the data for white miners between 55 and 59 years of age (see Table 7.2). From the data, we obtain the MLE $\widehat{\text{AR}}_{\text{apx}}(\doteq \widehat{\text{AR}}) = 0.639$, and

Table 7.2 Cumulative radiation exposure in WLM units for US uranium miners and mortality rates for lung cancer among US white males aged 55–59.

	Cumulative smoking (pack-years)			
	<20		≥20	
	WLM ≥120	WLM <120	WLM ≥120	WLM <120
Cases	5	2	34	5
Controls	41	35	64	44

Source: Whittemore (1982).

$\widehat{\text{Var}}(\hat{\phi}^*_{\text{conf}}) = 0.016\,96$. Applying interval estimators (7.14), (7.15), and (7.16), we obtain 95% confidence intervals for AR of [0.384, 0.894], [0.269, 0.822], and [0.370, 0.842]. Since all the resulting lower limits are above 0, the AR of lung cancer due to cumulative radiation exposure is significantly greater than 0 at the 5% level.

7.3 CASE–CONTROL STUDIES WITH MATCHED PAIRS

When studying a rare disease in the presence of strong nuisance confounders, we may often employ a case–control study design with matched pairs to increase efficiency. Suppose that a random sample of n cases is taken, and each is matched with a control with respect to certain nuisance confounders to form n matched pairs. Then each pair is classified according to the exposure status of the case and control as shown in the following table:

		Control: Exposed	Control: Unexposed	
Case	Exposed	π_{11}	π_{10}	$\pi_{1.}$
	Unexposed	π_{01}	π_{00}	$\pi_{0.}$
		$\pi_{.1}$	$\pi_{.0}$	1

where $0 < \pi_{ij} < 1$ denotes the corresponding cell probability, $\pi_{i.} = \pi_{i1} + \pi_{i0}$, and $\pi_{.j} = \pi_{1j} + \pi_{0j}$ for i and $j = 1, 0$. Recall that $\text{AR} = \text{P}(E|D)(\text{RR} - 1)/\text{RR}$ (**Exercise 7.6**). When the underlying disease is rare, we can approximate the RR by $\text{OR}(= \pi_{10}/\pi_{01})$ and thus approximate the AR by $\pi_{1.}(\pi_{10} - \pi_{01})/\pi_{10}$. In the following discussion we will assume that the underlying disease is so rare that AR and $\pi_{1.}(\pi_{10} - \pi_{01})/\pi_{10}$ are indistinguishable.

Let N_{ij} denote the random frequency of pairs falling into the cell with probability π_{ij}. The random vector $\mathbf{N}' = (N_{11}, N_{10}, N_{01}, N_{00})$ then follows the multinomial distribution (2.25) with parameters n and $\boldsymbol{\pi}' = (\pi_{11}, \pi_{10}, \pi_{01}, \pi_{00})$. Note that the sample proportion $\hat{\pi}_{ij} = N_{ij}/n$ is the MLE of π_{ij}, as are $\hat{\pi}_{i.} = N_{i.}/n$

and $\hat{\pi}_{.j} = N_{.j}/n$, where $N_{i.} = N_{i1} + N_{i0}$ and $N_{.j} = N_{1j} + N_{0j}$, for $\pi_{i.}$ and $\pi_{.j}$, respectively. Therefore, the MLE of AR is simply $\widehat{\text{AR}} = \hat{\pi}_{1.}(\hat{\pi}_{10} - \hat{\pi}_{01})/\hat{\pi}_{10}$. Define the random vector $\hat{\boldsymbol{\pi}}' = (\hat{\pi}_{11}, \hat{\pi}_{10}, \hat{\pi}_{01}, \hat{\pi}_{00})$. By the central limit theorem, we know that $\sqrt{n}(\hat{\boldsymbol{\pi}} - \boldsymbol{\pi})$ asymptotically follows the normal distribution with mean vector $\mathbf{0}$ and covariance matrix $\mathbf{D}(\boldsymbol{\pi}) - \boldsymbol{\pi}\boldsymbol{\pi}'$, where $\mathbf{0}'$ is a 1×4 vector of zeros, and $\mathbf{D}(\boldsymbol{\pi})$ is a 4×4 diagonal matrix with diagonal elements equal to $\pi_{11}, \pi_{10}, \pi_{01}$, and π_{00}. Using the delta method, the asymptotic variance of $\widehat{\text{AR}}$ is given by (**Exercise 7.15**) $\text{Var}(\widehat{\text{AR}}) = \{(\pi_{10} - \pi_{01})^2\pi_{11} + (\pi_{10}^2 + \pi_{01}\pi_{11})^2/\pi_{10} + \pi_{1.}^2\pi_{01} - [\pi_{1.}(\pi_{10} - \pi_{01})]^2\}/(n\pi_{10}^2)$, which we can estimate by substituting the MLEs $\hat{\pi}_{ij}$ for π_{ij}. We denote this estimated variance by $\widehat{\text{Var}}(\widehat{\text{AR}})$. We thus obtain an asymptotic $100(1-\alpha)$ percent confidence interval (Kuritz and Landis, 1987) for AR given by

$$[\widehat{\text{AR}} - Z_{\alpha/2}\sqrt{\widehat{\text{Var}}(\widehat{\text{AR}})}, \min\{\widehat{\text{AR}} + Z_{\alpha/2}\sqrt{\widehat{\text{Var}}(\widehat{\text{AR}})}, 1\}]. \tag{7.17}$$

In an effort to improve the normal approximation of $\widehat{\text{AR}}$, we may consider use of the logarithmic transformation (Katz *et al.*, 1978). Using the delta method, we obtain the estimated asymptotic variance $\widehat{\text{Var}}(\log(\widehat{\text{AR}})) = (\widehat{\text{AR}})^{-2}\widehat{\text{Var}}(\widehat{\text{AR}})$. Hence, an asymptotic $100(1-\alpha)$ percent confidence interval for the AR is

$$[\widehat{\text{AR}}\exp(-Z_{\alpha/2}\sqrt{\widehat{\text{Var}}(\log(\widehat{\text{AR}}))}), \min\{\widehat{\text{AR}}\exp(Z_{\alpha/2}\sqrt{\widehat{\text{Var}}(\log(\widehat{\text{AR}}))}), 1\}]. \tag{7.18}$$

Following Leung and Kupper (1981), we consider the logit transformation $\log(\widehat{\text{AR}}/(1 - \widehat{\text{AR}}))$. Because the estimated asymptotic variance $\widehat{\text{Var}}(\log(\widehat{\text{AR}}/(1 - \widehat{\text{AR}})))$ equals $(\widehat{\text{AR}}(1 - \widehat{\text{AR}}))^{-2}\widehat{\text{Var}}(\widehat{\text{AR}})$, an asymptotic $100(1-\alpha)$ percent confidence interval for the AR using the logit transformation is

$$[\{1 + ((1 - \widehat{\text{AR}})/\widehat{\text{AR}})\exp(Z_{\alpha/2}\sqrt{\widehat{\text{Var}}(\widehat{\text{AR}})}/(\widehat{\text{AR}}(1 - \widehat{\text{AR}})))\}^{-1},$$

$$\{1 + ((1 - \widehat{\text{AR}})/\widehat{\text{AR}})\exp(-Z_{\alpha/2}\sqrt{\widehat{\text{Var}}(\widehat{\text{AR}})}/(\text{AR}(1 - \widehat{\text{AR}})))\}^{-1}]. \tag{7.19}$$

Note that the logarithmic function $\log(x)$ is defined only for $x > 0$. When $\widehat{\text{AR}} < 0$, neither (7.18) nor (7.19) is applicable. Consider $\phi^{\dagger} = 1 - \text{AR} = (\pi_{10}\pi_{0.} + \pi_{1.}\pi_{01})/\pi_{10}$, which is always positive. Thus, we may consider use of the logarithmic transformation $\log(1 - \widehat{\text{AR}})$ rather than $\log(\widehat{\text{AR}})$ as used for deriving (7.18). Note that $\widehat{\text{Var}}(1 - \widehat{\text{AR}}) = \widehat{\text{Var}}(\widehat{\text{AR}})$. Using the delta method, we obtain the estimated asymptotic variance $\widehat{\text{Var}}(\log(\hat{\phi}^{\dagger}))$ equal to $\widehat{\text{Var}}(\widehat{\text{AR}})/(\hat{\phi}^{\dagger})^2$, where $\hat{\phi}^{\dagger} = 1 - \widehat{\text{AR}}$. Therefore, we obtain an asymptotic $100(1-\alpha)$ percent confidence interval for the AR given by

$$[1 - \hat{\phi}^{\dagger}\exp(Z_{\alpha/2}\sqrt{\widehat{\text{Var}}(\log(\hat{\phi}^{\dagger}))}), 1 - \hat{\phi}^{\dagger}\exp(-Z_{\alpha/2}\sqrt{\widehat{\text{Var}}(\log(\hat{\phi}^{\dagger}))})], \tag{7.20}$$

Recall that the asymptotic variance $\text{Var}(\widehat{\text{AR}})$ is equal to $\{(\pi_{10} - \pi_{01})^2\pi_{11} + (\pi_{10}^2 + \pi_{01}\pi_{11})^2/\pi_{10} + \pi_{1.}^2\pi_{01}\}/(n\pi_{10}^2) - \text{AR}^2/n$. Furthermore, if n is large, the probability $P((\widehat{AR} - AR)^2/\mathrm{Var}(\widehat{AR}) \leq Z^2_{\alpha/2}) \doteq 1 - \alpha$. These expressions lead to the following quadratic equation in AR: $A^\dagger AR^2 - 2B^\dagger AR + C^\dagger \leq 0$, where $A^\dagger = 1 + Z^2_{\alpha/2}/n$, $B^\dagger = \widehat{AR}$, and $C^\dagger = \widehat{AR}^2 - Z^2_{\alpha/2}\{(\hat{\pi}_{10} - \hat{\pi}_{01})^2\hat{\pi}_{11} + (\hat{\pi}^2_{10} + \hat{\pi}_{01}\hat{\pi}_{11})^2/\hat{\pi}_{10} + \hat{\pi}^2_{1.}\hat{\pi}_{01}\}/(n\hat{\pi}^2_{10})$ (**Exercise 7.16**). An asymptotic $100(1-\alpha)$ percent confidence interval for AR is then

$$[(B^\dagger - \sqrt{(B^\dagger)^2 - A^\dagger C^\dagger})/A^\dagger, \min\{(B^\dagger + \sqrt{(B^\dagger)^2 - A^\dagger C^\dagger})/A^\dagger, 1\}]. \qquad (7.21)$$

Note that because the coefficient $A^\dagger$ is positive, this equation is convex. Furthermore, with the commonly used adjustment procedure for sparse data of adding 0.5 to each cell frequency whenever any of the N_{ij} is 0, we can show that the inequality $(B^\dagger)^2 - A^\dagger C^\dagger > 0$ holds for all samples. The two distinct roots of the confidence limits in (7.21) thus always exist.

Lui (2001c) finds that, except for a few situations where the exposure prevalence in the case group is large ($\pi_{1.} = 0.80$), interval estimator (7.17) using Wald's statistic does perform reasonably well. While interval estimator (7.20) can improve the coverage probability over (7.17), using the former is likely to lead to a loss of efficiency as compared with the latter. By contrast, interval estimator (7.21) may generally not only improve the coverage probability of (7.17) but also increase the efficiency. Thus, interval estimator (7.21) is recommended for general use. When we know that both the underlying RR and $\pi_{1.}$ are not small ($RR \geq 4$ and $\pi_{1.} \geq 0.50$), say from some prior studies, however, we may wish to use interval estimator (7.19) as well, especially when n is not large. A discussion on estimation of the AR when there is more than one matched control per case can be found elsewhere (Kuritz and Landis, 1988b).

Example 7.6 Consider the data consisting of 183 pairs taken from a case–control study of oral conjugated estrogens and endometrial cancer (Antunes *et al.*, 1979; Schlesselman, 1982; Kuritz and Landis, 1987). We match each case with a control on race, age (within five years), date of admission (within 6 months), and hospital of admission. We then classify these 183 matched pairs according to their exposure status (ever versus never) with regard to use of the estrogens. From the data, we have $n_{11} = 12$, $n_{10} = 43$, $n_{01} = 7$, and $n_{00} = 121$. Suppose that we are interested in estimation of the AR of endometrial cancer due to the use of the estrogens. We obtain an estimate $\widehat{AR} = 0.252$. Using interval estimators (7.17)–(7.21), we obtain asymptotic 95% confidence intervals for AR of [0.172, 0.331], [0.183, 0.345], [0.181, 0.339], [0.168, 0.327], and [0.167, 0.325], respectively. The confidence intervals obtained from (7.18) and (7.19) are shifted slightly to the right as compared with the other three intervals, which are all similar to one another. In fact, applying Monte Carlo simulation to compare the performance of these estimators in the particular configuration given by the example, Lui (2001c) finds them all suitable for use; in each case the coverage probability is approximately equal to the desired confidence level. However, Lui notes that interval estimator (7.21) may be slightly more efficient than the other estimators in terms of average length.

7.4 MULTIPLE LEVELS OF EXPOSURE IN CASE–CONTROL STUDIES

There are situations in which the risk factor is polychotomous rather than dichotomous (Walter, 1976). For example, in a study of smoking and myocardial infarction (Shapiro *et al.*, 1979; Denman and Schlesselman, 1983), the risk factor of cigarettes smoked per day is categorized into several categories. Multiple levels of exposure can also be formed by a combination of several exposure variables. In this section, we discuss estimation of the AR due to each level of exposure. As noted elsewhere (Coughlin *et al.*, 1994), such estimates may have important policy implications for screening groups at highest risk of disease.

Suppose that we take an independent random sample of n_j subjects from the case ($j = 1$) and the control ($j = 0$) populations, respectively. We then retrospectively classify the subject according to the level of exposure E_k, where $k = 0, 1, 2, \ldots, K$. We define an exposure of E_0 as the baseline level. Let $N_{k|j}$ denote the number of subjects falling into the cell with probability $\pi_{k|j} (k = 0, 1, 2, \ldots, K)$ out of $n_j (= \sum_k N_{k|j})$ subjects. Thus, $\mathbf{N}_j = (N_{0|j}, N_{1|j}, \ldots, N_{K|j})'$ follows the multinomial distribution with parameters n_j and $\boldsymbol{\pi}_j = (\pi_{0|j}, \pi_{1|j}, \ldots, \pi_{K|j})'$. Therefore, the MLE of $\pi_{k|j}$ is $\hat{\pi}_{k|j} = N_{k|j}/n_j$. For $k \geq 1$, the attributable risk AR_k, denoting the proportional reduction of disease when we reduce the exposure level from E_k to E_0, is simply given by $[\mathrm{P}(D|E_k) - \mathrm{P}(D|E_0)]\mathrm{P}(E_k)/\mathrm{P}(D)$, where $\mathrm{P}(D|E_k)$ (for $k = 0, 1, 2, \ldots, K$) denotes the conditional probability of disease, given an exposure level equal to E_k, and $\mathrm{P}(E_k)$ and $\mathrm{P}(D)$ denote the exposure and the disease prevalence in the general population, respectively. We can show that AR_k is actually equal to (**Exercise 7.18**)

$$\mathrm{AR}_k = \mathrm{P}(E_k)(\mathrm{RR}_k - 1) \Big/ \left[1 + \sum_{k=1}^{K} \mathrm{P}(E_k)(\mathrm{RR}_k - 1)\right], \tag{7.22}$$

where $\mathrm{RR}_k = \mathrm{P}(D|E_k)/\mathrm{P}(D|E_0)$ is the RR between E_k and E_0. Note that, by definition, $\mathrm{RR}_0 = 1$ and $\mathrm{AR}_0 = 0$. When the underlying disease is rare, we can approximate the $\mathrm{P}(E_k)$ by the conditional probability $\mathrm{P}(E_k|\overline{D})$ of exposure level E_k in the control population and RR_k by $\mathrm{OR}_k = \mathrm{P}(E_k|D)\mathrm{P}(E_0|\overline{D})/[\mathrm{P}(E_0|D)\mathrm{P}(E_k|\overline{D})]$. Thus, the AR_k can be approximated by

$$\mathrm{P}(E_k|\overline{D})(\mathrm{OR}_k - 1) \Big/ \left[1 + \sum_{k} \mathrm{P}(E_k|\overline{D})(\mathrm{OR}_k - 1)\right]. \tag{7.23}$$

In terms of the $\pi_{k|j}$, we can express this approximation as (**Exercise 7.19**)

$$(\pi_{k|1}\pi_{0|0} - \pi_{0|1}\pi_{k|0})/\pi_{0|0}. \tag{7.24}$$

In the following discussion, we shall assume that the disease is so rare that the difference between AR_k and its approximation in (7.23) or (7.24) is negligible. From (7.24), the MLE of AR_k is then

$$\widehat{\text{AR}}_k = (\hat{\pi}_{k|1}\hat{\pi}_{0|0} - \hat{\pi}_{0|1}\hat{\pi}_{k|0})/\hat{\pi}_{0|0}. \tag{7.25}$$

Using the delta method, the estimated asymptotic variance of $\widehat{\text{AR}}_k$ is

$$\widehat{\text{Var}}(\widehat{\text{AR}}_k) = \left[\hat{\pi}_{k|1}(1-\hat{\pi}_{k|1}) + \left(\frac{\hat{\pi}_{k|0}}{\hat{\pi}_{0|0}}\right)^2 \hat{\pi}_{0|1}(1-\hat{\pi}_{0|1}) + 2\hat{\pi}_{k|1}\hat{\pi}_{0|1}\frac{\hat{\pi}_{k|0}}{\hat{\pi}_{0|0}}\right] \Big/ n_1$$
$$+ \left[(\hat{\pi}_{0|1})^2\left(\frac{\hat{\pi}_{k|0}}{\hat{\pi}^2_{0|0}} + \frac{\hat{\pi}^2_{k|0}}{\hat{\pi}^3_{0|0}}\right)\right] \Big/ n_0. \tag{7.26}$$

For $K = 1, \widehat{\text{AR}}_1$ (7.25) and variance (7.26) reduce to the MLE $\widehat{\text{AR}} = (\hat{\pi}_{1|1} - \hat{\pi}_{1|0})/\hat{\pi}_{0|0}$ and the variance $\text{Var}(\widehat{\text{AR}})$ (7.4) for the case of dichotomous levels of exposure discussed in Section 7.1.2 (**Exercise 7.20**). Note that the asymptotic variance (7.26) is slightly different from that obtained by Denman and Schlesselman (1983) by a term which decreases to 0 as either n_1 or n_0 goes to ∞. Thus, when the n_i are reasonably large, the difference between (7.26) and the variance formula developed by Denman and Schlesselman (1983) is negligible. From (7.26), we obtain an asymptotic $100(1-\alpha)$ percent confidence interval for AR_k given by

$$[\widehat{\text{AR}}_k - Z_{\alpha/2}\sqrt{\widehat{\text{Var}}(\widehat{\text{AR}}_k)},\ \min\{\widehat{\text{AR}}_k + Z_{\alpha/2}\sqrt{\widehat{\text{Var}}(\widehat{\text{AR}}_k)},\ 1\}]. \tag{7.27}$$

Similarly, to improve the normal approximation of $\widehat{\text{AR}}_k$, we consider use of the logarithmic transformation $\log(1-\widehat{\text{AR}})$. As noted before, $\widehat{\text{Var}}(1-\widehat{\text{AR}}_k) = \widehat{\text{Var}}(\widehat{\text{AR}}_k)$. Using the delta method, we obtain that the estimated asymptotic variance $\widehat{\text{Var}}(\log(1-\widehat{\text{AR}}_k)) = \widehat{\text{Var}}(\widehat{\text{AR}}_k)/(1-\widehat{\text{AR}}_k)^2$. Therefore, an asymptotic $100(1-\alpha)$ percent confidence interval for AR_k is given by

$$[1-(1-\widehat{\text{AR}}_k)\exp(Z_{\alpha/2}\sqrt{\widehat{\text{Var}}(\log(1-\widehat{\text{AR}}_k))}),$$
$$1-(1-\widehat{\text{AR}}_k)\exp(-Z_{\alpha/2}\sqrt{\widehat{\text{Var}}(\log(1-\widehat{\text{AR}}_k))})]. \tag{7.28}$$

Following the same idea as for deriving (7.21), we may rewrite the asymptotic variance of $\text{Var}(\widehat{\text{AR}})$ as

$$\left[\pi_{k|1} + \left(\frac{\pi_{k|0}}{\pi_{0|0}}\right)^2 \pi_{0|1} - \text{AR}_k^2\right] \Big/ n_1 + \left[(\pi_{0|1})^2\left(\frac{\pi_{k|0}}{\pi^2_{0|0}} + \frac{\pi^2_{k|0}}{\pi^3_{0|0}}\right)\right] \Big/ n_0. \tag{7.29}$$

Asymptotically in n, the probability $P((\widehat{\text{AR}}_k - \text{AR}_k)^2/\text{Var}(\widehat{\text{AR}}_k) \le Z^2_{\alpha/2}) \doteq 1-\alpha$. These considerations lead to the following quadratic equation in AR_k:

$$\mathcal{A}_k\text{AR}_k^2 - 2\mathcal{B}_k\text{AR}_k + \mathcal{C}_k \le 0,$$

where $\mathcal{A}_k = 1 + Z^2_{\alpha/2}/n_1$, $\mathcal{B}_k = \widehat{\mathrm{AR}}_k$, and $\mathcal{C}_k = \widehat{\mathrm{AR}}^2_k - Z^2_{\alpha/2}[\hat{\pi}_{k|1} + (\hat{\pi}_{k|0}/\hat{\pi}_{0|0})^2 \hat{\pi}_{0|1}]/n_1 - Z^2_{\alpha/2}[(\hat{\pi}_{0|1})^2(\hat{\pi}_{k|0}/\hat{\pi}^2_{0|0} + \hat{\pi}^2_{k|0}/\hat{\pi}^3_{0|0})]/n_0$ (**Exercise 7.21**). An asymptotic $100(1-\alpha)$ percent confidence interval for AR_k is then given by

$$[(\mathcal{B}_k - \sqrt{\mathcal{B}^2_k - \mathcal{A}_k\mathcal{C}_k})/\mathcal{A}_k, \min\{(\mathcal{B}_k + \sqrt{\mathcal{B}^2_k - \mathcal{A}_k\mathcal{C}_k})/\mathcal{A}_k, 1\}]. \tag{7.30}$$

Note that we can easily show that $\mathrm{AR}_k = \mathrm{P}(E_k|D)(\mathrm{RR}_k - 1)/\mathrm{RR}_k$ (see **Exercise 7.6**). Thus, the AR due to the K mutually exclusive levels ($k = 1, 2, \ldots, K$) of exposure is simply equal to $\Sigma^K_{k=1}\mathrm{AR}_k = 1 - \Sigma^K_{k=0}\mathrm{P}(E_k|D)/\mathrm{RR}_k = 1 - \mathrm{P}(D|E_0)/\mathrm{P}(D)$ (**Exercise 7.22**). In fact, this is the AR for the case of dichotomous exposure and hence all the results presented in Section 7.1.2. can be applied. Similarly, when there are confounders that form S distinct levels $C_1, C_2, \ldots, C_S$, the AR due to all K exposure levels is then equal to $\Sigma^S_{s=1}\Sigma^K_{k=1}[\mathrm{P}(D|E_k, C_s) - \mathrm{P}(D|E_0, C_s)]\mathrm{P}(E_k|C_s)\mathrm{P}(C_s)/\mathrm{P}(D) = 1 - \Sigma^S_{s=1}\Sigma^K_{k=0}\mathrm{P}(E_k, C_s|D)/\mathrm{RR}_{k|s}$, where $\mathrm{RR}_{k|s} = \mathrm{P}(D|E_k, C_s)/\mathrm{P}(D|E_0, C_s)$ (Lui, 2003; **Exercise 7.23**). Note that if $\mathrm{RR}_{k|s}$ does not depend on the level C_s of confounders (i.e. $\mathrm{RR}_{k|s} = \mathrm{RR}^*_k$), the above formula can be simplified to $1 - \Sigma^K_{k=0}\mathrm{P}(E_k|D)/\mathrm{RR}^*_k$. However, we should not interpret this as allowing us to ignore the confounders in estimation of the AR. When estimating the RR^*_k, we still need to control the confounders to avoid bias. The above formulae also indicate that once we have estimates $\widehat{\mathrm{RR}}_{k|s}$ (or $\widehat{\mathrm{RR}}^*_k$ when $\mathrm{RR}_{k|s}$ is constant across the levels of C_s), we can estimate the AR from the distribution of exposure among the cases only. When the underlying disease is rare, the $\mathrm{RR}_{k|s}$ can be approximated by $\mathrm{OR}_{k|s} = \mathrm{P}(E_k|D, C_s)\mathrm{P}(E_0|\bar{D}, C_s)/[(\mathrm{P}(E_0|D, C_s)\mathrm{P}(E_k|\bar{D}, C_s)]$. Furthermore, because $\mathrm{P}(E_k|D, C_s) = \mathrm{P}(E_k, C_s|D)/\mathrm{P}(C_s|D)$ and $\mathrm{P}(E_k|\bar{D}, C_s) = \mathrm{P}(E_k, C_s|\bar{D})/\mathrm{P}(C_s|\bar{D})$, we can apply the corresponding empirical estimators to estimate these parameters from a case–control study. When the number $K \times S$ of cells cross-classified by the exposure variables and confounders is large relative to the number of subjects, however, this approach may not be appropriate (Lui, 2003). This is because in this case the number of subjects falling into each cell is likely to be small and the MLE of AR_k can suffer a serious underestimation bias (Lui, 2003; Whittemore, 1982). This leads us to consider use of a multivariate model-based approach for estimation of the AR in the next section.

Example 7.7 Consider the data consisting of women aged 35–39 in a study of the relation between oral contraceptive use and myocardial infarction (MI) (Shapiro, *et al.*, 1979; Denman and Schlesselman, 1983). Information on the number of cigarettes smoked per day is collected as part of the study. The observed frequencies from the case and control groups (in parentheses) falling in the categories 'none', '1–24', and '25+' are: 3 (161), 12 (130), and 22 (65), respectively. From these data, the MLEs $\widehat{\mathrm{AR}}_1$ and $\widehat{\mathrm{AR}}_2$ are 0.259 and 0.562, respectively. This suggests that an estimated 26% of the cases of MI are attributable to smoking 1–24 cigarettes per day, and 56% are attributable to smoking 25 or more cigarettes per day, all other factors being equal between the case and the control groups. Applying interval estimators (7.27), (7.28), and (7.30), we obtain the asymptotic 95% confidence intervals for AR_1 to be [0.079, 0.439], [0.055, 0.419], and [0.061, 0.408]. Similarly, we obtain the asymptotic 95% confidence intervals for AR_2 to be [0.388, 0.736], [0.348, 0.706], and [0.335, 0.683] for using (7.27), (7.28), and (7.30). When comparing these interval estimators, we may note that interval estimator (7.28) seems to produce a slightly longer confidence interval for AR_k than the other two estimators.

7.5 LOGISTIC MODELING IN CASE–CONTROL STUDIES

When the confounding effects can be controlled through stratification and the number of subjects is large relative to the number of combined levels determined by all variables under consideration, we may apply the model-free approach as discussed in Section 7.2 or recent results of (Lui, 2003) to estimate the AR. On the other hand, if the number of subjects is not large, the estimator of the AR based on a function of the $\widehat{OR}_{k|s}$, which themselves are subject to large variation, may be questionable. The multivariate model-based approach is an appealing and logical approach to solving this practical difficulty. Based on the fact that the AR can be written as $1 - \sum_{s=1}^{S}\sum_{k=0}^{K} P(E_k, C_s|D)/RR_{k|s}$, Bruzzi *et al.* (1985) suggest that one can estimate the AR by simply using the empirical joint distribution of exposure and confounder variables from the sample of cases and the resulting estimate $\widehat{RR}_{k|s}$ based on the assumed logistic regression model. Benichou and Gail (1990) further derive the asymptotic variance of the estimator proposed by Bruzzi *et al.* (1985). Greenland and Drescher (1993) discuss the MLE of the AR under logistic regression models. Drescher and Schill (1991) propose a simple innovative approach by observing that the intercept parameters in the logistic regression are actually a function of the AR. Thus, we can easily apply standard statistical software to obtain both point and interval estimates of AR. Because Drescher and Schill's approach is easily understood and simple to use, here we concentrate the following discussion on using this approach.

7.5.1 Logistic model containing only the exposure variables of interest

We begin by considering the situation where the vector $\mathbf{z}$ consists of only the exposure variables of interest (i.e., there are no confounders). Let $\mathbf{z}_0$ represent the baseline level for $\mathbf{z}$. Let D denote the random variable of disease status (1 for a case, and 0 otherwise). The AR can then be defined as

$$1 - P(D = 1|\mathbf{z}_0)/P(D = 1), \tag{7.31}$$

where $P(D = 1)$ and $P(D = 1|\mathbf{z}_0)$ denote the disease rate in the general population and the reference population with $\mathbf{z} = \mathbf{z}_0$, respectively. Suppose we take an independent random sample of size n_j from the case ($j = 1$) and the control ($j = 0$) populations, respectively. Following Drescher and Schill (1991), we consider the logistic regression model $P(D = 1|\mathbf{z}) = \exp(\beta_0 + \boldsymbol{\beta}'(\mathbf{z} - \mathbf{z}_0))/(1 + \exp(\beta_0 + \boldsymbol{\beta}'(\mathbf{z} - \mathbf{z}_0)))$. We can easily see that the intercept β_0 is simply equal to $\log(P(D = 1|\mathbf{z}_0)/P(D = 0|\mathbf{z}_0))$. Furthermore, as shown elsewhere (Anderson, 1972; Farewell, 1979; Prentice and Pyke, 1979; **Exercise 5.19**), when deriving the MLE of $\boldsymbol{\beta}$ from retrospective data, we can focus attention on the likelihood

$$\prod_{i=1}^{n}\left(\frac{\exp(\beta_0^* + \boldsymbol{\beta}'(\mathbf{z}_i - \mathbf{z}_0))}{1 + \exp(\beta_0^* + \boldsymbol{\beta}'(\mathbf{z}_i - \mathbf{z}_0))}\right)^{D_i}\left(\frac{1}{1 + \exp(\beta_0^* + \boldsymbol{\beta}'(\mathbf{z}_i - \mathbf{z}_0))}\right)^{1-D_i}, \quad (7.32)$$

where $\beta_0^* = \beta_0 + \log(\tau_1/\tau_0)$, τ_j being the sampling fraction of population j. Note that $\tau_1/\tau_0 = (n_1/n_0)[P(D = 0)/P(D = 1)]$. When the underlying disease is so rare that $P(D = 0|\mathbf{z}_0) = P(D = 0) \doteq 1$, Drescher and Schill (1991) note that

$$\beta_0^* \doteq \log(P(D = 1|\mathbf{z}_0)/P(D = 1)) + \log(n_1/n_0) = \phi^* + \log(n_1/n_0), \quad (7.33)$$

where $\phi^* = \log(1 - AR)$. Assuming that $n_j/n \to \rho_j > 0$ as $n \to \infty$, where $n = n_1 + n_0$, Prentice and Pyke (1979) show that $\sqrt{n}[(\hat{\phi}^*, \hat{\boldsymbol{\beta}}')' - (\phi^*, \boldsymbol{\beta}')']$ asymptotically follows the normal distribution with an estimated covariance matrix given by

$$[\mathbf{I}_o(\hat{\beta}_0^*, \hat{\boldsymbol{\beta}})/n]^{-1} - \begin{bmatrix} (\hat{\rho}_1\hat{\rho}_0)^{-1} & \mathbf{0} \\ \mathbf{0} & \mathbf{0} \end{bmatrix}, \quad (7.34)$$

where $\hat{\beta}_0^*$ and $\hat{\boldsymbol{\beta}}$ are the MLEs of β_0^* and $\boldsymbol{\beta}$, $\mathbf{I}_o(\hat{\beta}_0^*, \hat{\boldsymbol{\beta}})$ is the observed information matrix, and $\hat{\rho}_j = n_j/n$. We can easily obtain these estimates by using PROC LOGISTIC in SAS (1990). Thus, the estimated asymptotic variance $\widehat{\mathrm{Var}}(\hat{\phi}^*)$ is given by $(\mathbf{I}_o(\hat{\beta}_0^*, \hat{\boldsymbol{\beta}})^{-1})_{11} - (1/n_1 + 1/n_0)$, where $(\mathbf{I}_o(\hat{\beta}_0^*, \hat{\boldsymbol{\beta}})^{-1})_{11}$ is the (1, 1)th element of the inverse of the observed information matrix. Based on these results, the MLE of the AR is

$$\widehat{AR} = 1 - \exp(\hat{\beta}_0^* - \log(n_1/n_0)). \quad (7.35)$$

Furthermore, the corresponding asymptotic $100(1 - \alpha)$ percent confidence interval for the AR is given by

$$\begin{aligned}&[1 - \exp(\hat{\beta}_0^* - \log(n_1/n_0) + Z_{\alpha/2}\sqrt{\widehat{\mathrm{Var}}(\hat{\phi}^*)}),\\ &1 - \exp(\hat{\beta}_0^* - \log(n_1/n_0) - Z_{\alpha/2}\sqrt{\widehat{\mathrm{Var}}(\hat{\phi}^*)})]. \end{aligned} \quad (7.36)$$

Example 7.8 To illustrate the approach of Drescher and Schill (1991) and allow readers to easily compare the results between the model-free approach discussed previously and the logistic regression discussed in this section, we consider the simplest case, where the vector $\mathbf{z}$ contains only a single variable of dichotomous exposure. Consider the data on the association between smoking ($z = 1$ for smoking, and $z = 0$ for non-smoking) and bladder cancer ($D = 1$ for a case, and

$D = 0$ for a non-case) from the case–control study discussed in Example 7.3 (Cole *et al.*, 1971). Recall that $X_1 = 39$, $n_1 - X_1 = 14$, $X_0 = 32$, and $n_0 - X_0 = 20$. When applying PROC LOGISTIC in SAS (1990) to the likelihood (7.32), we obtain $\hat{\beta}_0^* = -0.3567$, with estimated standard error $\widehat{\mathrm{SD}}(\hat{\beta}_0^*) = 0.3485$. This leads to $\widehat{\mathrm{AR}} = 0.313$ from (7.35), the same as obtained in Example 7.3. Applying (7.36), we obtain an asymptotic 95% confidence interval for the AR of $[-0.209, 0.610]$, again identical to the interval obtained from (7.6) using the $\log(1 - x)$ transformation. In fact, as noted by Drescher and Schill (1991), we can show that when the logistic regression model contains only a single dichotomous exposure variable, the MLE (7.35) of the AR is identical to $(\hat{\pi}_{1|1} - \hat{\pi}_{1|0})/(1 - \hat{\pi}_{1|0})$. Furthermore, we can show that the estimated asymptotic variance $\widehat{\mathrm{Var}}(\hat{\phi}^*)$ is equivalent to $\hat{\pi}_{1|1}/[n_1(1 - \hat{\pi}_{1|1})] + \hat{\pi}_{1|0}/[n_0(1 - \hat{\pi}_{1|0})]$ (**Exercise 7.25**).

7.5.2 Logistic regression model containing both exposure and confounding variables

When estimating the AR, we may encounter situations in which there are confounders. For simplicity, we will restrict the following discussion to the situation in which the confounders can be controlled by stratified analysis. A discussion extending this to accommodate a more general situation can be found elsewhere (Greenland and Drescher, 1993).

Suppose that the combinations of all confounders form S levels, denoted by C_s, $s = 1, 2, \ldots, S$. Suppose further that from each stratum $s (s = 1, 2, \ldots, S)$ we take an independent random sample of n_{sj} subjects from the case $(j = 1)$ and the control $(j = 0)$ populations, respectively. Let $n_{s.} = n_{s1} + n_{s0}$ denote the total number of sampled subjects in stratum s. The AR can be defined as (**Exercise 7.10**)

$$\sum_s \mathrm{P}(C_s|D = 1)\mathrm{AR}_s, \tag{7.37}$$

where $\mathrm{AR}_s = 1 - \mathrm{P}(D = 1|\mathbf{z}_0, C_s)/\mathrm{P}(D = 1|C_s)$ and $\mathbf{z}_0$ is the desired baseline level. For each stratum, we assume that the OR of the disease rate for a person with exposure covariate $\mathbf{z}$ to a person with $\mathbf{z}_0$ is homogeneous and is given by $\exp(\boldsymbol{\beta}'(\mathbf{z} - \mathbf{z}_0))$. Following Drescher and Schill (1991), we consider the logistic regression model $\mathrm{P}(D = 1|\mathbf{z}, C_s) = \exp(\beta_{0s} + \boldsymbol{\beta}'(\mathbf{z} - \mathbf{z}_0))/(1 + \exp(\beta_{0s} + \boldsymbol{\beta}'(\mathbf{z} - \mathbf{z}_0)))$. The intercept β_{0s} is simply equal to $\log(\mathrm{P}(D = 1|\mathbf{z}_0, C_s)/\mathrm{P}(D = 0|\mathbf{z}_0, C_s))$. Furthermore, when deriving the MLE of $\boldsymbol{\beta}$, we can focus attention on the likelihood

$$\prod_{s=1}^{S}\prod_{i=1}^{n_{s.}} \left(\frac{\exp(\beta_{0s}^* + \boldsymbol{\beta}'(\mathbf{z}_{is} - \mathbf{z}_0))}{1 + \exp(\beta_{0s}^* + \boldsymbol{\beta}'(\mathbf{z}_{is} - \mathbf{z}_0))}\right)^{D_{is}} \left(\frac{1}{1 + \exp(\beta_{0s}^* + \boldsymbol{\beta}'(\mathbf{z}_{is} - \mathbf{z}_0))}\right)^{1-D_{is}}, \tag{7.38}$$

where D_{is} is 1 if the ith subject in stratum s is a case and 0 otherwise, $\beta_{0s}^* = \beta_{0s} + \log(\tau_{s1}/\tau_{s0})$, τ_{sj} being the sampling fraction of population j in stratum C_s, and $\mathbf{z}_{is}$ is the value of $\mathbf{z}$ on the ith subject from stratum s. Since $\tau_{s1}/\tau_{s0} =$

$(n_{s1}/n_{s0})[\mathrm{P}(D=0|C_s)/\mathrm{P}(D=1|C_s)]$, under the assumption that the disease is rare for each stratum, we have

$$\beta_{0s}^* \doteq \log(\mathrm{P}(D=1|\mathbf{z}_0, C_s)/\mathrm{P}(D=1|C_s)) + \log(n_{s1}/n_{s0}) = \phi_s^* + \log(n_{s1}/n_{s0}), \tag{7.39}$$

where $\phi_s^* = \log(1-\mathrm{AR}_s)$. Define $\boldsymbol{\beta}_0^* = (\beta_{01}^*, \beta_{02}^*, \ldots, \beta_{0s}^*, \ldots, \beta_{0S}^*)'$ and $\phi^* = (\phi_1^*, \phi_2^*, \ldots, \phi_s^*, \ldots, \phi_S^*)'$. Assume that $n_{sj}/n \to \rho_{sj} > 0$, where $n = \sum_s (n_{s1}+n_{s0})$, $(j=1,0)$. Following Prentice and Pyke (1979), we may claim that $\sqrt{n}[(\hat{\boldsymbol{\phi}}^*, \hat{\boldsymbol{\beta}}')' - (\boldsymbol{\phi}^*, \boldsymbol{\beta}')']$, where $\hat{\boldsymbol{\phi}}^* = (\hat{\beta}_{01}^* - \log(n_{11}/n_{10}), \hat{\beta}_{02}^* - \log(n_{21}/n_{20}), \ldots, \hat{\beta}_{0s}^* - \log(n_{s1}/n_{s0}), \ldots, \hat{\beta}_{0S}^* - \log(n_{S1}/n_{S0}))'$, asymptotically follows the normal distribution with estimated covariance matrix given by

$$[\mathbf{I}_o(\hat{\boldsymbol{\beta}}_0^*, \hat{\boldsymbol{\beta}})/n]^{-1} - \begin{bmatrix} \boldsymbol{\Gamma} & \mathbf{0} \\ \mathbf{0} & \mathbf{0} \end{bmatrix}, \tag{7.40}$$

where $\hat{\boldsymbol{\beta}}_0^*$, and $\hat{\boldsymbol{\beta}}$ are the MLEs of $\boldsymbol{\beta}_0^*$ and $\boldsymbol{\beta}, \mathbf{I}_o(\hat{\boldsymbol{\beta}}_0^*, \hat{\boldsymbol{\beta}})$ is the observed information matrix, and $\boldsymbol{\Gamma}$ is given by the diagonal matrix diag $(1/\hat{\rho}_{11} + 1/\hat{\rho}_{10}, 1/\hat{\rho}_{21} + 1/\hat{\rho}_{20}, \ldots, 1/\hat{\rho}_{S1} + 1/\hat{\rho}_{S0})$, with $\hat{\rho}_{sj} = n_{sj}/n$. Thus, the estimated asymptotic covariance matrix $\widehat{\mathrm{Cov}}(\hat{\boldsymbol{\phi}}^*)$ is given by $(\mathbf{I}_o(\hat{\boldsymbol{\beta}}_0^*, \hat{\boldsymbol{\beta}})^{-1})_{11} - \mathrm{diag}(1/n_{11} + 1/n_{10}, 1/n_{21} + 1/n_{20}, \ldots, 1/n_{S1} + 1/n_{S0})$, where $(\mathbf{I}_o(\hat{\boldsymbol{\beta}}_0^*, \hat{\boldsymbol{\beta}})^{-1})_{11}$ is the upper left $S \times S$ matrix of the inverse of the observed information matrix. Based on these results, the MLE of AR_s is

$$\widehat{\mathrm{AR}}_s = 1 - \exp(\hat{\beta}_{0s}^* - \log(n_{s1}/n_{s0})). \tag{7.41}$$

The asymptotic $100(1-\alpha)$ percent confidence interval for AR_s is then given by

$$\begin{aligned}&[1 - \exp(\hat{\beta}_{0s}^* - \log(n_{s1}/n_{s0}) + Z_{\alpha/2}\sqrt{\widehat{\mathrm{Var}}(\hat{\phi}_s^*)}),\\ &\quad 1 - \exp(\hat{\beta}_{0s}^* - \log(n_{s1}/n_{s0}) - Z_{\alpha/2}\sqrt{\widehat{\mathrm{Var}}(\hat{\phi}_s^*)})].\end{aligned} \tag{7.42}$$

To estimate the AR due to reducing from $\mathbf{z}$ to $\mathbf{z}_0$, we may use

$$\widehat{\mathrm{AR}} = \sum_s \hat{p}_{s|1} \widehat{\mathrm{AR}}_s, \tag{7.43}$$

where $\hat{p}_{s|1} = n_{s|1}/n_1$. Define $\widehat{\mathbf{AR}}' = (\widehat{\mathrm{AR}}_1, \widehat{\mathrm{AR}}_2, \ldots, \widehat{\mathrm{AR}}_S)$, $\hat{\mathbf{p}}' = (\hat{p}_{1|1}, \hat{p}_{2|1}, \ldots, \hat{p}_{S|1})$, and $\hat{\mathbf{A}}' = ((1-\widehat{\mathrm{AR}}_1)\hat{p}_{1|1}, (1-\widehat{\mathrm{AR}}_2)\hat{p}_{2|1}, \ldots, (1-\widehat{\mathrm{AR}}_S)\hat{p}_{S|1})$. Drescher and Schill (1991) note that $\widehat{\mathbf{AR}}$ is asymptotically independent of $\hat{\mathbf{p}}$. Thus, using the delta method, we obtain the estimated asymptotic variance (Drescher and Schill, 1991)

$$\widehat{\mathrm{Var}}(\widehat{\mathrm{AR}}) = \widehat{\mathbf{AR}}'\widehat{\mathrm{Cov}}(\hat{\mathbf{p}})\widehat{\mathbf{AR}} + \hat{\mathbf{A}}'\widehat{\mathrm{Cov}}(\hat{\boldsymbol{\phi}}^*)\hat{\mathbf{A}}, \tag{7.44}$$

where $\widehat{\mathrm{Cov}}(\hat{\mathbf{p}})$ and $\widehat{\mathrm{Cov}}(\hat{\boldsymbol{\phi}}^*)$ are the estimated covariance matrices of $\hat{\mathbf{p}}$ and $\hat{\boldsymbol{\phi}}^*$, respectively. To improve the normal approximation of $\widehat{\mathrm{AR}}$, we may use the logarithmic transformation of $1-\widehat{\mathrm{AR}}$. This leads to an asymptotic $100(1-\alpha)$

percent confidence interval for AR given by

$$\left[1-\exp\left(\log(1-\widehat{\mathrm{AR}})+Z_{\alpha/2}\frac{\sqrt{\widehat{\mathrm{Var}}(\widehat{\mathrm{AR}})}}{1-\widehat{\mathrm{AR}}}\right),\right.$$
$$\left.1-\exp\left(\log(1-\widehat{\mathrm{AR}})-Z_{\alpha/2}\frac{\sqrt{\widehat{\mathrm{Var}}(\widehat{\mathrm{AR}})}}{1-\widehat{\mathrm{AR}}}\right)\right]. \tag{7.45}$$

Example 7.9 To allow readers to compare the results using different approaches to estimating the AR, we first consider the same data (Table 7.2) as discussed in Example 7.5. Suppose that we want to estimate AR due to radon-daughter exposure measured in working level months (WLM) when reducing 'WLM $\geq$ 120' to 'WLM $<$ 120' based on the logistic regression model $P(D=1|z_1, C_s) = \exp(\beta_{0s}+\beta_1 z_1)/(1+\exp(\beta_{0s}+\beta_1 z_1))$, where $z_1 = 1$ for WLM ≥ 120, and $z_1 = 0$ otherwise; $S = 2$ for the two strata formed by the cumulative smoking levels (pack-years) less than 20 and 20 or more. Using (7.41) and (7.42), we obtain stratum-specific estimates $\widehat{\mathrm{AR}}_s$ (and 95% confidence intervals) of 0.606 ([0.203, 0.805]) and 0.635 ([0.256, 0.821]), respectively. Thus, using (7.43), we obtain a summary estimate $\widehat{\mathrm{AR}}$ of 0.631, which is similar to the estimate $\widehat{\mathrm{AR}}_{\mathrm{apx}} = 0.639$ obtained in Example 7.5. Using (7.45), we obtain a 95% confidence interval for the AR of [0.248, 0.819], again similar to that obtained using (7.15).

Example 7.10 To illustrate the use of the logistic regression model when there are multiple variables of exposure, we consider the retrospective data (Table 7.4) relating myocardial infarction to recent oral contraceptive use and cigarette smoking with strata formed by different age categories (Shapiro *et al.*, 1979). Suppose that we are interested in estimating the AR due to the joint effects of recent oral contraceptive use and number of cigarettes smoked per day while controlling the confounding effect due to age. We assume the logistic

Table 7.3 Cumulative radiation exposure in WLM units for US uranium miners and mortality rates for lung cancer among US white males aged 50–54.

	Cumulative smoking (pack-years)			
	<20		≥20	
	WLM ≥120	WLM <120	WLM ≥120	WLM <120
Cases	10	4	12	2
Controls	26	29	30	27

Source: Whittemore (1982).

Table 7.4 Frequency distribution of women with myocardial infarction versus recent oral contraceptive use, cigarette smoking level (number of cigarettes smoked per day) and age.

Age	Smoking	Oral contraceptive	Cases	Controls
25–34	0	No	1	281
		Yes	0	38
	1–24	No	5	221
		Yes	2	35
	≥25	No	8	112
		Yes	11	22
35–44	0	No	13	318
		Yes	1	12
	1–24	No	32	249
		Yes	1	15
	≥25	No	53	125
		Yes	8	8
45–49	0	No	20	155
		Yes	3	2
	1–24	No	42	96
		Yes	0	1
	≥25	No	31	50
		Yes	3	2

Source: Shapiro *et al.* (1979).

regression model $P(D = 1|z_1, z_2, z_3, C_s) = \exp(\beta_{0s} + \beta_1 z_1 + \beta_2 z_2 + \beta_3 z_3)/\ (1 + \exp(\beta_{0s} + \beta_1 z_1 + \beta_2 z_2 + \beta_3 z_3))$, where $z_1 = 0$ and $z_2 = 0$ for patients with zero cigarettes smoked per day, $z_1 = 1$ and $z_2 = 0$ for patients with 1–24 cigarettes smoked per day, and $z_1 = 0$ and $z_2 = 1$ for patients with 25 or more cigarettes smoked per day; z_3 is 1 for patients with recent oral contraceptive use, and 0 otherwise; and the number S of strata equals 3, for the three age categories 25–34, 35–44, and 45+. Using (7.41) and (7.42), we obtain stratum-specific estimates $\widehat{AR}_s$ (and 95% confidence interval) of 0.758 ([0.650, 0.833]), 0.708 ([0.598, 0.788]), and 0.645 ([0.525, 0.734]), respectively. Using (7.43) and (7.45), we obtain a summary estimate $\widehat{AR}$ of 0.687, with asymptotic 95% confidence interval [0.577, 0.769].

7.6 CASE–CONTROL STUDIES UNDER INVERSE SAMPLING

Suppose that we employ independent inverse sampling, in which we continue sampling subjects until we obtain the predetermined number $x_j(> 0)$ of subjects with exposure from the case ($j = 1$) and the control ($j = 0$) populations, respectively. Let Y_j denote the number of subjects with non-exposure collected before obtaining exactly the desired x_j from group j. Then the random variable

Y_j follows the negative binomial distribution (1.13) with parameters x_j and $\pi_{1|j}$, where $\pi_{1|j}$ denotes the conditional probability of being exposed, given the subject is a case ($j = 1$) or a control ($j = 0$). Recall that the AR can be approximated by $P(E|D)(\text{OR} - 1)/\text{OR} = 1 - \phi^*$, where $\phi^* = (1 - \pi_{1|1})/(1 - \pi_{1|0})$ when the disease is rare. Define $N_j = x_j + Y_j$. The MLE of $\pi_{1|j}$ is $\hat{\pi}_{1|j} = x_j/N_j$ under (1.13), with estimated asymptotic variance $\hat{\pi}^2_{1|j}(1 - \hat{\pi}_{1|j})/x_j$ (**Exercise 1.11**). Thus, the MLE of the AR is $\widehat{\text{AR}} = 1 - \hat{\phi}^*$, where $\hat{\phi}^* = (1 - \hat{\pi}_{1|1})/(1 - \hat{\pi}_{1|0})$. It is easy to show that an estimated asymptotic variance $\widehat{\text{Var}}(\widehat{\text{AR}})$ is $(\hat{\phi}^*)^2\{\hat{\pi}^2_{1|1}/[x_1(1 - \hat{\pi}_{1|1})] + \hat{\pi}^2_{1|0}/[x_0(1 - \hat{\pi}_{1|0})]\}$. Thus, an asymptotic $100(1 - \alpha)$ percent confidence interval for AR is given by

$$[\widehat{\text{AR}} - Z_{\alpha/2}\sqrt{\widehat{\text{Var}}(\widehat{\text{AR}})},\ \min\{\widehat{\text{AR}} + Z_{\alpha/2}\sqrt{\widehat{\text{Var}}(\widehat{\text{AR}})},\ 1\}]. \tag{7.46}$$

Since the sampling distribution of $\widehat{\text{AR}}$ is likely skewed when x_j is not large and $\pi_{1|j}$ is small, we may consider use of a logarithmic transformation to improve the normal approximation. We obtain an asymptotic $100(1 - \alpha)$ percent confidence interval for the AR given by

$$[1 - \hat{\phi}^* \exp(Z_{\alpha/2}\sqrt{\widehat{\text{Var}}(\log(\widehat{\text{AR}}))}),\ 1 - \hat{\phi}^* \exp(-Z_{\alpha/2}\sqrt{\widehat{\text{Var}}(\log(\hat{\phi}^*))})], \tag{7.47}$$

where $\widehat{\text{Var}}(\log(\widehat{\text{AR}})) = \hat{\pi}^2_{1|1}/[x_1(1 - \hat{\pi}_{1|1})] + \hat{\pi}^2_{1|0}/[x_0(1 - \hat{\pi}_{1|0})]$.

Recall that for $x_i > 1$, the unbiased estimator of $\pi_{1|j}$ under (1.13) is $\hat{\pi}^{(u)}_{1|j} = (x_j - 1)/(N_j - 1)$. Therefore, if both x_j are large, we have $P(\{[(1 - \hat{\pi}^{(u)}_{1|1}) - \phi^*(1 - \hat{\pi}^{(u)}_{1|0})]/\sqrt{\text{Var}((1 - \hat{\pi}^{(u)}_{1|1}) - \phi^*(1 - \hat{\pi}^{(u)}_{1|0}))}\}^2 \leq Z^2_{\alpha/2}) \doteq 1 - \alpha$. Because we can estimate the variance $\text{Var}(1 - \hat{\pi}^{(u)}_{1|j})$ by the unbiased estimator $\hat{\pi}^{(u)}_{1|j}(1 - \hat{\pi}^{(u)}_{1|j})/(N_j - 2)$ (1.18), we arrive at the following quadratic equation in ϕ^*:

$$A^{\ddagger}\phi^{*2} - 2B^{\ddagger}\phi^* + C^{\ddagger} \leq 0, \tag{7.48}$$

where $A^{\ddagger} = (1 - \hat{\pi}^{(u)}_{1|0})^2 - Z^2_{\alpha/2}\hat{\pi}^{(u)}_{1|0}(1 - \hat{\pi}^{(u)}_{1|0})/(N_0 - 2)$, $B^{\ddagger} = (1 - \hat{\pi}^{(u)}_{1|1})(1 - \hat{\pi}^{(u)}_{1|0})$, and $C^{\ddagger} = (1 - \hat{\pi}^{(u)}_{1|1})^2 - Z^2_{\alpha/2}\hat{\pi}^{(u)}_{1|1}(1 - \hat{\pi}^{(u)}_{1|1})/(N_1 - 2)$. If $A^{\ddagger} > 0$ and $B^{\ddagger 2} - A^{\ddagger}C^{\ddagger} > 0$, then an asymptotic $100(1 - \alpha)$ percent confidence interval for the AR is given by

$$[1 - (B^{\ddagger} + \sqrt{B^{\ddagger 2} - A^{\ddagger}C^{\ddagger}})/A^{\ddagger},\ 1 - \max\{(B^{\ddagger} - \sqrt{B^{\ddagger 2} - A^{\ddagger}C^{\ddagger}})/A^{\ddagger},\ 0\}]. \tag{7.49}$$

Note that (7.46) (7.47), and (7.49) are derived on the basis of large-sample theory. When x_j is small, these interval estimators may not be valid. However, we can derive a $100(1 - \alpha)$ percent confidence interval on the basis of the exact conditional distribution, given that the marginal $Y_1 + Y_0 = y_.$ is fixed. This exact conditional confidence interval is can be used even when the number of subjects x_j with exposure is as small as 1.

As shown elsewhere (Lui, 1995), the conditional distribution of Y_1, given a fixed total number of subjects with exposure $y_{.} = Y_1 + Y_0$, is then (**Exercise 3.11**)

$$P(Y_1 = y_1 | y_{.}, x_1, x_0, \phi^*) = \frac{\binom{y_1 + x_1 - 1}{y_1}\binom{y_{.} - y_1 + x_0 - 1}{y_{.} - y_1}(\phi^*)^{y_1}}{\sum_{y=0}^{y_{.}}\binom{y + x_1 - 1}{y}\binom{y_{.} - y + x_0 - 1}{y_{.} - y}(\phi^*)^{y}}, \quad (7.50)$$

where $y_1 = 0, 1, \ldots, y_{.}$. On the basis of conditional distribution (7.50), note that the conditional MLE $\widehat{AR}_{cond}$ of the AR is $1 - \hat{\phi}^*_{cond}$, where $\hat{\phi}^*_{cond}$ is the conditional MLE of ϕ^* and is obtained by solving the equation $y_1 = E(Y_1 | y_{.}, x_1, x_0, \phi^*)$ for ϕ^* (**Exercise 3.13**). Furthermore, we may obtain the estimated asymptotic conditional variance $\widehat{Var}(\hat{\phi}^*_{cond}) = (\hat{\phi}^*_{cond})^2 / Var(Y_1 | y_{.}, x_1, x_0, \hat{\phi}^*_{cond})$ from the inverse Fisher information matrix. The sufficient and necessary conditions for the unique existence of the conditional MLE of ϕ are given in **Exercise 3.14**. An asymptotic $100(1-\alpha)$ percent conditional confidence interval for the AR is

$$[1 - \hat{\phi}^*_{cond} - Z_{\alpha/2}\sqrt{\widehat{Var}(\hat{\phi}^*_{cond})},\ 1 - \max\{\hat{\phi}^*_{cond} - Z_{\alpha/2}\sqrt{\widehat{Var}(\hat{\phi}^*_{cond})},\ 0\}]. \quad (7.51)$$

Note that $\sum_{y=0}^{y_1} P(Y = y | y_{.}, x_1, x_0, \phi^*)$ is a decreasing function of ϕ^*. Thus, we can obtain an exact $100(1-\alpha)$ percent confidence interval $[\phi^*_l, \phi^*_u]$ by solving the following two equations (Casella and Berger, 1990) for ϕ^*_l and ϕ^*_u:

$$\sum_{y=y_1}^{y_{.}} P(Y = y | y_{.}, x_1, x_0, \phi^*_l) = \alpha/2,$$

$$\sum_{y=0}^{y_1} P(Y = y | y_{.}, x_1, x_0, \phi^*_u) = \alpha/2. \quad (7.52)$$

If y_1 were 0, then we would define the lower limit ϕ^*_l to be 0. Similarly, if y_0 were 0 (or equivalently, $y_1 = y_{.}$), we would define the upper limit $\phi^*_u = \infty$. An exact $100(1-\alpha)$ percent confidence interval for the AR is then given by

$$[1 - \phi^*_u, 1 - \phi^*_l]. \quad (7.53)$$

Example 7.11 For a rare disease in a case–control study, suppose that we wish to obtain an approximate 95% confidence interval for the AR. Suppose further that we employ inverse sampling and collect ($y_1 =$)5 and ($y_0 =$)25 subjects before we obtain 50 subjects ($= x_1 = x_0$) with exposure from the case and the control groups, respectively. From these data, application of an iterative numerical procedure to solve $\Sigma_{y=5}^{y_{.}} P(Y = y | y_{.} = 30, x_1 = x_0 = 50, \phi^*_l) = 0.025$ and $\Sigma_{y=0}^{5} P(Y = y | y_{.} = 30, x_1 = x_0 = 50, \phi^*_u) = 0.025$ leads to $\phi^*_l = 0.086$ and $\phi^*_u = 0.657$. Therefore, an approximate 95% confidence interval for AR is [0.343, 0.914].

EXERCISES

7.1. Show that we can rewrite the AR as $[P(D) - P(D|\overline{E})]/P(D)$, where $P(D|\overline{E})$ and $P(D)$ represent the prevalence of disease in the non-exposed and the general populations, respectively.

7.2. Show that we can express the AR as $P(E)(RR - 1)/[P(E)(RR - 1) + 1]$, where $RR = P(D|E)/P(D|\overline{E})$ denotes the risk ratio between the exposed and non-exposed populations. Thus, the AR depends on both the prevalence $P(E)$ and RR. A risk factor with a large RR does not necessarily lead to a large value of AR if the prevalence of this risk factor $P(E)$ is small.

7.3. Suppose that the prevalence of smoking and the prevalence of drinking in a population are 0.02 and 0.30, respectively. Suppose further that the RR of possessing coronary artery disease for smoking is 10, while the RR of possessing coronary artery disease for drinking is only 2. What are the corresponding ARs for smoking and drinking? From the public health point of view, which risk factor, if eliminated, would achieve a higher proportional reduction in coronary artery disease?

7.4. In the notation of Section 7.1.1, when n is large, show that $\sqrt{n}(\log(\hat{\phi}) - \log(\phi))$, where $\hat{\phi} = \hat{\pi}_{01}/(\hat{\pi}_{0.}\hat{\pi}_{.1})$, has the asymptotic normal distribution with mean 0 and variance $(1 - \pi_{01})/\pi_{01} - (\pi_{0.} + \pi_{.1} - 2\pi_{01})/(\pi_{0.}\pi_{.1})$. (Hint: When n is large, $\sqrt{n}[(\hat{\pi}_{01}, \hat{\pi}_{0.}, \hat{\pi}_{.1})' - (\pi_{01}, \pi_{0.}, \pi_{.1})']$ has asymptotic normal distribution with mean vector $(0, 0, 0)'$ and covariance matrix $\mathbf{\Sigma}$ with diagonal terms equal to $\pi_{01}(1 - \pi_{01})$, $\pi_{0.}(1 - \pi_{0.})$, and $\pi_{.1}(1 - \pi_{.1})$, and with off-diagonal terms given by the covariances $\text{Cov}(\hat{\pi}_{01}, \hat{\pi}_{0.}) = \pi_{01}(1 - \pi_{0.})$, $\text{Cov}(\hat{\pi}_{01}, \hat{\pi}_{.1}) = \pi_{01}(1 - \pi_{.1})$, and $\text{Cov}(\hat{\pi}_{0.}, \hat{\pi}_{.1}) = -(\pi_{11}\pi_{00} - \pi_{10}\pi_{01})$. We then use the delta method and the function $f(X_1, X_2, X_3) = \log(X_1/(X_2X_3))$.)

7.5. Consider the situation discussed in Section 7.1.1. We define $Z = \hat{\pi}_{01} - f\phi\hat{\pi}_{0.}\hat{\pi}_{.1}$, where $f = n/(n-1)$. (a) Show that the expectation E(Z) converges to 0 as n increases to ∞. Thus, when n is large, we have the probability $P((\hat{\pi}_{01} - f\phi\hat{\pi}_{0.}\hat{\pi}_{.1})^2/\text{Var}(\hat{\pi}_{01} - f\phi\hat{\pi}_{0.}\hat{\pi}_{.1})) \doteq 1 - \alpha$, where the asymptotic variance $\text{Var}(\hat{\pi}_{01} - f\phi\hat{\pi}_{0.}\hat{\pi}_{.1}) = \text{Var}(\hat{\pi}_{01}) + f^2\phi^2\text{Var}(\hat{\pi}_{0.}\hat{\pi}_{.1}) - 2f\phi\text{Cov}(\hat{\pi}_{01}, \hat{\pi}_{0.}\hat{\pi}_{.1})$. (b) Based on this result, derive the following quadratic equation in $\phi : \mathfrak{A}\phi^2 - 2\mathfrak{B}\phi + \mathfrak{C} \leq 0$, where $\mathfrak{A} = f^2[(\hat{\pi}_{0.}\hat{\pi}_{.1})^2 - Z^2_{\alpha/2}\hat{\pi}_{0.}\hat{\pi}_{.1}(\hat{\pi}_{11} + \hat{\pi}_{00} + 4(\hat{\pi}_{01} - \hat{\pi}_{0.}\hat{\pi}_{.1}))/n]$, $\mathfrak{B} = f[\hat{\pi}_{0.}\hat{\pi}_{.1}\hat{\pi}_{01} - Z^2_{\alpha/2}\hat{\pi}_{01}(\hat{\pi}_{.1}(1 - \hat{\pi}_{0.}) + \hat{\pi}_{0.}(1 - \hat{\pi}_{.1}))/n]$, and $\mathfrak{C} = \hat{\pi}^2_{01} - Z^2_{\alpha/2}\hat{\pi}_{01}(1 - \hat{\pi}_{01})/n$. If both $\mathfrak{A} > 0$ and $\mathfrak{B}^2 - \mathfrak{A}\mathfrak{C} > 0$, then an asymptotic $100(1 - \alpha)$ percent confidence interval for AR will be $[1 - (\mathfrak{B} + \sqrt{\mathfrak{B}^2 - \mathfrak{A}\mathfrak{C}})/\mathfrak{A}, 1 - \max\{(\mathfrak{B} - \sqrt{\mathfrak{B}^2 - \mathfrak{A}\mathfrak{C}})/\mathfrak{A}, 0\}]$.

7.6. When there are no confounders, show that the attributable risk AR can be expressed as $P(E|D)(RR - 1)/RR$, where $P(E|D)$ denotes the exposure prevalence in the case population and $RR = P(D|E)/P(D|\overline{E})$.

7.7. For a case–control study with no confounders, show that $P(E|D)(OR - 1)/OR = (\pi_{1|1} - \pi_{1|0})/(1 - \pi_{1|0})$, where the $\pi_{i|j}$ are defined in Section 7.1.2.

7.8. Show that the estimated asymptotic variance for $\widehat{\text{AR}}$ is $\widehat{\text{Var}}(\widehat{\text{AR}}) = (\hat{\phi}^*)^2 \{\hat{\pi}_{1|1}/[n_1(1-\hat{\pi}_{1|1})] + \hat{\pi}_{1|0}/[n_0(1-\hat{\pi}_{1|0})]\}$, where $\widehat{\text{AR}}$, $\hat{\phi}^*$, and $\hat{\pi}_{i|j}$ are defined in Section 7.1.2.

7.9. Consider the data consisting of subjects aged 75–79 taken from the case–control study (Cole *et al.*, 1971; Schlesselman, 1982, p. 49) described in Example 7.3. Exposure is defined according to whether the number of cigarettes smoked during a patient's lifetime is at least 100. As shown elsewhere, we have $X_1 = 46$, $n_1 - X_1 = 7$, $X_0 = 42$, and $n_0 - X_0 = 15$. What is the MLE $\widehat{\text{AR}}$? What are the 95% confidence intervals for AR using interval estimators (7.5) and (7.6)?

7.10. Show that the AR in the presence of confounders $C_s (s = 1, 2, \ldots, S)$ can be expressed as $\sum_s \text{P}(C_s|D)\text{AR}_s$, where $\text{AR}_s = 1 - \text{P}(D|\overline{E}, C_s)/\text{P}(D|C_s) = [\text{P}(D|E, C_s) - \text{P}(D|\overline{E}, C_s)]\text{P}(E|C_s)]/\text{P}(D|C_s) = \text{P}(E|D, C_s)(\text{RR}_s - 1)/\text{RR}_s = \text{P}(E|C_s)(\text{RR}_s - 1)/[\text{P}(E|C_s)(\text{RR}_s - 1) + 1]$, and where $\text{RR}_s = \text{P}(D|E, C_s)/\text{P}(D|\overline{E}, C_s)$, the risk ratio at the confounder level C_s.

7.11. Show that when the underlying risk ratio RR_s is constant and equal to RR_0 for $s = 1, 2, \ldots, S$, the AR in the presence of confounders can be simplified to $\text{P}(E|D)(\text{RR}_0 - 1)/\text{RR}_0$, where $\text{P}(E|D)$ is the exposure prevalence in the case population.

7.12. Show that AR $\left(= \sum_s \text{P}(C_s|D)\text{AR}_s\right)$ in the presence of confounders $C_s (s = 1, 2, \ldots, S)$ can be expressed as $1 - \sum_s \text{P}(C_s|D)\text{P}(\overline{E}|C_s, D)/\text{P}(\overline{E}|C_s)$. When the underlying disease is rare, this can be further approximated by $1 - \sum_s \text{P}(C_s|D)$ $\text{P}(\overline{E}|C_s, D)/\text{P}(\overline{E}|C_s, \overline{D})$.

7.13. Show that the estimated asymptotic variance (Whittemore, 1982) of $\hat{\phi}^*_{\text{conf}}$ is given by

$$\widehat{\text{Var}}(\hat{\phi}^*_{\text{conf}}) = \frac{1}{n_1}\left\{\sum_s \left(\frac{\hat{\pi}^2_{0s|1}\hat{\pi}^2_{.s|0}}{\hat{\pi}^2_{0s|0}}\right)\left(\frac{1}{\hat{\pi}_{0s|1}} + \frac{n_1\hat{\pi}_{1s|0}}{n_0\hat{\pi}_{.s|0}\hat{\pi}_{0s|0}}\right) - (\hat{\phi}^*_{\text{conf}})^2\right\},$$

where $\hat{\phi}^*_{\text{conf}}$ is as defined in Section 7.2.2.

7.14. Consider the data (Table 7.3) consisting of white miners aged 50–54 (Whittemore, 1982, p. 234). Suppose that, as in Example 7.5, we are interested in estimation of the AR due to radon-daughter exposure measured in working level months (WLM) by reducing 'WLM $\geq$ 120' to 'WLM $<$ 120', while controlling the confounder of cumulative smoking (pack-years) categorized as less than 20 or at least 20. What is the MLE $\widehat{\text{AR}}$? What are the 95% confidence intervals for the AR using interval estimators (7.14), (7.15), and (7.16)?

7.15. Show that the asymptotic variance of $\widehat{\text{AR}}(= \hat{\pi}_{1.}(\hat{\pi}_{10} - \hat{\pi}_{01})/\hat{\pi}_{10})$ defined in Section 7.3 is given by $\text{Var}(\widehat{\text{AR}}) = \{(\pi_{10} - \pi_{01})^2\pi_{11} + (\pi^2_{10} + \pi_{01}\pi_{11})^2/\pi_{10} + \pi^2_{1.}\pi_{01} - [\pi_{1.}(\pi_{10} - \pi_{01})]^2\}/(n\pi^2_{10})$.

7.16. Show that from the probability $\text{P}((\widehat{\text{AR}} - \text{AR})^2/\text{Var}(\widehat{\text{AR}}) \leq Z^2_{\alpha/2}) \doteq 1 - \alpha$, as discussed in Section 7.3, where $\text{Var}(\widehat{\text{AR}}) = \{(\pi_{10} - \pi_{01})^2\pi_{11} + (\pi^2_{10} +$

$\pi_{01}\pi_{11})^2/\pi_{10} + \pi_{1.}^2\pi_{01}\}/(n\pi_{10}^2) - \mathrm{AR}^2/n$, we can derive asymptotic confidence limits based on the following quadratic equation in AR: $A^\dagger \mathrm{AR}^2 - 2B^\dagger \mathrm{AR} + C^\dagger \leq 0$, where $A^\dagger = 1 + Z_{\alpha/2}^2/n$, $B^\dagger = \widehat{\mathrm{AR}}$, and $C^\dagger = \widehat{\mathrm{AR}}^2 - Z_{\alpha/2}^2\{(\hat{\pi}_{10} - \hat{\pi}_{01})^2\hat{\pi}_{11} + (\hat{\pi}_{10}^2 + \hat{\pi}_{01}\hat{\pi}_{11})^2/\hat{\pi}_{10} + \hat{\pi}_{1.}^2\hat{\pi}_{01}\}/(n\hat{\pi}_{10}^2)$.

7.17. Consider the data consisting of 80 matched pairs in a study of the association between the number of beverages drunk 'burning hot' and esophageal cancer (Breslow, 1982, p. 665). For the purpose of illustration only, suppose we define exposure as 'at least one beverage drunk "burning hot" ' and non-exposure as 'no beverages drunk "burning hot" '. In the notation of Section 7.3, we have $n_{11} = 7$, $n_{10} = 32$, $n_{01} = 10$, and $n_{00} = 31$. What is the MLE $\widehat{\mathrm{AR}}$? What are the 95% confidence intervals for the AR using interval estimators (7.17)–(7.21)?

7.18. Show that $[\mathrm{P}(D|E_k) - \mathrm{P}(D|E_0)]\mathrm{P}(E_k)/\mathrm{P}(D) = \mathrm{P}(E_k)(\mathrm{RR}_k - 1)/\left[1 + \sum_k \mathrm{P}(E_k)(\mathrm{RR}_k - 1)\right]$, where $\mathrm{RR}_k = \mathrm{P}(D|E_k)/\mathrm{P}(D|E_0)$.

7.19. Show that the approximation of AR_k (7.23) can be expressed as $(\pi_{k|1}\pi_{0|0} - \pi_{0|1}\pi_{k|0})/\pi_{0|0}$ (see (7.24)), where the $\pi_{k|j}$ are as defined in Section 7.4.

7.20. For $K = 1$, show that $\widehat{\mathrm{AR}}_1$(7.25) reduces to the MLE $\widehat{\mathrm{AR}} = (\hat{\pi}_{1|1} - \hat{\pi}_{1|0})/\hat{\pi}_{0|0}$ for the case of dichotomous exposure and that variance (7.26) simplifies to variance $\mathrm{Var}(\widehat{\mathrm{AR}})$ (7.4).

7.21. Using formula (7.29), derive the quadratic equation: $\mathcal{A}\mathrm{AR}_k^2 - 2\mathcal{B}\mathrm{AR}_k + \mathcal{C} \leq 0$, where $\mathcal{A} = 1 + Z_{\alpha/2}^2/n_1$, $\mathcal{B} = \widehat{\mathrm{AR}}_k$, and

$$\mathcal{C} = \widehat{\mathrm{AR}}_k^2 - Z_{\alpha/2}^2\left[\hat{\pi}_{k|1} + \left(\frac{\hat{\pi}_{k|0}}{\hat{\pi}_{0|0}}\right)^2\hat{\pi}_{0|1}\right]\Big/ n_1$$
$$- Z_{\alpha/2}^2\left[(\hat{\pi}_{0|1})^2\left(\frac{\hat{\pi}_{k|0}}{\hat{\pi}_{0|0}^2} + \frac{\hat{\pi}_{k|0}^2}{\hat{\pi}_{0|0}^3}\right)\right]\Big/ n_0.$$

7.22. Show that when there are multiple levels $E_k (k = 0, 1, \ldots, K)$ of exposure, the AR due to all exposure levels $k \geq 1$, as discussed in Section 7.4, can be expressed as $1 - \sum_{k=0}^{K}\mathrm{P}(E_k|D)/\mathrm{RR}_k$, where $\mathrm{RR}_k = \mathrm{P}(D|E_k)/\mathrm{P}(D|E_0)$.

7.23. Show that

$$\sum_{s=1}^{S}\sum_{k=1}^{K}[\mathrm{P}(D|E_k, C_s) - \mathrm{P}(D|E_0, C_s)]\mathrm{P}(E_k|C_s)\mathrm{P}(C_s)/\mathrm{P}(D)$$
$$= 1 - \sum_{s=1}^{S}\sum_{k=0}^{K}\mathrm{P}(E_k, C_s|D)/\mathrm{RR}_{k|s},$$

where $\mathrm{RR}_{k|s} = \mathrm{P}(D|E_k, C_s)/\mathrm{P}(D_0|E_0, C_s)$ (Hint: First, show that $[\mathrm{P}(D|E_k, C_s) - \mathrm{P}(D|E_0, C_s)]\mathrm{P}(E_k|C_s)]/\mathrm{P}(D|C_s) = \mathrm{P}(E_k|D, C_s)(\mathrm{RR}_{k|s} - 1)/\mathrm{RR}_{k|s}$.)

7.24. Consider the myocardial infarction (MI) data for women aged 40–44 (Shapiro, *et al.*, 1979). The observed frequencies from the case and control groups (in parentheses) falling in the categories 'none', '1–24', and '25+' are: 11 (169), 21 (134), and 39 (68), respectively. What are the MLEs $\widehat{\text{AR}}_1$ and $\widehat{\text{AR}}_2$ due to the '1–24' and '25+' levels? What are the corresponding 95% confidence intervals for $\text{AR}_k (k = 1, 2)$ when we apply interval estimators (7.27), (7.28), and (7.30)?

7.25. Show that for the logistic regression model containing only a single dichotomous exposure variable ($z = 1$ for being exposed and $z = 0$ for being non-exposed), the MLE (7.35) is given by $(\hat{\pi}_{1|1} - \hat{\pi}_{1|0})/(1 - \hat{\pi}_{1|0})$. Furthermore, the estimated asymptotic variance $\widehat{\text{Var}}(\hat{\phi}^*)$ is given by $\hat{\pi}_{1|1}/[n_1(1 - \hat{\pi}_{1|1})] + \hat{\pi}_{1|0}/[n_0(1 - \hat{\pi}_{1|0})]$.

7.26. Consider the data in Table 7.5 taken from a case–control study on esophageal cancer (Tuyns *et al.*, 1977; Breslow and Day, 1980). There were 200 males diagnosed with esophageal cancer in one of the regional hospitals in France between January 1972 and April 1974. The controls were a sample of 778 adult males drawn from electoral lists in each commune, of whom 775 provided sufficient data for analysis. Suppose that we are interested in estimating the AR due to the joint effects of alcohol and tobacco, while controlling the confounder of age. We assume the logistic regression model $P(Y = 1|\mathbf{z}, C_s) = \exp(\beta_{0s} + \boldsymbol{\beta}'\mathbf{z})/(1 + \exp(\beta_{0s} + \boldsymbol{\beta}'\mathbf{z}))$ where $s = 1, 2, \ldots, 6$, for the six age categories, and the covariate vector $\mathbf{z}' = (z_1, z_2, z_3, \ldots, z_6)$ consisting of six index variables. We define $z_1 = z_2 = z_3 = 0$ for the baseline alcohol level of 0–39 g/day; $z_1 = 1, z_2 = z_3 = 0$ for an alcohol level of 40–79; $z_2 = 1, z_1 = z_3 = 0$ for an alcohol level of 80–119; and $z_3 = 1, z_1 = z_2 = 0$ for an alcohol level of over 120. Similarly, we use the other three index variables z_4, z_5, and z_6 to define the four distinct levels of tobacco. For example, we define $z_4 = z_5 = z_6 = 0$ for the tobacco baseline level of 0–9 g/day; $z_4 = 1, z_5 = z_6 = 0$ for a tobacco level of 10–19, etc.
(a) Based on the above logistic regression model, what are the stratum-specific estimates $\widehat{\text{AR}}_s$ (7.41) and the 95% confidence intervals using (7.42) for $s = 1, 2, \ldots, 6$?
(b) What is the summary estimate $\widehat{\text{AR}}$ (7.43) over all strata?
(c) Using (7.45), what is the 95% confidence interval for AR?

Table 7.5 Grouped data from the Ille-et-Vilaine study of esophageal cancer.

Age	Alcohol (g/day)	Tobacco (g/day)	Cases	Controls
25–34	0–39	0–9	0	40
		10–19	0	10
		20–29	0	6
		30+	0	5
	40–79	0–9	0	27

Table 7.5 (*continued*)

Age	Alcohol (g/day)	Tobacco (g/day)	Cases	Controls
		10–19	0	7
		20–29	0	4
		30+	0	7
	80–119	0–9	0	2
		10–19	0	1
		20–29	0	0
		30+	0	2
	120+	0–9	0	1
		10–19	1	0
		20–29	0	1
		30+	0	2
35–44	0–39	0–9	0	60
		10–19	1	13
		20–29	0	7
		30+	0	8
	40–79	0–9	0	35
		10–19	3	20
		20–29	1	13
		30+	0	8
	80–119	0–9	0	11
		10–19	0	6
		20–29	0	2
		30+	0	1
	120+	0–9	2	1
		10–19	0	3
		20–29	2	2
		30+	0	0
45–54	0–39	0–9	1	45
		10–19	0	18
		20–29	0	10
		30+	0	4
	40–79	0–9	6	32
		10–19	4	17
		20–29	5	10
		30+	5	2
	80–119	0–9	3	13
		10–19	6	8
		20–29	1	4
		30+	2	2
	120+	0–9	4	0
		10–19	3	1
		20–29	2	1
		30+	4	0
55–64	0–39	0–9	2	47
		10–19	3	19
		20–29	3	9

(*continued overleaf*)

Table 7.5 (*continued*)

Age	Alcohol (g/day)	Tobacco (g/day)	Cases	Controls
		30+	4	2
	40–79	0–9	9	31
		10–19	6	15
		20–29	4	13
		30+	3	3
	80–119	0–9	9	9
		10–19	8	7
		20–29	3	3
		30+	4	0
	120+	0–9	5	5
		10–19	6	1
		20–29	2	1
		30+	5	1
65–74	0–39	0–9	5	43
		10–19	4	10
		20–29	2	5
		30+	0	2
	40–79	0–9	17	17
		10–19	3	7
		20–29	5	4
		30+	0	0
	80–119	0–9	6	7
		10–19	4	8
		20–29	2	1
		30+	1	0
	120+	0–9	3	1
		10–19	1	1
		20–29	1	0
		30+	1	0
75+	0–39	0–9	1	17
		10–19	2	4
		20–29	0	0
		30+	1	2
	40–79	0–9	2	3
		10–19	1	2
		20–29	0	3
		30+	1	0
	80–119	0–9	1	0
		10–19	1	0
		20–29	0	0
		30+	0	0
	120+	0–9	2	0
		10–19	1	0
		20–29	0	0
		30+	0	0

Source: Breslow and Day (1980).

REFERENCES

Anderson, J. A. (1972) Separate sample logistic discrimination. *Biometrika*, **59**, 19–35.

Anderson, T. W. (1958) *An Introduction to Multivariate Statistical Analysis*. Wiley, New York.

Antunes, C. M. F., Stolley, P. D., Rosenshein, N. B., Davies, J. L., Tonascia, J. A., Brown, C., Burnett, L., Rutledge, A., Pokempner, M. and Garcia, R. (1979) Endometrial cancer and estrogen use: report of a large case–control study. *New England Journal of Medicine*, **300**, 9–13.

Basu, S. and Landis, J. R. (1995) Model-based estimation of population attributable risk under cross-sectional sampling. *American Journal of Epidemiology*, **142**, 1338–1343.

Benichou, J. (1991) Methods of adjustment for estimating the attributable risk in case–control studies: a review. *Statistics in Medicine*, **10**, 1753–1773.

Benichou, J. and Gail, M. H. (1990) Variance calculations and confidence intervals for estimates of the attributable risk based on logistic models. *Biometrics*, **46**, 991–1003.

Breslow, N. (1982) Covariance adjustment of relative risk estimates in matched studies. *Biometrics*, **38**, 661–672.

Breslow, N. E. and Day, N. E. (1980) *Statistical Methods in Cancer Research*. International Agency for Research on Cancer, Lyon.

Bruzzi, P., Green, S. B., Byar, D. P., Brinton, L. A. and Schairer, D. (1985) Estimating the population attributable risk for multiple risk factors using case–control data. *American Journal of Epidemiology*, **122**, 904–914.

Casella, G. and Berger, R. L. (1990) *Statistical Inference*. Duxbury, Belmont, CA.

Cole, P. and MacMahon, B. (1971) Attributable risk percent in case–control studies. *British Journal of Preventive and Social Medicine*, **25**, 242–244.

Cole, P., Monson, R. R., Haning, H. and Friedell, G. H. (1971) Smoking and cancer of the lower urinary tract. *New England Journal of Medicine*, **284**, 129–134.

Coughlin, S. S., Benichou, J. and Weed, D. L. (1994) Attributable risk estimation in case–control studies. *Epidemiologic Reviews*, **16**, 51–64.

Denman, D. W. and Schlesselman, J. J. (1983) Interval estimation of the attributable risk for multiple exposure levels in case–control studies. *Biometrics*, **39**, 185–192.

Deubner, D. C., Wilkinson, W. E., Helms, M. J., *et al.* (1980) Logistic model estimation of death attributable to risk factors for cardiovascular disease in Evans County, Georgia. *American Journal of Epidemiology*, **112**, 135–143.

Drescher, K. and Schill, W. (1991) Attributable risk estimation from case–control data via logistic regression. *Biometrics*, **47**, 1247–1256.

Farewell, V. T. (1979) Some results on the estimation of logistic models based on retrospective data. *Biometrika*, **66**, 27–32.

Fleiss, J. L. (1979) Inference about population attributable risk from cross-sectional studies. *American Journal of Epidemiology*, **110**, 103–104.

Fleiss, J. L. (1981) *Statistical Methods for Rates and Proportions*, 2nd edition. Wiley, New York.

Gefeller, O. (1990) Theory and application of attributable risk estimation in cross-sectional studies. *Statistica Applicata*, **2**, 323–330.

Gefeller, O. (1992a) An annotated bibliography on the attributable risk. *Biometrical Journal*, **8**, 1007–1012.

Gefeller, O. (1992b) Comparison of adjusted attributable risk estimators. *Statistics in Medicine*, **11**, 2083–2091.

Greenland, S. (1987) Variance estimators for attributable fraction estimates consistent in both large strata and sparse data. *Statistics in Medicine*, **6**, 701–708.

Greenland, S. and Drescher, K. (1993) Maximum likelihood estimation of the attributable fraction from logistic models. *Biometrics*, **49**, 865–872.

Greenland, S. and Robins, J. M. (1988) Conceptual problems in the definition and interpretation of attributable fractions. *American Journal of Epidemiology*, **128**, 1185–1197.

Katz, D., Baptista, J., Azen, S. P. and Pike, M. C. (1978) Obtaining confidence intervals for the risk ratio in cohort studies. *Biometrics*, **34**, 469–474.

Kleinbaum, D. G., Kupper, L. L. and Morgenstern, H. (1982) *Epidemiological Research: Principles and Quantitative Methods*. Lifetime Learning, Belmont, CA.

Kuritz, S. J. and Landis, J. R. (1987) Attributable risk ratio estimation from matched-pairs case–control data. *American Journal of Epidemiology*, **125**, 324–328.

Kuritz, S. J. and Landis, J. R. (1988a) Summary attributable risk estimation from unmatched case–control data. *Statistics in Medicine*, **7**, 507–517.

Kuritz, S. J. and Landis, J. R. (1988b) Attributable risk estimation from matched case–control data. *Biometrics*, **44**, 355–367.

Last, J. M. (1983) *A Dictionary of Epidemiology*. Oxford University Press, New York.

Leung, H. M. and Kupper, L. L. (1981) Comparisons of confidence intervals for attributable risk. *Biometrics*, **37**, 293–302.

Levin, M. L. (1953) The occurrence of lung cancer in man. *Acta Unio Internationalis contra Cancrum*, **9**, 531–541.

Lui, K.-J. (1995) Notes on conditional confidence limits under inverse sampling. *Statistics in Medicine*, **14**, 2051–2056.

Lui, K.-J. (2001a) Notes on interval estimation of the attributable risk in cross-sectional sampling. *Statistics in Medicine*, **20**, 1797–1809.

Lui, K.-J. (2001b) Confidence intervals of the attributable risk under cross-sectional sampling with confounders. *Biometrical Journal*, **43**, 767–779.

Lui, K.-J. (2001c) Interval estimation of the attributable risk in case control studies with matched pairs. *Journal of Epidemiology and Community Health*, **55**, 885–890.

Lui, K.-J. (2003) Interval estimation of the attributable risk for multiple exposure levels in case–control studies with confounders. *Statistics in Medicine*, **22**, 2443–2457.

Miettinen, O. S. (1974) Proportion of disease caused or prevented by a given exposure, trait or intervention. *American Journal of Epidemiology*, **99**, 325–332.

Miettinen, O. S. (1976) Estimability and estimation in case-referent studies. *American Journal of Epidemiology*, **103**, 226–235.

Prentice, R. L. and Pyke, R. (1979) Logistic disease incidence models and case–control studies. *Biometrika*, **66**, 403–411.

SAS Institute, Inc. (1990) *SAS/STAT User's Guide, Volume 2, Version 6*. 4th edition. SAS Institute, Inc., Cary, NC.

Schlesselman, J. J. (1982). *Case–Control Studies*. Oxford University Press, New York.

Shapiro, S., Slone, D., Rosenberg, L., Kaufman, D. W., Stolley, P. D. and Miettinen, O.S. (1979) Oral contraceptive use in relation to myocardial infarction. *Lancet*, **1**, 743–747.

Tarone, R. E. (1981) On summary estimators of relative risk. *Journal of Chronic Diseases*, **34**, 463–468.

Tuyns, A. J., Pequignot, G. and Jensen, O. M. (1977) Le cancer de l'oesophage en Ille-et-Vilaine en fonction des niveaux de consommation d'alcool et tabac. *Bulletin of Cancer*, **64**, 45–60.

Walter, S. D. (1976) The estimation and interpretation of attributable risk in health research. *Biometrics*, **32**, 829–849.

Wermuth, N. (1976) Exploratory analyses of multidimensional contingency tables. In *Proceedings of the 9th International Biometric Conference*, Vol. 1. American Statistical Association and Biometric Society, Washington, DC, pp. 279–295.

Whittemore, A. S. (1982) Statistical methods for estimating attributable risk from retrospective data. *Statistics in Medicine*, **1**, 229–243.

Whittemore, A. S. (1983) Estimating attributable risk from case–control studies. *American Journal of Epidemiology*, **117**, 76–85.

8

Number Needed to Treat

The 'number needed to treat' (NNT), defined as the average number of patients who need to be treated in order to prevent one patient in the placebo group suffering an adverse event, was first proposed by Laupacis *et al.* (1988) to measure the discrepancy in adverse event rates between the treatment and placebo groups in randomized controlled clinical trials. To clarify the meaning of this measure, consider the numerical example given in the Veterans Administration Cooperative Study on hypertension (Veterans Administration Cooperative Study Group on Antihypertensive Agents, 1972). As given elsewhere (Laupacis *et al.*, 1988), the proportions of patients with no target-organ damage experiencing adverse event rates in the treatment and the control groups are 4.0% and 9.8%, respectively. If 100 patients were treated with antihypertensive agents, then we would expect to save six patients ($\doteq 100(0.098 - 0.04)$) with adverse events. Thus, the NNT, calculated as the number of patients treated divided by the expected number of patients saved from an adverse event, is equal to $17 (\doteq 100/6)$. This means that we need to treat 17 patients on average in order to prevent one patient with an adverse event. Similarly, in epidemiologic studies, consider the example of all-cause mortality rates between smokers and non-smokers given by Sheps (1958). The all-cause mortality rates were 1325 and 1884 per 100 000 person-years for non-smokers and heavy smokers, respectively. From these data, we obtain an NNT of $179 (\doteq 100\,000/[100\,000(0.01884 - 0.01325)]$. In other words, if 179 heavy smokers were persuaded to quit smoking every year, we would expect to save the life of one heavy smoker. Because clinicians find it easy to understand and interpret the NNT, this measure has recently become popular in reporting clinical findings in randomized controlled trials and in evidence-based medicine (Cook and Sackett, 1995; Tramèr *et al.*, 1995; Sackett *et al.*, 1996, 2000; McQuay and Moore, 1997; Chatellier *et al.*, 1996; Elferink and Van Zwieten-Boot, 1997). Note that the average number of subjects needed to treat in order to prevent a subject with adverse events is, by definition, always positive. To avoid the unrealistic case where we may obtain a negative NNT, as for defining the relative difference discussed in Chapter 3, we may wish to employ the NNT only when we have some prior knowledge to rank the order of the expected response rates

Statistical Estimation of Epidemiological Risk K-J. Lui
 ISBN: 0-470-85071-X (HB)

between the two groups under comparison (Lui, 2003). As noted elsewhere (Fleiss, 1981), this assumption is generally tenable if the control group is given an inert placebo, or if the control group is given an active drug and the treatment is that drug plus another active compound or that drug at a greater dosage level. Although Altman (1998) attempts to give an interpretation of a negative NNT, we do not allow negative values here so as to avoid obtaining the absurd results and escape criticisms of the use of NNT noted elsewhere (Altman, 1998; Hutton, 2000; Lesaffre and Pledger, 1999; Lui, 2003).

In this chapter, we begin by discussing the estimation of the NNT in the simplest situation – independent binomial sampling. We then extend this to cover the situation in which we employ pre-stratified sampling in multicenter studies. As noted in the previous chapters, these results are useful for meta-analysis. Also, we discuss how to incorporate the intraclass correlation between responses within clusters into estimation of the NNT in cluster randomization trials (Cornfield, 1978; Donner *et al.*, 1981; Klar and Donner, 2001; Lui *et al.*, 2000). Finally, because the matched-pair design is often used to increase efficiency in randomized clinical trials or in epidemiological studies, we consider estimation of NNT for paired-sample data as well.

8.1 INDEPENDENT BINOMIAL SAMPLING

Suppose that we are comparing the response rates between the treatment and control groups in a randomized controlled trial. Assume that a higher response rate is a better outcome. Without loss of generality, we assume that the response rate π_1 for the treatment is larger than the rate π_0 for the control. Note that this is equivalent to assuming that the non-response (or the adverse event) rate $\pi_0^*(= 1 - \pi_0)$ for the control is higher than the rate $\pi_1^*(= 1 - \pi_1)$ for the treatment. By definition, $\pi_1 - \pi_0 = \pi_0^* - \pi_1^*$. Suppose that we randomly assign n_1 patients to the treatment group and n_0 patients to the control group. In the treatment group, the expected number of subjects who responded, but who would not have done so if they had been assigned to the control group, is equal to $n_1(\pi_1 - \pi_0)$. Thus, we define the NNT as

$$\tau = n_1/[n_1(\pi_1 - \pi_0)] = 1/(\pi_1 - \pi_0), \tag{8.1}$$

which is simply the reciprocal of the risk difference (RD) discussed in Chapter 2. The range of NNT is, by definition, $\tau > 1$. In the extreme case where $\pi_1 = 1$ and $\pi_0 = 0$, we have NNT $= 1$. When the response rates between the two comparison groups are equal (i.e., $\pi_1 = \pi_0$), NNT $= \infty$. Thus, it is not convenient to base the NNT on testing whether the response rates are equal between two comparison groups. Note that we can rewrite (8.1) as $\tau = 1/[(\mathrm{RR} - 1)\pi_0]$, where $\mathrm{RR} = \pi_1/\pi_0$ is the risk ratio discussed in Chapter 4. Thus, given RR fixed, the lower the baseline response rate (i.e., π_0 is small), the larger is the NNT. Thus, it will be more effective to treat a patient with a high response-rate profile than one with a low response-rate profile. Note further that the NNT is very sensitive to rounding error when the underlying RD is extremely small – a small change in the scale of the RD may result in a large difference in the scale of NNT. For example, suppose that the true value of RD is 0.005, but that it is rounded to two decimal places and recorded as 0.01 for convenience. The NNT corresponding to the former is 200, and to the latter is 100. Thus, when the resulting estimate of NNT is large, we should treat this index with caution.

Let X_i denote the number of patients who respond among n_i patients assigned to the treatment ($i = 1$) and to the placebo ($i = 0$). Then X_i follows the binomial distribution (1.1) with parameters n_i and π_i. Using the functional invariance property (Casella and Berger, 1990; see the Appendix), the maximum likelihood estimator (MLE) of τ for $\hat{\pi}_1 > \hat{\pi}_0$ is simply

$$\hat{\tau} = 1/(\hat{\pi}_1 - \hat{\pi}_0), \tag{8.2}$$

where $\hat{\pi}_i = X_i/n_i$. Note that because $\tau > 1$, we define $\hat{\tau}$ (8.2) to be ∞ if $\hat{\pi}_1 - \hat{\pi}_0 \leq 0$. Note also that because $\hat{\tau}$ can be ∞ with positive probability, the point estimator $\hat{\tau}$ is biased and has no finite expectation. To obtain an interval estimator of τ, we may employ the monotonic transformation $\tau = 1/\Delta$, where $\Delta = \pi_1 - \pi_0$, over the range (0, 1), and interval estimators discussed in Chapter 2 for the RD. For brevity, we will present only a couple of simple interval estimators for the NNT here. Bender (2001) discusses using (2.7) to derive a confidence interval for the NNT. We refer readers to Chapter 2 for details.

First, applying the simplest, naive interval estimator (2.3) using Wald's statistic, we obtain an asymptotic $100(1 - \alpha)\%$ confidence interval for τ given by (**Exercise 8.2**)

$$\left[\frac{1}{\min\{\hat{\Delta} + Z_{\alpha/2}\sqrt{\hat{\pi}_1(1-\hat{\pi}_1)/n_1 + \hat{\pi}_0(1-\hat{\pi}_0)/n_0},\, 1\}},\right.$$
$$\left.\frac{1}{\max\{\hat{\Delta} - Z_{\alpha/2}\sqrt{\hat{\pi}_1(1-\hat{\pi}_1)/n_1 + \hat{\pi}_0(1-\hat{\pi}_0)/n_0}),\, 0\}}\right], \tag{8.3}$$

where $\hat{\Delta} = \hat{\pi}_1 - \hat{\pi}_0$ and Z_α is the upper 100αth percentile of the standard normal distribution. If $\max\{\hat{\Delta} - Z_{\alpha/2}\sqrt{\hat{\pi}_1(1-\hat{\pi}_1)/n_1 + \hat{\pi}_0(1-\hat{\pi}_0)/n_0}), 0\}$ were 0, we would define the upper limit (8.3) to be ∞. In this case, (8.3) becomes a one-sided half-open confidence interval. Thus, when there is no statistically significant difference in the response rates π_1 and π_0 at (two-sided) level α, we can only obtain the lower limit of a $100(1 - \alpha)$ percent confidence interval unless we reduce the desired confidence level, use other more efficient statistics, or increase the sample size. The non-existence of the upper limit suggests that the data are not sufficiently large to provide an accurate estimate of the upper limit for the NNT. As noted by McQuay and Moore (1997), this point is also relevant to clinicians and should be taken into account in their clinical decisions.

Recall that the probability $P([(\hat{\Delta} - \Delta)/\sqrt{\mathrm{Var}(\hat{\Delta})}]^2 \leq Z^2_{\alpha/2}) \doteq 1 - \alpha$ as n_i is large. Define $\mathcal{T} = \pi_1 + \pi_0$. Then we have $\pi_1 = (\mathcal{T} + \Delta)/2$ and $\pi_0 = (\mathcal{T} - \Delta)/2$. Thus, we can express the variance $\mathrm{Var}(\hat{\Delta}) \left(= \sum_{i=0}^{1} \pi_i(1 - \pi_i)/n_i\right)$ in terms of parameters $\mathcal{T}$ and Δ. This leads to the following asymptotic $100(1 - \alpha)$ percent confidence interval for the NNT:

$$[1/\Delta_u(\hat{\mathcal{T}}_1), 1/\Delta_l(\hat{\mathcal{T}}_1)], \tag{8.4}$$

where $\Delta_l(\hat{\mathcal{T}}_1) = \max\{(B - \sqrt{B^2 - AC})/A, 0\}$, $\Delta_u(\hat{\mathcal{T}}_1) = \min\{(B + \sqrt{B^2 - AC})/A, 1\}$, $A = 1 + [1/n_1 + 1/n_0]Z^2_{\alpha/2}/4$, $B = \hat{\pi}_1 - \hat{\pi}_0 + (1 - \hat{\mathcal{T}}_1)[1/n_1 - 1/n_0]Z^2_{\alpha/2}/4$, $C = (\hat{\pi}_1 - \hat{\pi}_0)^2 - \hat{\mathcal{T}}_1(2 - \hat{\mathcal{T}}_1)[1/n_1 + 1/n_0]Z^2_{\alpha/2}/4$, and $\hat{\mathcal{T}}_1 = \hat{\pi}_1 + \hat{\pi}_0$. When either of the $\hat{\pi}_i$ is 0 or 1, we recommend use of the *ad hoc* adjustment procedure of adding 0.5 for sparse data, substituting $\hat{\pi}_i^*(1 - \hat{\pi}_i^*)/n_i$ for $\hat{\pi}_i(1 - \hat{\pi}_i)/n_i$ in (8.3) and using $\hat{\pi}_i^*$ instead of $\hat{\pi}_i$ in (8.4), where $\hat{\pi}_i^* = (X_i + 0.5)/(n_i + 1)$.

Example 8.1 Consider the study of the efficacy and safety of taking topiramate 400 mg/day as adjunct therapy to the traditional anti-epileptic drugs for partial onset seizures with or without secondary generalization (Sharief *et al.*, 1996; Lesaffre and Pledger, 1999). The parameters π_1 and π_0 here denote the proportions of patients who show at least 50% reduction in seizures for the topiramate and placebo groups, respectively. Patients enrolled have previously had seizure rates of at least one per week during an 8-week baseline period. There are 23 patients randomly assigned to topiramate treatment and 24 patients to the placebo for a 3 week titration period followed by an 8-week stabilization period. We find eight patients in the topiramate treatment group and two patients in the placebo group who show at least 50% reduction. Given these data, the point estimate of the NNT is $4(\doteq 3.78 = 1/(\hat{\pi}_1 - \hat{\pi}_0))$. Applying interval estimators (8.3) and (8.4), we obtain the 95% confidence intervals of [2, 25] and [2, 34], respectively. Since the number of patients is a positive integer, confidence limits are rounded to the nearest integer. Note that the upper limit of the 95% confidence interval obtained from (8.3) is much lower than that from (8.4). This is because the number of patients in the placebo group who showed at least 50% reduction in seizures is small (2); it may thus be inappropriate to use (8.3) here. This example also implies that when the expected number of subjects in either of the two comparison groups is small, using (8.3) tends to shift to the left compared to (8.4), and hence likely overestimates the efficacy of the treatment under investigation.

Example 8.2 Consider the placebo-controlled multicenter randomized trial of interferon β-1b in treatment of secondary progressive multiple sclerosis (European Study Group on Interferon β-1b in Secondary Progressive MS, 1998). Outpatients in the secondary progressive phase of MS having scores of 3.0–6.5 on the Expanded Disability Status Scale are randomly allocated to receive either 8 million IU interferon β-1 subcutaneously every other day or placebo, for up to 3 years. As given elsewhere (Sackett *et al.*, 2000, p. 130), among the 358 patients assigned to placebo 49.8% exhibit confirmed progression of disability, while among the 360 patients assigned to interferon β-1b the proportion is 38.9%. From these data, the point estimate of the NNT is 9. Both (8.3) and (8.4) give a 95% confidence interval for the NNT of [6, 28]. Because the expected numbers of subjects with progression of disability in the two comparison groups are reasonably large, interval estimators (8.3) and (8.4) are expected to produce similar results.

8.2 A SERIES OF INDEPENDENT BINOMIAL SAMPLING PROCEDURES

Consider a multicenter study with S centers. From center $s(s = 1, 2, \ldots, S)$, suppose that we sample n_{is} subjects from the treatment ($i = 1$) and the placebo groups ($i = 0$), respectively. Suppose further that we obtain X_{is} subjects with response. Let π_{is} denote the response probability for a randomly selected subject from treatment i in the sth stratum. Define $\Delta_s = \pi_{1s} - \pi_{0s}$ and $T_s = \pi_{1s} + \pi_{0s}$. Therefore, $\pi_{1s} = (\Delta_s + T_s)/2$ and $\pi_{0s} = (T_s - \Delta_s)/2$. The joint probability mass function of the random vector $\mathbf{X}' = (\mathbf{X}'_1, \mathbf{X}'_0)$, where $\mathbf{X}'_i = (X_{i1}, X_{i2}, \ldots, X_{iS})$, is then given by (2.8).

The MLEs of Δ_s and T_s are simply $\hat{\Delta}_s = \hat{\pi}_{1s} - \hat{\pi}_{0s}$ and $\hat{T}_s = \hat{\pi}_{1s} + \hat{\pi}_{0s}$, respectively, where $\hat{\pi}_{is} = X_{is}/n_{is}$. Furthermore, we can easily show that the variance $\mathrm{Var}(\hat{\Delta}_s)$ is $\pi_{1s}(1 - \pi_{1s})/n_{1s} + \pi_{0s}(1 - \pi_{0s})/n_{0s}$. Note that the NNT τ_s in stratum s is, by definition, equal to $1/\Delta_s$. In this section, we assume that τ_s is constant across all strata, and we denote this common value of τ_s by τ_c.

First, consider the situation in which we take a reasonably large sample size from each center s. To estimate τ_c, if the variance $\mathrm{Var}(\hat{\Delta}_s)$ were known, we would employ the reciprocal of the weighted least-squares (WLS) estimator of Δ_s : $1/[\sum_s W_s\hat{\Delta}_s/\sum_s W_s]$, where $W_s = 1/\mathrm{Var}(\hat{\Delta}_s)$ (Lesaffre and Pledger, 1999). If $\mathrm{Var}(\hat{\Delta}_s)$ were unknown, we would substitute the unbiased estimator $\widehat{\mathrm{Var}}(\hat{\Delta}_s) = \hat{\pi}_{1s}(1 - \hat{\pi}_{1s})/(n_{1s} - 1) + \hat{\pi}_{0s}(1 - \hat{\pi}_{0s})/(n_{0s} - 1)$ for $\mathrm{Var}(\hat{\Delta}_s)$, and obtain the following WLS estimator of τ_c:

$$\hat{\tau}_{\mathrm{WLS}} = 1/\left(\sum_s \hat{W}_s\hat{\Delta}_s/\sum_s \hat{W}_s\right), \tag{8.5}$$

where $\hat{W}_s = 1/\widehat{\mathrm{Var}}(\hat{\Delta}_s)$. On the basis of simulations, Lesaffre and Pledger (1999) note that using (8.5) is preferable to using the estimator $\sum_s \hat{\mathcal{W}}_s\hat{\tau}_s/\sum_s \hat{\mathcal{W}}_s$, where $\hat{\mathcal{W}}_s$ is proportional to the reciprocal of the estimated asymptotic variance $\widehat{\mathrm{Var}}(\hat{\tau}_s)$. In fact, we can show that under the assumption that $\tau_1 = \tau_2 = \ldots = \tau_S$, the WLS estimator $\sum_s \hat{W}_s\hat{\tau}_s/\sum_s \hat{W}_s$ may be even preferable to $\sum_s \hat{\mathcal{W}}_s\hat{\tau}_s/\sum_s \hat{\mathcal{W}}_s$ (**Exercise 8.5**). Furthermore, note that the variance $\mathrm{Var}((\sum_s W_s\hat{\Delta}_s)/(\sum_s W_s)) = 1/\sum_s W_s$. This leads to an asymptotic $100(1 - \alpha)\%$ confidence interval for τ_c

given by

$$\left[1/\min\left\{\frac{\sum_s \hat{W}_s\hat{\Delta}_s}{\sum_s \hat{W}_s}+\frac{Z_{\alpha/2}}{\sqrt{\sum_s \hat{W}_s}},1\right\},\quad 1/\max\left\{\frac{\sum_s \hat{W}_s\hat{\Delta}_s}{\sum_s \hat{W}_s}-\frac{Z_{\alpha/2}}{\sqrt{\sum_s \hat{W}_s}},0\right\}\right]. \tag{8.6}$$

When the data are sparse, we may consider using estimator (2.11) (Greenland and Robins, 1985) and obtain the Mantel–Haenszel type estimator of τ_s given by

$$\hat{\tau}_{\text{MH}} = 1/\hat{\Delta}_{\text{MH}}, \tag{8.7}$$

where

$$\hat{\Delta}_{\text{MH}} = \frac{\sum_s (X_{1s}n_{0s} - X_{0s}n_{1s})/n_{.s}}{\sum_s n_{1s}n_{0s}/n_{.s}}.$$

Furthermore, when using the estimated asymptotic variance $\widehat{\text{Var}}(\hat{\Delta}_{\text{MH}})$ (2.12) (Sato, 1989), we obtain an asymptotic $100(1-\alpha)$ percent confidence interval for τ_c given by

$$[1/\min\{\hat{\Delta}_{\text{MH}} + Z_{\alpha/2}\sqrt{\widehat{\text{Var}}(\hat{\Delta}_{\text{MH}})}, 1\},\ 1/\max\{\hat{\Delta}_{\text{MH}} - Z_{\alpha/2}\sqrt{\widehat{\text{Var}}(\hat{\Delta}_{\text{MH}})}, 0\}]. \tag{8.8}$$

Note that, by definition, $\tau_s = 1/\Delta_s$. Thus, τ_s is constant over s if and only if Δ_s is constant. Therefore, to examine the assumption that τ_s is constant, we may employ test procedures described in Section 2.2.2, for testing the homogeneity of the RD (Lipsitz *et al.*, 1998; Lui and Kelly, 2000).

Example 8.3 Marson *et al.* (1996) provide a systematic review of the efficacy and tolerability of new anti-epileptic drugs. Here, we concentrate our attention on comparing the proportion of patients who show 50% or greater reduction (50% responders) in the frequency of seizures between taking gabapentin 1200 mg per day and the placebo. In Table 8.1, we summarize the data obtained from ($S =$)4 parallel studies (Anhut *et al.*, 1994; Sivenius *et al.*, 1991; UK Gabapentin Study Group, 1990; US Gabapentin Study Group No. 5, 1993). Given these data, we obtain point estimates $\hat{\tau}_s$ of 6, 6, 8, and 13, respectively. Apart from the last estimate for the US Study Gabapentin Group, the estimates are similar to one another. Applying test procedures (2.14)–(2.16) to test the homogeneity of the NNT, we obtain p-values of 0.654, 0.713, and 0.635. These suggest that there is no significant evidence at the 5% level against the assumption that the NNT for taking gabapentin is constant over these four studies. Using estimators (8.5)–(8.8), we obtain estimates $\hat{\Delta}_{\text{WLS}}$ and $\hat{\Delta}_{\text{MH}}$, together with their 95% confidence intervals (in parentheses), of 9 ([6, 20]) and 8 ([5, 17]), respectively; these resulting estimates are similar to each other.

Table 8.1 Observed number of 50% responders/the total number of patients between the group of taking gabapentin 1200 mg per day and the group of taking placebo in four parallel studies.

	Anhut *et al.*	Sivenius *et al.*	UK Study Group	US Study Group
Gabapentin	14/52	3/9	13/61	16/101
Placebo	10/109	3/18	6/66	8/98

Source: Marson *et al.* (1996).

Example 8.4 Consider the all-cause mortality data from the multicenter trials comparing aspirin ($i = 0$) with placebo ($i = 1$) in post-myocardial infarction patients (Table 2.1). As noted in Example 2.3, because of the baseline imbalance of medical conditions between the aspirin and placebo groups in the sixth trial ($s = 6$) (Canner, 1987), we exclude this trial's data from consideration here. For the first five trials, we obtain NNT estimates $\hat{\tau}_s$ of 36, 39, 40, 54, and 43. Applying test procedures using (2.14)–(2.16) for testing the homogeneity of the NNT, we obtain p-values of 0.998, 0.915, and 1.00. Therefore, the hypothesis that the NNT for the all-cause mortality rate between placebo and aspirin is constant over these five trials seems reasonable. Applying estimators (8.5)–(8.8), we obtain $\hat{\tau}_{\text{WLS}} = 40$ with 95% confidence interval [25, 101], and $\hat{\tau}_{\text{MH}} = 40$ with 95% confidence interval [25, 105]. Because the numbers of subjects in these trials are reasonably large, we obtain similar findings using (8.5) and (8.7), suggesting that for every 40 post-myocardial infarction patients, we would expect to save one life if they all took aspirin.

8.3 INDEPENDENT CLUSTER SAMPLING

In randomized intervention trials, we may randomly assign clusters of subjects to treatments rather than individuals for administrative convenience or to reduce the effect of treatment contamination (Donner *et al.*, 1981; Klar and Donner, 2001; Lui *et al.*, 2000). For example, in the randomized trial studying the effect of vitamin A supplementation on childhood mortality, the unit of randomization is the household (Herrera *et al.*, 1992). Because subject responses within clusters are likely correlated, the confidence interval for the NNT without taking the intraclass correlation between responses within clusters into account can be misleading.

Suppose that we randomly assign n_i classes, of which each has m_{ij} subjects, to the treatment ($i = 1$) and the placebo ($i = 0$) groups, respectively. We define the random variable $X_{ijk} = 1$ if the kth ($k = 1, 2, \ldots, m_{ij}$) subject in the jth ($j = 1, \ldots, n_i$) cluster from treatment i is positive, and $X_{ijk} = 0$ otherwise. We let the probability $P(X_{ijk} = 1) = p_{ij}$ and $P(X_{ijk} = 0) = 1 - p_{ij}$, where $0 < p_{ij} < 1$. Because subject responses within clusters are likely correlated, we assume that the p_{ij} independently identically follow the beta distribution beta(α_i, β_i) with mean $\pi_i = \alpha_i/T_i$ and variance $\pi_i(1-\pi_i)/(T_i+1)$, where $T_i = \alpha_i + \beta_i$ (Johnson and Kotz, 1970). Note that the beta-binomial model has been frequently applied to model dependent binary data (Lui, 1991, 2001a; Lui *et al.*, 1996, 2000) and encompasses a wide variety of shapes (Johnson and Kotz, 1970). On the basis of the model assumptions, the intraclass correlation between X_{ijk} and $X_{ijk'}$ for $k \neq k'$ is $\rho_i = 1/(T_i+1)$ (**Exercise 1.7**). Note that the probability of a randomly selected subject being positive from treatment i is equal to $E(X_{ijk}) = \pi_i$. We assume that $\pi_1 > \pi_0$ and define the NNT as $\tau = 1/\Delta$, where $\Delta = \pi_1 - \pi_0$.

Define $\hat{\pi}_i = \sum_j \sum_k X_{ijk}/m_{i.}$, where $m_{i.} = \sum_j m_{ij}$ is the total number of subjects assigned to treatment i. We can easily show that $\hat{\pi}_i$ is an unbiased estimator of π_i with variance $\text{Var}(\hat{\pi}_i) = \pi_i(1-\pi_i)f(\mathbf{m}_i, \rho_i)/m_{i.}$, where $\mathbf{m}_i' = (m_{i1}, m_{i2}, \ldots, m_{in_i})$ and $f(\mathbf{m}_i, \rho_i) = \sum_j m_{ij}[1+(m_{ij}-1)\rho_i]/m_{i.}$ is the variance inflation factor due to the intraclass correlation ρ_i (**Exercise 1.8**).

First, note that

$$\hat{\Delta} = \hat{\pi}_1 - \hat{\pi}_0, \tag{8.9}$$

is an unbiased estimator of Δ under the beta-binomial distribution with variance

$$\text{Var}(\hat{\Delta}) = \pi_1(1-\pi_1)f(\mathbf{m}_1, \rho_1)/m_{1.} + \pi_0(1-\pi_0)f(\mathbf{m}_0, \rho_0)/m_{0.}. \tag{8.10}$$

We can simply substitute $\hat{\pi}_i$ for π_i and the traditional intraclass correlation estimator $\hat{\rho}_i$ (2.19) for ρ_i in (8.10) to obtain the estimated variance $\widehat{\text{Var}}(\hat{\Delta})$ (Fleiss, 1986; Elston, 1977; Lui *et al.*, 1996). Therefore, an asymptotic $100(1-\alpha)\%$ confidence interval for τ is given by

$$[1/\min\{\hat{\Delta} + Z_{\alpha/2}\sqrt{\widehat{\text{Var}}(\hat{\Delta})}, 1\}, 1/\max\{\hat{\Delta} - Z_{\alpha/2}\sqrt{\widehat{\text{Var}}(\hat{\Delta})}, 0\}], \tag{8.11}$$

where $\widehat{\text{Var}}(\hat{\Delta}) = \hat{\pi}_1(1-\hat{\pi}_1)f(\mathbf{m}_1, \hat{\rho}_1)/m_{1.} + \hat{\pi}_0(1-\hat{\pi}_0)f(\mathbf{m}_0, \hat{\rho}_0)/m_0.$ Note that since the sampling distribution of $\hat{\tau}$ may be skewed, interval estimator (8.11) may not perform well, especially when the expected number of responses in either of the two comparison groups is moderate or small. Thus, may we consider generalizing interval estimator (8.4) to cover cluster sampling.

First, recall that the probability $P([(\hat{\Delta} - \Delta)/\sqrt{\text{Var}(\hat{\Delta})}]^2 \leq Z^2_{\alpha/2}) \doteq 1-\alpha$ as both $m_{i.}$ is large. This leads to the following quadratic equation in Δ:

$$A^\dagger\Delta^2 - 2B^\dagger\Delta + C^\dagger \leq 0, \tag{8.12}$$

where

$$A^\dagger = 1 + [f(\mathbf{m}_1, \hat{\rho}_1)/m_{1.} + f(\mathbf{m}_0, \hat{\rho}_0)/m_{0.}]Z^2_{\alpha/2}/4,$$

$$B^\dagger = \hat{\Delta} + (1-\hat{T})[f(\mathbf{m}_1, \hat{\rho}_1)/m_{1.} - f(\mathbf{m}_0, \hat{\rho}_0)/m_{0.}]Z^2_{\alpha/2}/4,$$

$$C^\dagger = \hat{\Delta}^2 - \hat{T}(2-\hat{T})[f(\mathbf{m}_1, \hat{\rho}_1)/m_{1.} + f(\mathbf{m}_0, \hat{\rho}_0)/m_{0.}]Z^2_{\alpha/2}/4,$$

with $\hat{T} = \hat{\pi}_1 + \hat{\pi}_0$. Because $A^\dagger > 0$, (8.12) is always convex. If $B^{\dagger 2} - A^\dagger C^\dagger > 0$, then an asymptotic $100(1-\alpha)$ percent confidence interval for Δ would be given by $[\Delta_l(\hat{T}_1), \Delta_u(\hat{T}_1)]$, where $\Delta_l(\hat{T}_1) = \max\{(B^\dagger - \sqrt{B^{\dagger 2} - A^\dagger C^\dagger})/A^\dagger, 0\}$ and $\Delta_u(\hat{T}_1) = \min\{(B^\dagger + \sqrt{B^{\dagger 2} - A^\dagger C^\dagger})/A^\dagger, 1\}$ are the two distinct roots of (8.12) subject to Δ lying in the range (0, 1). Thus, we obtain an asymptotic $100(1-\alpha)$ percent confidence interval for τ given by

$$[1/\Delta_u(\hat{T}_1), 1/\Delta_l(\hat{T}_1)]. \tag{8.13}$$

Note that if either $\hat{\pi}_i$ is 0 or 1, we recommend using $\hat{\pi}_i^*(1-\hat{\pi}_i^*)f(\mathbf{m}_i, \hat{\rho}_i)/m_{i.}$ to estimate $\mathrm{Var}(\hat{\pi}_i)$ in (8.10), or substituting $\hat{T}^* = \hat{\pi}_1^* + \hat{\pi}_0^*$ for $\hat{T}$ in (8.12), where $\hat{\pi}_i^* = (X_{i..} + 0.5)/(m_{i.} + 1)$.

Example 8.5 Consider the data in Table 1.1 from the study of the effect of an education intervention on behavior change (Mayer *et al.*, 1997). We randomly assign 29 classes, of size ranging from 1 to 6, to the intervention group and 29 to the control group and wish to study the effect of the education intervention program on the possession of an adequate level of solar protection. Based on the data (Table 1.1), the point estimate $\hat{\tau}$ (8.9) of the NNT is 5. This suggests that when we apply the education intervention program to the children in the control group we can expect that for every five children, one additional child will employ an adequate level of solar protection. Using either (8.11) or (8.13) gives a 95% confidence interval of $[3, \infty)$. In other words, the data considered here are not precise enough to produce an accurate estimate of the upper limit of the NNT.

8.4 PAIRED-SAMPLE DATA

Suppose that to increase the efficiency of a randomized trial, we match patients with respect to some strong nuisance confounders to form n matched pairs. Suppose further that for each matched pair we randomly assign one patient to the new treatment and the other to the standard treatment. For clarity, we use the following table to summarize the data structure:

			Standard treatment Response status		
			Yes	No	
New treatment	Response status	Yes	π_{11}	π_{10}	$\pi_{1.}$
		No	π_{01}	π_{00}	$\pi_{0.}$
			$\pi_{.1}$	$\pi_{.0}$	

where π_{ij} denotes the corresponding cell probability, $\pi_{i.} = \pi_{i1} + \pi_{i0}$, and $\pi_{.j} = \pi_{1j} + \pi_{0j}$ for i and $j = 0, 1$. Without loss of generality, we assume that $\pi_{1.} > \pi_{.1}$ and define the NNT as $\tau = 1/\Delta$, where $\Delta = \pi_{1.} - \pi_{.1} (= \pi_{10} - \pi_{01})$. Let $\mathfrak{T} = \pi_{10} + \pi_{01}$ represent the probability of discordance between responses within a given pair. Thus, we have $\pi_{10} = (\mathfrak{T} + \Delta)/2$ and $\pi_{01} = (\mathfrak{T} - \Delta)/2$.

Let N_{ij} denote the observed frequency of matched pairs falling in cell (i, j) with probability π_{ij}. Then, the random vector $\mathbf{N}' = (N_{11}, N_{10}, N_{01}, N_{00})$ follows the multinomial distribution (2.25) with parameters n and $\boldsymbol{\pi}' = (\pi_{11}, \pi_{10}, \pi_{01}, \pi_{00})$. Therefore, the MLE $\hat{\pi}_{ij}$ of π_{ij} is N_{ij}/n for i and $j = 0, 1$. We thus have that the MLE of τ is $1/\hat{\Delta}$ when $\hat{\Delta} = \hat{\pi}_{10} - \hat{\pi}_{01} > 0$. If $\hat{\Delta} \leq 0$, we would define $\hat{\tau}$ to be ∞. We can easily show that $\text{Var}(\hat{\Delta}) = [\pi_{10} + \pi_{01} - (\pi_{10} - \pi_{01})^2]/n$ (**Exercise 2.7**). Following ideas similar to those for deriving interval estimator (2.26), we obtain an asymptotic $100(1 - \alpha)$ percent confidence interval for τ given by

$$[1/\min\{\hat{\pi}_{10} - \hat{\pi}_{01} + Z_{\alpha/2}\sqrt{\widehat{\text{Var}}(\hat{\Delta})}, 1\}, 1/\max\{\hat{\pi}_{10} - \hat{\pi}_{01} - Z_{\alpha/2}\sqrt{\widehat{\text{Var}}(\hat{\Delta})}, 0\}], \quad (8.14)$$

where $\widehat{\text{Var}}(\hat{\Delta}) = [\hat{\pi}_{10} + \hat{\pi}_{01} - (\hat{\pi}_{10} - \hat{\pi}_{01})^2]/\text{n}$.
When either n_{10} or n_{01} is small, interval estimator (8.14) may not perform well.

To improve the performance of (8.14) when n is not large, we may consider interval estimator (2.32) subject to the condition $\pi_1 > \pi_0$. This leads to an asymptotic $100(1 - \alpha)$ percent confidence interval for τ given by (Lui, 1998; May and Johnson, 1997)

$$[1/\min\{(B^{\ddagger} + \sqrt{B^{\ddagger 2} - A^{\ddagger}C^{\ddagger}})/A^{\ddagger}, 1\}, 1/\max\{(B^{\ddagger} - \sqrt{B^{\ddagger 2} - A^{\ddagger}C^{\ddagger}})/A^{\ddagger}, 0\}], \quad (8.15)$$

where $\min\{(B^{\ddagger} + \sqrt{B^{\ddagger 2} - A^{\ddagger}C^{\ddagger}})/A^{\ddagger}, 1\}$ and $\max\{(B^{\ddagger} - \sqrt{B^{\ddagger 2} - A^{\ddagger}C^{\ddagger}})/A^{\ddagger}, 0\}$ are the two distinct real roots of the quadratic equation

$$A^{\ddagger}\Delta^2 - 2B^{\ddagger}\Delta + C^{\ddagger} \leq 0, \quad (8.16)$$

where $A^{\ddagger} = (1 + Z^2_{\alpha/2}/n)$, $B^{\ddagger} = \hat{\Delta}$, $C^{\ddagger} = \hat{\Delta}^2 - Z^2_{\alpha/2}\hat{\Sigma}/n$ and $\hat{\Sigma} = \hat{\pi}_{10} + \hat{\pi}_{01}$ a root lying below the interval (0, 1) will be deemed to be 0, while a root lying above the interval will be replaced by 1.

Following Edwardes (1995), we may consider use of the $\tanh^{-1}(x)$ transformation to improve the normal approximation of $\hat{\Delta}$. This leads to an asymptotic $100(1 - \alpha)$ percent confidence interval for τ given by

$$[1/\tanh(\tanh^{-1}(\hat{\Delta}) + Z_{\alpha/2}\sqrt{\widehat{\text{Var}}(\hat{\Delta})/(1 - \hat{\Delta}^2)}),$$

$$1/\max\{\tanh(\tanh^{-1}(\hat{\Delta}) - Z_{\alpha/2}\sqrt{\widehat{\text{Var}}(\hat{\Delta})/(1 - \hat{\Delta}^2)}), 0\}]. \quad (8.17)$$

Whenever $n_{10} = 0$, $n_{01} = 0$, or $n - n_{10} - n_{01} = 0$, we adjust for sparse data by adding 0.50 to each cell when using the above interval estimators. As also noted in Chapter 2, when π_1 is large (at least 0.50) and π_0 is small (at most 0.025) or when π_1 is small (at most 0.025) and π_0 is large (at least 0.50), (8.17) is likely preferable to (8.15), especially when n is not large. On the other hand, the latter is preferable to the former in the other situations. When collecting matched pairs under cluster sampling, we can easily apply results published elsewhere (Lui, 2001b) to derive interval estimators for the NNT.

Although the NNT may be useful for summarizing findings in clinical trials, as noted previously, we may not want to employ this measure unless we can rank from prior knowledge the order of response rates between the two treatments under comparison. When testing whether there is a difference in the response rates of two treatments, note that the NNT is ∞ under the null hypothesis of no difference. Thus, statistics directly based on the NNT are awkward to use in detecting whether there is a difference between treatments. Thus, if detecting a difference in the response rates is one of our main goals, we may wish to use test statistics related to other indices such as the risk difference (Chapter 2), relative difference (Chapter 3), risk ratio (Chapter 4), odds ratio (Chapter 5), or generalized odd ratio (Chapter 6).

Example 8.6 Consider the numerical example given by Rosner (1990, pp. 342–343), in which two treatments for a rare form of cancer are compared. Within each pair, we randomly assign patients to receive either chemotherapy or surgery, and determine the vital status, survival or death, at the end of a 5-year follow-up. There are ($n =$)621 pairs of patients matched with respect to age, sex, and clinical condition. We obtain $n_{10} = 16$ (pairs where the patient receiving chemotherapy survives but the patient receiving surgery dies), and $n_{01} = 5$ (pairs where the patient receiving surgery survives but the patient receiving chemotherapy dies). Given these data, we estimate the NNT to be $\hat{\tau}(= 1/\hat{\Delta}) = 57$. This means that if we gave chemotherapy instead of surgery to patients, we would expect to save one life for every 57 patients. Applying (8.14), (8.15), and (8.17), we obtain 95% confidence intervals for NNT of [31, 302], [31, 308], and [31, 302]. All these confidence intervals are wide, although they are similar to one another.

EXERCISES

8.1. (a) Show that NNT $= 1/[(\mathrm{RR} - 1)\pi_0]$, where RR $= \pi_1/\pi_0$ is the risk ratio between the two treatments.
(b) Show that NNT $= 1/[(\mathrm{OR} - 1)\pi_0] + \mathrm{OR}/[(\mathrm{OR} - 1)(1 - \pi_0)]$, where OR $= \pi_1(1 - \pi_0)/[\pi_0(1 - \pi_1)]$ is the odds ratio of the response probability between the two treatments.

8.2. Show that if both the numbers of subjects n_i are large, interval estimator (8.3) is an asymptotic $100(1 - \alpha)$ percent confidence interval for $\tau = 1/(\pi_1 - \pi_0)$, where $\pi_1 > \pi_0$.

8.3. In the study the efficacy and safety of topiramate 400 mg/day (Example 8.1), suppose that we obtain 7 (out of 23) patients in the topiramate treatment and 14 (out of 24) patients in the placebo indicating 'poor patient rating of medication' (Sharief *et al.*, 1996). What is the point estimate of the NNT based on these data? What are the 95% confidence intervals for the NNT using (8.3) and (8.4)?

Table 8.2 Observed numbers of 50% responders/the total number of patients in the group taking tiagabine and the group taking placebo in three parallel studies.

	TIA-106	TIA-107	TIA-109
Tiagabine	17/88	11/77	53/210
Placebo	4/91	5/77	8/108

Source: Marson *et al.* (1996).

8.4. Suppose that we obtain a 95% confidence interval of (−5%, 25%) for $\Delta = \pi_1 - \pi_0$. Discuss what would be wrong if we claimed that the corresponding 95% confidence interval for the NNT is simply (−20, 4) (Altman, 1998).

8.5. Using the delta method, show that in **Section 8.2**, the asymptotic variance $\mathrm{Var}(\hat{\tau}_s)$ of the NNT is given by $\tau_s^4 \mathrm{Var}(\hat{\Delta}_s)$. Thus, under the assumption that $\tau_1 = \tau_2 = \ldots = \tau_S$, the asymptotic variance $\mathrm{Var}(\hat{\tau}_s)$ is proportional to $\mathrm{Var}(\hat{\Delta}_s)$.

8.6. Consider the data (Table 8.2) on the number of 50% responders (as defined in Example 8.3) for tiagabine (Marson *et al.*, 1996) and placebo groups. (a) What are the point estimates of the NNT for each stratum? (b) What are the summary estimates using (8.5) and (8.7)? What are the 95% confidence intervals for the NNT when we apply (8.6) and (8.8)?

8.7. Consider a randomized controlled clinical trial with 100 matched pairs. Suppose that we expect the response rate for the new treatment to be higher than that for the standard treatment. Suppose that we obtain (n_{10} =)20 matched pairs in which the patient receiving the new treatment shows positive but the patient receiving the standard treatment shows negative, and (n_{01} =)5 matched pairs in which the patient receiving the new treatment shows negative but the patient receiving the standard treatment shows positive. What is the point estimate of the NNT? What are the 95% confidence intervals for the NNT using (8.14), (8.15), and (8.17)?

REFERENCES

Altman, D. G. (1998) Confidence intervals for the number needed to treat. *British Medical Journal*, **317**, 1309–1312.

Anhut, H., Ashman, P., Feuerstein, T. J., Sauermann, W., Saunders, M. and Schmidt, B. (1994) Gabapentin (Neurontin) as add-on therapy in patients with partial seizures: a double-blind, placebo-controlled study. The International Gabapentin Study Group. *Epilepsia*, **35**, 795–801.

Bender, R. (2001) Calculating confidence intervals for the number needed to treat. *Controlled Clinical Trials*, **22**, 102–110.

Canner, P. L. (1987) An overview of six clinical trials of aspirin in coronary heart disease. *Statistics in Medicine*, **6**, 255–263.

Casella, G. and Berger, R. L. (1990) *Statistical Inference*. Duxbury, Belmont, CA.

Chatellier, G., Zapletal, E., Lemaitre, D., Menard, J. and Degoulet, P. (1996) The number needed to treat: a clinically useful nomogram in its proper context. *British Medical Journal*, **312**, 426–429.

Cook, R. J. and Sackett, D. L. (1995) The number needed to treat: a clinically useful measure of treatment effect. *British Medical Journal*, **310**, 452–454.

Cornfield, J. (1978) Randomization by group: a formal analysis. *American Journal of Epidemiology*, **108**, 100–102.

Donner, A., Birkett, N., and Buck, C. (1981) Randomization by cluster sample size requirements and analysis. *American Journal of Epidemiology*, **114**, 906–914.

Edwardes, M. D. (1995) A confidence interval for $\Pr(X < Y) - \Pr(X > Y)$ estimated from simple cluster samples. *Biometrics*, **41**, 571–578.

Elferink, A. J. A. and Van Zwieten-Boot, B. J. (1997) Analysis based on number needed to treat shows differences between drugs studied. *British Medical Journal*, **314**, 603.

Elston, R. C. (1977) Response to query: estimating 'inheritability' of a dichotomous trait. *Biometrics* **33**, 232–233.

European Study Group on Interferon β-lb in Secondary Progressive MS (1998) Placebo-controlled multicenter randomized trial of interferon β-lb in treatment of secondary progressive multiple sclerosis. *Lancet*, **352**, 1491–1497.

Fleiss, J. L. (1981) *Statistical Methods for Rates and Proportions*, 2nd edition. John Wiley & Sons, Inc., New York.

Fleiss, J. L. (1986) *The Design and Analysis of Clinical Experiments*. Wiley, New York.

Greenland, S. and Robins, J. M. (1985) Estimation of a common effect parameter from sparse follow-up data. *Biometrics*, **41**, 55–68.

Herrera, M. G., Nestel, P., El Amin, A., Fawzi, W. W., Mohamed, K. A. and Weld, L. (1992) Vitamin A supplementation and child survival. *Lancet*, **340**, 267–271.

Hutton, J. L. (2000) Number needed to treat: properties and problems. *Journal of the Royal Statistical Society A*, **163**, 403–419.

Johnson, N. L. and Kotz, S. (1970) *Distributions in Statistics: Continuous Univariate Distributions*, Vol. 2. Wiley, New York.

Klar, N. and Donner, A. (2001) Current and future challenges in the design and analysis of cluster randomization trials. *Statistics in Medicine*, **20**, 3729–3740.

Laupacis, A., Sackett, D. L. and Roberts, R. S. (1988) An assessment of clinically useful measures of the consequences of treatment. *New England Journal of Medicine*, **318**, 1728–1733.

Lesaffre, E. and Pledger, G. (1999) A note on the number needed to treat. *Controlled Clinical Trials*, **20**, 439–447.

Lipsitz, S. R., Dear, K. B. G., Laird, N. M. and Molenberghs, G. (1998) Tests for homogeneity of the risk difference when data are sparse. *Biometrics*, **54**, 148–160.

Lui, K.-J. (1991) Sample size for repeated measurements in dichotomous data. *Statistics in Medicine*, **10**, 463–472.

Lui, K.-J. (1998) Confidence intervals for differences in correlated binary proportions. *Statistics in Medicine*, **17**, 2017–2021.

Lui, K.-J. (2001a) Interval estimation of simple difference in dichotomous data with repeated measurements. *Biometrical Journal*, **43**, 845–861.

Lui, K.-J. (2001b) A note on interval estimation of the simple difference in data with correlated matched pairs. *Biometrical Journal*, **43**, 235–247.

Lui, K.-J. (2003) When would it be appropriate to use the number needed to treat: a simple logical solution. Unpublished manuscript, Department of Mathematics and Statistics, San Diego State University.

Lui, K.-J. and Kelly, C. (2000) A revisit of tests for homogeneity of the risk difference. *Biometrics*, **56**, 309–315.

Lui, K.-J., Cumberland, W. G. and Kuo, L. (1996) An interval estimate for the intraclass correlation in beta-binomial sampling. *Biometrics*, **52**, 412–425.

Lui, K.-J., Mayer, J. A. and Eckhardt, L. (2000) Confidence intervals for the risk ratio under cluster sampling based on the beta-binomial model. *Statistics in Medicine*, **19**, 2933–2942.

Marson, A. G., Kadir, Z. A. and Chadwick, D. W. (1996) New antiepileptic drugs: a systematic review of their efficacy and tolerability. *British Medical Journal*, **313**, 1169–1174.

May, W. L. and Johnson, W. D. (1997) Confidence intervals for differences in correlated binary proportions. *Statistics in Medicine*, **16**, 2127–2136.

McQuay, H. J. and Moore, R. A. (1997) Using numerical results from systematic reviews in clinical practice. *Annals of Internal Medicine*, **126**, 712–720.

Rosner, B. (1990). *Fundamentals of Biostatistics*. PWS-Kent, Boston.

Sackett, D. L., Deeks, J. J. and Altman, D. G. (1996) Down with odds ratios!. *Evidence-Based Medicine*, **1**, 164–166.

Sackett, D. L, Straus, S. E., Richardson, W. S., Rosenberg, W. and Haynes, R. B. (2000) *Evidence-Based Medicine: How to Practice and Teach EBM*. Churchill Livingstone, Edinburgh.

Sato, T. (1989) On the variance estimator for the Mantel–Haenszel risk difference. *Biometrics*, **45**, 1323–1324.

Sharief, M., Viteri, C., Ben-Menachem, E., Weber, M., Reife, R., Pledger, G. and Karim, R. (1996) Double-blind, placebo-controlled study of topiramate in patients with refractory partial epilepsy. *Epilepsy Research*, **25**, 217–224.

Sheps, M. C. (1958) Shall we count the living or the dead? *New England Journal of Medicine*, **259**, 1210–1214.

Sivenius, J., Kalviainen, R., Ylinen, A. and Riekkinen, P. (1991). Double-blind study of gabapentin in the treatment of partial seizures. *Epilepsia*, **32**, 539–542.

Tramèr, M. R., Moore, A. and McQuay, H. (1995) Prevention of vomiting after paediatric strabismus surgery: a systematic review using the numbers-needed-to-treat method. *British Journal of Anaesthesia*, **75**, 556–561.

UK Gabapentin Study Group (1990) Gabapentin in partial epilepsy. *Lancet*, **335**, 1114–1117.

US Gabapentin Study Group No 5 (1993) Gabapentin as add-on therapy in refractory partial epilepsy: a double-blind, placebo-controlled, parallel-group study. *Neurology*, **43**, 2292–2298.

Veterans Administration Cooperative Study Group on Antihypertensive Agents (1972) Effects on treatment on morbidity in hypertension III. Influence of age, diastolic pressure, and prior cardiovascular disease; further analysis of side effects. *Circulation*, **45**, 991–1004.

Appendix

Maximum Likelihood Estimator and Large-Sample Theory

In this appendix, we present the definition of the maximum likelihood estimator (MLE) and its asymptotic statistical properties. We describe Wald's test, the score test, and the asymptotic likelihood ratio test and discuss how to apply them to find an asymptotic $100(1-\alpha)$ percent confidence region. We also present a brief description of the delta method. This method is most useful for interval estimation.

A.1 THE MAXIMUM LIKELIHOOD ESTIMATOR, WALD'S TEST, THE SCORE TEST, AND THE ASYMPTOTIC LIKELIHOOD RATIO TEST

Suppose that we obtain n random observations, $x_1, x_2, x_3, \ldots$, and x_n, that are assumed to be independently and identically distributed from a population with a probability density function (pdf) (or a probability mass function (pmf)) $f(x|\boldsymbol{\mu})$, where $\boldsymbol{\mu} = (\mu_1, \mu_2, \ldots, \mu_S)'$ is an $S \times 1$ vector of parameters from a known parameter space $\boldsymbol{\Theta}$. By definition, the likelihood is then given by

$$L(\boldsymbol{\mu}|\mathbf{x}) = \prod_{i=1}^{n} f(x_i|\boldsymbol{\mu}), \tag{A.1}$$

where $\mathbf{x}' = (x_1, x_2, x_3, \ldots, x_n)$. The MLE of $\boldsymbol{\mu}$ is defined as the value $\hat{\boldsymbol{\mu}}(\in \boldsymbol{\Theta})$ that maximizes (A.1) over $\boldsymbol{\Theta}$. That is,

$$L(\hat{\boldsymbol{\mu}}|\mathbf{x}) = \max_{\boldsymbol{\mu}\in\boldsymbol{\Theta}} L(\boldsymbol{\mu}|\mathbf{x}).$$

Note that when the observations x_i are discrete random variables, the MLE $\hat{\boldsymbol{\mu}}$ actually maximizes the probability of obtaining the particular sample $\mathbf{x}$ that

Statistical Estimation of Epidemiological Risk K-J. Lui
 ISBN: 0-470-85071-X (HB)

we have obtained. Because the function $\log(x)$ is a monotonically increasing function for $x > 0$, the value $\hat{\boldsymbol{\mu}}$ that maximizes the log-likelihood $\log(L(\boldsymbol{\mu}|\mathbf{x}))$ will also maximize the likelihood $L(\boldsymbol{\mu}|\mathbf{x})$. Furthermore, because it is usually more convenient to work with the former than the latter, we commonly find the MLE based on the log-likelihood. In practice, we can generally obtain the MLE by finding the roots of the following equations:

$$\frac{\partial \log(L(\boldsymbol{\mu}|\mathbf{x}))}{\partial \mu_i} = 0, \qquad i = 1, 2, \ldots, S. \tag{A.2}$$

Note that if $\hat{\boldsymbol{\mu}}$ is the MLE of $\boldsymbol{\mu}$, then $f(\hat{\boldsymbol{\mu}})$ will be the MLE of $f(\boldsymbol{\mu})$ for a given function $f(\mathbf{x})$ (Casella and Berger, 1990). This property is called the functional invariance. Under mild conditions (Casella and Berger, 1990), if $\hat{\boldsymbol{\mu}}$ is the MLE, then as long as n is large, one may assume that $\hat{\boldsymbol{\mu}} - \boldsymbol{\mu}$ approximately follows the multivariate normal distribution with mean $\mathbf{0} = (0, 0, \ldots, 0)'$, and estimated covariance matrix $\mathbf{I}^{-1}(\hat{\boldsymbol{\mu}})$, where $\mathbf{I}(\boldsymbol{\mu})$ denotes the $S \times S$ Fisher's information matrix with (i, j)th element $-\mathrm{E}(\partial^2 \log(L(\boldsymbol{\mu}|\mathbf{x}))/\partial \mu_i \partial \mu_j)$. This leads to Wald's statistic $(\hat{\boldsymbol{\mu}} - \boldsymbol{\mu}_0)'\mathbf{I}(\hat{\boldsymbol{\mu}})(\hat{\boldsymbol{\mu}} - \boldsymbol{\mu}_0)$ for testing the hypothesis $H_0 : \boldsymbol{\mu} = \boldsymbol{\mu}_0$. When $(\hat{\boldsymbol{\mu}} - \boldsymbol{\mu}_0)'\mathbf{I}(\hat{\boldsymbol{\mu}})(\hat{\boldsymbol{\mu}} - \boldsymbol{\mu}_0) > \chi^2_{S,\alpha}$, we reject $H_0 : \boldsymbol{\mu} = \boldsymbol{\mu}_0$ at level-α; $\chi^2_{f,\alpha}$ is the upper 100αth percentile of the central chi-squared distribution with f degrees of freedom. Furthermore, since the observed information matrix $\mathbf{I}_o(\hat{\boldsymbol{\mu}})$, defined as the $S \times S$ matrix with (i, j)th element $\partial^2 \log(L(\boldsymbol{\mu}|\mathbf{x}))/\partial \mu_i \partial \mu_j|_{\boldsymbol{\mu}=\hat{\boldsymbol{\mu}}}$, is a consistent estimator of $\mathbf{I}(\boldsymbol{\mu})$, we may often substitute $\mathbf{I}_o(\hat{\boldsymbol{\mu}})$ for $\mathbf{I}(\hat{\boldsymbol{\mu}})$ in the above test statistic and obtain $(\hat{\boldsymbol{\mu}} - \boldsymbol{\mu}_0)'\mathbf{I}_o(\hat{\boldsymbol{\mu}})(\hat{\boldsymbol{\mu}} - \boldsymbol{\mu}_0)$. We reject $H_0 : \boldsymbol{\mu} = \boldsymbol{\mu}_0$ at level α when $(\hat{\boldsymbol{\mu}} - \boldsymbol{\mu}_0)'\mathbf{I}_o(\hat{\boldsymbol{\mu}})(\hat{\boldsymbol{\mu}} - \boldsymbol{\mu}_0) > \chi^2_{S,\alpha}$. On the other hand, by inverting the acceptance region for Wald's statistic at level-α (Casella and Berger, 1990), we can obtain an asymptotic $100(1-\alpha)$ percent confidence region $\{\boldsymbol{\mu}|(\hat{\boldsymbol{\mu}} - \boldsymbol{\mu})'\mathbf{I}(\hat{\boldsymbol{\mu}})(\hat{\boldsymbol{\mu}} - \boldsymbol{\mu}) \leq \chi^2_{S,\alpha}\}$ or $\{\boldsymbol{\mu}|(\hat{\boldsymbol{\mu}} - \boldsymbol{\mu})'\mathbf{I}_o(\hat{\boldsymbol{\mu}})(\hat{\boldsymbol{\mu}} - \boldsymbol{\mu}) \leq \chi^2_{S,\alpha}\}$ for $\boldsymbol{\mu}$. When $S = 1$, we call this confidence region, which reduces to an asymptotic $100(1-\alpha)$ percent confidence interval, the interval estimator for μ_1 using Wald's statistic.

Define $\mathbf{U}(\boldsymbol{\mu}) = (U_1(\boldsymbol{\mu}), U_2(\boldsymbol{\mu}), \ldots, U_S(\boldsymbol{\mu}))'$, where $U_i(\boldsymbol{\mu}) = \partial \log(L(\boldsymbol{\mu}|\mathbf{x}))/\partial \mu_i$. As noted elsewhere (Lawless, 1982; Cox and Hinkley, 1974; Casella and Berger, 1990), the score vector $\mathbf{U}(\boldsymbol{\mu})$ asymptotically follows the multivariate normal distribution with mean $\mathbf{0}$ and covariance matrix $\mathbf{I}(\boldsymbol{\mu})$. Thus, we may apply the score statistic $\mathbf{U}(\boldsymbol{\mu}_0)'\mathbf{I}(\boldsymbol{\mu}_0)^{-1}\mathbf{U}(\boldsymbol{\mu}_0)$ to test the hypothesis $H_0 : \boldsymbol{\mu} = \boldsymbol{\mu}_0$ if n is large. When $\mathbf{U}(\boldsymbol{\mu}_0)'\mathbf{I}(\boldsymbol{\mu}_0)^{-1}\mathbf{U}(\boldsymbol{\mu}_0) > \chi^2_{S,\alpha}$, we reject $H_0 : \boldsymbol{\mu} = \boldsymbol{\mu}_0$ at level α. By inverting the acceptance region of the score test, we obtain an asymptotic $100(1-\alpha)$ percent confidence region for $\boldsymbol{\mu}$ given by $\{\boldsymbol{\mu}|\mathbf{U}(\boldsymbol{\mu})'\mathbf{I}(\boldsymbol{\mu})^{-1}\mathbf{U}(\boldsymbol{\mu}) \leq \chi^2_{S,\alpha}\}$. Note that when using the score test procedure, we do not need to obtain the MLE $\hat{\boldsymbol{\mu}}$, which often involves an iterative numerical procedure. If we are interested in testing $H_0 : \boldsymbol{\mu}_1 = \boldsymbol{\mu}_{10}$ (where $\boldsymbol{\mu} = (\boldsymbol{\mu}_1', \boldsymbol{\mu}_2')'$) in the presence of an $(S-r) \times 1$ subvector of nuisance parameters $\boldsymbol{\mu}_2$, where $\boldsymbol{\mu}_{10}$ is a specified value for the $r \times 1$ subvector $\boldsymbol{\mu}_1$ of interest, we may apply the partial score statistic as follows.

We partition $\mathbf{U}(\boldsymbol{\mu})$, $\mathbf{I}(\boldsymbol{\mu})$ and $\mathbf{I}(\boldsymbol{\mu})^{-1}$ in an analogous way as for $\boldsymbol{\mu} = (\boldsymbol{\mu}_1, \boldsymbol{\mu}_2)'$:

$$\mathbf{U}(\boldsymbol{\mu}) = \begin{pmatrix} \mathbf{U}_1(\boldsymbol{\mu}) \\ \mathbf{U}_2(\boldsymbol{\mu}) \end{pmatrix}, \qquad \mathbf{I}(\boldsymbol{\mu}) = \begin{pmatrix} \mathbf{I}_{11}(\boldsymbol{\mu}) & \mathbf{I}_{12}(\boldsymbol{\mu}) \\ \mathbf{I}_{21}(\boldsymbol{\mu}) & \mathbf{I}_{22}(\boldsymbol{\mu}) \end{pmatrix},$$

$$\mathbf{I}(\boldsymbol{\mu})^{-1} = \begin{pmatrix} \mathbf{I}^*_{11}(\boldsymbol{\mu}) & \mathbf{I}^*_{12}(\boldsymbol{\mu}) \\ \mathbf{I}^*_{21}(\boldsymbol{\mu}) & \mathbf{I}^*_{22}(\boldsymbol{\mu}) \end{pmatrix}, \tag{A.3}$$

where $\mathbf{I}^*_{11}(\boldsymbol{\mu}) = (\mathbf{I}_{11}(\boldsymbol{\mu}) - \mathbf{I}_{12}(\boldsymbol{\mu})\mathbf{I}_{22}(\boldsymbol{\mu})^{-1}\mathbf{I}_{21}(\boldsymbol{\mu}))^{-1}$ (Graybill, 1976). For a given $\boldsymbol{\mu}_1 = \boldsymbol{\mu}_{10}$, let $\hat{\boldsymbol{\mu}}_2(\boldsymbol{\mu}_{10})$ be the restricted MLE of $\boldsymbol{\mu}_2$ obtained by maximizing the likelihood $L(\boldsymbol{\mu}_1 = \boldsymbol{\mu}_{10}, \boldsymbol{\mu}_2|\mathbf{x}) = \prod_{i=1}^n f(x_i|\boldsymbol{\mu}_1 = \boldsymbol{\mu}_{10}, \boldsymbol{\mu}_2)$. We denote $\tilde{\boldsymbol{\mu}} = (\boldsymbol{\mu}'_{10}, \hat{\boldsymbol{\mu}}'_2(\boldsymbol{\mu}_{10}))'$. Then, the statistic $\mathbf{U}_1(\tilde{\boldsymbol{\mu}})'\mathbf{I}^*_{11}(\tilde{\boldsymbol{\mu}})\mathbf{U}_1(\tilde{\boldsymbol{\mu}})$ asymptotic follows the χ^2 distribution with r degrees of freedom when $\boldsymbol{\mu}_1 = \boldsymbol{\mu}_{10}$. Thus, if $\mathbf{U}_1(\tilde{\boldsymbol{\mu}})'\mathbf{I}^*_{11}(\tilde{\boldsymbol{\mu}})\mathbf{U}_1(\tilde{\boldsymbol{\mu}}) > \chi^2_{r,\alpha}$, we would reject the null hypothesis $H_0 : \boldsymbol{\mu}_1 = \boldsymbol{\mu}_{10}$ at level α. Again, by inverting the acceptance region of this partial score test, we obtain an asymptotic $100(1 - \alpha)$ percent confidence region for $\boldsymbol{\mu}_1$ given by $\{\boldsymbol{\mu}_1|\mathbf{U}_1(\tilde{\boldsymbol{\mu}})'\mathbf{I}^*_{11}(\tilde{\boldsymbol{\mu}})\mathbf{U}_1(\tilde{\boldsymbol{\mu}}) \le \chi^2_{r,\alpha}\}$.

A third method for testing $H_0 : \boldsymbol{\mu} = \boldsymbol{\mu}_0$ is to consider the asymptotic likelihood ratio test. We will reject H_0 at level α if $-2\log(L(\boldsymbol{\mu}_0|\boldsymbol{x})/(L(\hat{\boldsymbol{\mu}}|\boldsymbol{x}))) > \chi^2_{S,\alpha}$. Furthermore, when the parameters of interest form a subvector $\boldsymbol{\mu}_1$ (rather than $\boldsymbol{\mu}$), we may reject $H_0 : \boldsymbol{\mu}_1 = \boldsymbol{\mu}_{10}$ at level α if $-2\log(L(\tilde{\boldsymbol{\mu}}|\boldsymbol{x})/L(\hat{\boldsymbol{\mu}}|\boldsymbol{x})) > \chi^2_{r,\alpha}$. Similarly, we can obtain an asymptotic $100(1 - \alpha)$ percent confidence region for $\boldsymbol{\mu}$ or $\boldsymbol{\mu}_1$ by simply inverting the acceptance region of the test.

A.2 THE DELTA METHOD AND ITS APPLICATIONS

Suppose that $\hat{\boldsymbol{\mu}}' = (\hat{\mu}_1, \hat{\mu}_2, \ldots, \hat{\mu}_S)$ asymptotically follows the multivariate normal distribution with mean $\boldsymbol{\mu}' = (\mu_1, \mu_2, \ldots, \mu_S)$, and covariance matrix $\boldsymbol{\Sigma}/n$. Suppose further that $g(\mathbf{x})$ has a continuous non-zero differential $\partial g(\mathbf{x})/\partial x_i$ at $\mathbf{x} = \boldsymbol{\mu}$. Define $\partial g/\partial \boldsymbol{x}|_{\mathbf{x}=\boldsymbol{\mu}}$ as the vector $(\partial g(\mathbf{x})/\partial x_1, \partial g(\mathbf{x})/\partial x_2, \ldots, \partial g(\mathbf{x})/\partial x_S)'$ evaluated at $\mathbf{x} = \boldsymbol{\mu}$. Then, $\sqrt{n}(g(\hat{\boldsymbol{\mu}}) - g(\boldsymbol{\mu}))$ asymptotically follows the multivariate normal distribution with mean $\mathbf{0}$ and variance $(\partial g/\partial \boldsymbol{x})'|_{\mathbf{x}=\boldsymbol{\mu}}\boldsymbol{\Sigma}(\partial g/\partial \boldsymbol{x})|_{\mathbf{x}=\boldsymbol{\mu}}$. This result is referred to as the delta method (Agresti, 1990).

To illustrate the use of the delta method, consider the random vector $(Y_1, Y_2, \ldots, Y_S)'$, for example, following the multinomial distribution with parameters n and $\boldsymbol{\pi} = (\pi_1, \pi_2, \ldots, \pi_S)'$. By the central limit theorem, if n is large, the random vector $(\hat{\pi}_1, \hat{\pi}_2, \ldots, \hat{\pi}_S)'$ asymptotically has the multivariate normal distribution with mean $\boldsymbol{\pi} = (\pi_1, \pi_2, \ldots, \pi_S)'$ and covariance matrix $[\text{diag}(\boldsymbol{\pi}) - \boldsymbol{\pi}\boldsymbol{\pi}']/n$, where $\hat{\pi}_i = Y_i/n$, and $\text{diag}(\boldsymbol{\pi})$ is a diagonal matrix with diagonal elements equal to π_i. Thus, using the delta method, we may claim that $\sqrt{n}(g(\hat{\boldsymbol{\mu}}) - g(\boldsymbol{\mu}))$ asymptotically follows the multivariate normal distribution with mean $\mathbf{0}$ and variance $\sum_i \pi_i(\partial g/\partial x_i)^2|_{\mathbf{x}=\boldsymbol{\pi}} - (\sum_i \pi_i(\partial g/\partial x_i)|_{\mathbf{x}=\boldsymbol{\pi}})^2$.

REFERENCES

Agresti, A. (1990) *Categorical Data Analysis*. Wiley, New York.
Casella, G. and Berger, R. L. (1990) *Statistical Inference*. Duxbury Press, Belmont, CA.
Cox, D. R. and Hinkley, D. V. (1974) *Theoretical Statistics*. Chapman & Hall, London.
Graybill, F. A. (1976) *Theory and Application of the Linear Model*. Duxbury Press, Belmont, CA.
Lawless, J. F. (1982) *Statistical Models and Methods for Lifetime Data*. Wiley, New York.

Answers to Selected Exercises

CHAPTER 1

1.10 (a) $\hat{\pi} = 0.618$; (b) [0.471, 0.764], [0.467, 0.748], and [0.487, 0.783].
1.18 [0.001, 0.009], [0.002, 0.012], and [0.002, 0.012].
1.19 [9, 62].
1.22 (a) 0.20; (b) [0.076, 0.324]; (c) [0.109, 0.368]; (d) [0.096, 0.368].

CHAPTER 2

2.17 [−0.000, 0.175], [−0.003, 0.171], [−0.005, 0.172], and [−0.007, 0.178].
2.18 (a) $\hat{\Delta} = 0.264$; (b) [0.107, 0.420], [0.107, 0.415], [0.106, 0.415], and [0.110, 0.414].
2.19 (a) 0.155 and 0.166; (b) [0.009, 0.301] and [0.017, 0.315]; (c) 0.644, 0.712, and 0.626.
2.20 [−0.0336, 0.1479], [−0.0346, 0.1505], [−0.0341, 0.1443], and [−0.0339, 0.1472].
2.21 (a) $\hat{\Delta} = 0.150$ and $\hat{\Delta}^{(u)} = 0.145$; (b) [0.046, 0.254], [0.042, 0.248], [0.052, 0.259].
2.24 (a) 0.00012; (b) [−0.00102, 0.00126]; (c) [−0.00114, 0.00116].
2.26 [−0.00021, 0.00071], [−0.00022, 0.00072], [−0.00022, 0.00072].

CHAPTER 3

3.4 (a) 0.0464; (b) [0.030, 0.063].
3.15 (a) 0.146 and 0.227; (b) [0.000, 0.317] and [0.013, 0.394]; (c) 0.363, 0.495, and 0.304.
3.16 (a) 0.429; (b) [0.217, 0.640] and [0.172, 0.605].

Statistical Estimation of Epidemiological Risk K-J. Lui
 ISBN: 0-470-85071-X (HB)

3.17 (a) $\hat{\delta} = 0.111$; $\hat{\delta}_{\text{cond}} = 0.115$; (b) [0.000, 0.247], [0.000, 0.237], and [0.000, 0.232]; (c) [0.000, 0.252] and [0.000, 0.258].

CHAPTER 4

4.7 (a) 0.95, 0.88, and 0.98; (b) 0.74, 0.83, 0.70, 0.82, and 0.82; (c) 0.79; (d) [0.680, 0.913] and [0.679, 0.911].
4.13 (a) 1.073; (b) [1.038, 1.109].
4.14 (a) 10 and 9.018; (b) [4.182, 23.190] and [4.058, 24.645].
4.19 (a) p-value is 0.999; (b) $\hat{\theta}_{\text{c}}^*$ and $\hat{\theta}_{\text{MH}}^*$ both equal 1.161; (c) [0.885, 1.524], [0.885, 1.523], [0.885, 1.523], and [0.886, 1.522].

CHAPTER 5

5.12 2.64; [1.005, 7.316].
5.13 (a) 0.96; (b) 0.766 and 0.767; (c) [0.648, 0.902], [0.647, 0.901], and [0.648, 0.901]; (d) approximately 0.08.
5.20 (a) 0.96; (b) $\hat{\beta}_1 = -0.2692$; $\widehat{\text{OR}} = \exp(\hat{\beta}_1) = 0.764$; [0.648, 0.901]; (c) 0.08.

CHAPTER 7

7.26 (a) 0.884 [0.738, 0.948], 0.845 [0.735, 0.909], 0.840 [0.740, 0.901], 0.831 [0.736, 0.892], 0.739 [0.614, 0.823], and 0.699 [0.516, 0.812]); (b) 0.80; (c) [0.700, 0.867].

Index

Statistical Estimation of Epidemiological Risk K-J. Lui

ISBN: 0-470-85071-X (HB)

STATISTICS IN PRACTICE

Human and Biological Sciences

Brown and Prescott – Applied Mixed Models in Medicine
Ellenberg, Fleming and DeMets – Data Monitoring Committees in Clinical Trials: A Practical Perspective
Lawson, Browne and Vidal Rodeiro – Disease Mapping with WinBUGS and MLwiN
Lui – Statistical Estimation of Epidemiological Risk
Marubini and Valsecchi – Analysing Survival Data from Clinical Trials and Observation Studies
Parmigiani – Modeling in Medical Decision Making: A Bayesian Approach
Senn – Cross-over Trials in Clinical Research, Second Edition
Senn – Statistical Issues in Drug Development
Spiegelhalter, Abrams and Myles – Bayesian Approaches to Clinical Trials and Health-Care Evaluation
Whitehead – Design and Analysis of Sequential Clinical Trials, Revised Second Edition
Whitehead – Meta-Analysis of Controlled Clinical Trials

Earth and Environmental Sciences

Buck, Cavanagh and Litton – Bayesian Approach to Interpreting Archaeological Data
Glasbey and Horgan – Image Analysis for the Biological Sciences
Webster and Oliver – Geostatistics for Environmental Scientists

Industry, Commerce and Finance

Aitken – Statistics and the Evaluation of Evidence for Forensic Scientists, Second Edition
Lehtonen and Pahkinen – Practical Methods for Design and Analysis of Complex Surveys, Second Edition
Ohser and Mücklich – Statistical Analysis of Microstructures in Materials Science

CPSIA information can be obtained at www.ICGtesting.com
Printed in the USA
BVOW06*0009220915

419014BV00010B/139/P

9 780470 850718